Principles of Optics for Engineers

Uniting historically different approaches by presenting optical analyses as solutions of Maxwell's equations, this unique book enables students and practicing engineers to fully understand the similarities and differences between the various methods.

The book begins with a thorough discussion of plane wave analysis, which provides a clear understanding of optics without considering boundary condition or device configuration. It then goes on to cover diffraction analysis, including a rigorous analysis of TEM waves using Maxwell's equations, and the use of Gaussian beams to analyze different applications. Modes of simple waveguides and fibers are also covered, as well as several approximation methods including the perturbation technique, the coupled mode analysis, and the super mode analysis. Analysis and characterization of guided wave devices, such as power dividers, modulators, and switches, are presented via these approximation methods.

With theory linked to practical examples throughout, it provides a clear understanding of the interplay between plane wave, diffraction, and modal analysis, and how the different techniques can be applied to various areas such as imaging, spectral analysis, signal processing, and optoelectronic devices.

William S. C. Chang is an Emeritus Professor of the Department of Electrical and Computer Engineering, University of California, San Diego (UCSD). After receiving his Ph.D. from Brown University in 1957, he pioneered maser and laser research at Stanford University, and he has been involved in guided-wave teaching and research at Washington University and UCSD since 1971. He has published over 200 technical papers and several books, including *Fundamentals of Guided-Wave Optoelectronic Devices* (Cambridge, 2009), *Principles of Lasers and Optics* (Cambridge, 2005) and *RF Photonic Technology in Optical Fiber Links* (Cambridge, 2002).

Principles of Optics for Engineers

Diffraction and Modal Analysis

BY WILLIAM S. C. CHANG

University of California, San Diego

CAMBRIDGE
UNIVERSITY PRESS

CAMBRIDGE
UNIVERSITY PRESS

University Printing House, Cambridge CB2 8BS, United Kingdom

Cambridge University Press is part of the University of Cambridge.

It furthers the University's mission by disseminating knowledge in the pursuit of education, learning and research at the highest international levels of excellence.

www.cambridge.org
Information on this title: www.cambridge.org/9781107074903

First published 2015

A catalog record for this publication is available from the British Library

Library of Congress Cataloging in Publication data
Chang, William S. C. (William Shen-chie), 1931–
Principles of optics for engineers: diffraction and modal analysis / by William S. C. Chang, University of California, San Diego.
pages cm
ISBN 978-1-107-07490-3
1. Optical engineering. 2. Diffraction. 3. Modal analysis. I. Title.
TA1520.C47 2015
535–dc23

2014042202

ISBN 978-1-107-07490-3 Hardback

Additional resources for this publication at www.cambridge.org/9781107074903

Contents

Introduction

Optics is a very old field of science. It has been taught traditionally as propagation, imaging, and diffraction of polychromatic natural light, then as interference, diffraction, and propagation of monochromatic light. Books like *Principles of Optics* by E. Wolf in 1952 gave a comprehensive and extensive in-depth discussion of properties of polychromatic and monochromatic light. Topics such as optical waveguide, fiber optics, optical signal processing, and holograms for laser light have been presented separately in more recent books. There appears to be no need for any new book in optics. However, there are several reasons to present optics differently, such as is done in this book.

Many contemporary optics books are concerned with components and instruments such as lenses, microscopes, interferometers, gratings, etc. Reflection, refraction, and diffraction of optical radiation are emphasized in these books. Other books are concerned with the propagation of laser light in devices and systems such as optical fibers, optical waveguides, and lasers, where they are analyzed more like microwave devices and systems. The mathematical techniques used in the two approaches are very different. In one case, diffraction integrals and their analysis are important. In the other case, modal analysis is important. Students usually learn optical analysis in two separate ways and then reconcile, if they can, the similarities and differences between them. Practicing engineers are also not fully aware of the interplay of these two different approaches. These difficulties can be resolved if optical analyses are presented from the beginning as solutions of Maxwell's equations and then applied to various applications using different techniques, such as diffraction or modal analysis.

The major difficulty to present optics from the solutions of Maxwell's equations is the complexity of the mathematics. Complex mathematical analyses often obscure the basic differences and similarities of the mathematical techniques and mask the understanding of basic concepts.

Optical device configurations vary from simple mirrors to complex waveguide devices. How to solve Maxwell's equations depends very much on the configuration of the components to be analyzed. The more complex the configuration, the more difficult the solution. Optics is presented in this book in the order of the complexity of the configuration in which the analysis is carried out. In this manner, the reasons for using different analytical techniques can be easily understood, and basic principles are not masked by any unnecessary mathematical complexity.

Optics in unbounded media is first presented in this book in the form of plane wave analysis. A plane wave is the simplest solution of Maxwell's equations. Propagation, refraction, diffraction, and focusing of optical radiation, even optical resonators and planar waveguides, can be analyzed and understood by plane wave analysis. It leads directly to ray optics, which is the basis of traditional optics. It provides a clear demonstration and understanding of optics without considering boundary condition or device configuration. Even sophisticated concepts such as modal expansion can also be introduced using plane waves. Plane wave analysis is the focus of the first two chapters.

Realistically, wave propagation in bulk optical components involves a finite boundary such as a lens that has a finite aperture. Plane wave analysis can no longer be used in this configuration. However, in these situations, the waves are still transverse electric and magnetic (TEM). Therefore, TEM waves are rigorously analyzed using Maxwell's equations in Chapter 3. The diffraction analysis presented in Chapter 3 is identical to traditional optical analysis. Since applications of diffraction analysis are already covered extensively in existing optics books, only a few basic applications of diffraction theory are presented here. The distinct features of our presentation here are: (1) Both the TEM assumption of the Kirchoff's integral analysis and the relation between diffraction theory and Maxwell's equations are clearly presented. (2) Modern engineering concepts such as convolution, unit impulse response, and spatial filtering are introduced.

Diffraction integrals are again used to analyze laser cavities in the first part of Chapter 4, for three reasons: (1) Laser modes are used in many applications. (2) The diffraction analysis leads directly to the concept of modes. It is instructive to recognize that they are inter-related. (3) An important consequence of laser cavity analysis is that laser modes are Gaussian. A Gaussian mode retains its functional form not only inside, but also outside of the cavity.

The second part of Chapter 4 is focused on Gaussian beams and how different applications can be analyzed using Gaussian beams. Gaussian modes are also natural solutions of the Maxwell's equations. It constitutes a complete set. Just like any other set of modes, such as plane waves, any radiation can be represented as summation of Gaussian modes. When the diffraction integral is used in Chapter 3 to analyze waves propagating through components with finite apertures, the diffraction loss needs to be calculated by the Kirchoff's integral for each aperture. In comparison, the diffraction loss of a Gaussian beam propagating through an aperture can be calculated without any integration. Therefore, a Gaussian beam is used to represent TEM waves in many engineering applications.

Although TEM modes exist in solid-state and gas laser cavities, waves propagating in waveguides and fibers are no longer transverse electric and magnetic. Microwave-like modal analysis needs to be used to analyze optical devices that have dimensions of the order of optical wavelength.

Optical waveguides and fibers are dielectric devices. They are different from microwave devices. Microwave waveguides have closed metallic boundaries. The mathematical complexity of finding microwave waveguide modes is much simpler than that of optical waveguides.

The distinct features in the analysis of dielectric waveguides are: (1) There are analytical solutions for very few basic device configurations because of the complex boundary conditions. Analyses of practical devices need to be carried out by approximation techniques. (2) There is a continuous set of radiation modes in addition to the discrete guided-wave modes. Any abrupt discontinuity will excite radiation modes. (3) The evanescent tail of the guided-wave modes not only reduces propagation loss, but also provides access to excite the modes by coupling through evanescent fields. (4) Multiple modes are often excited in devices. The performance of the device depends on what modes have been excited.

Because of the complexity of modal analysis of optical waveguides and fibers, it is presented here in four parts.

In the first part, modes of simple waveguides and fibers are discussed in Chapter 5. Analytical solutions for planar waveguides and step–index fiber are presented. Although these are not realistic devices, they are the only solutions that can be obtained from Maxwell's equations. Modes of these simple basic devices are very useful for demonstrating various properties of the guided waves. Approximation methods are then presented to discuss modes of realistic devices. For example, the effective index method is used here to analyze channel waveguides.

Guided-wave devices operate by mutual interactions among modes. These interactions need to be analyzed in the absence of exact solutions. Therefore, several approximation methods, the perturbation technique, the coupled mode analysis, and the super mode analysis, are presented in Chapter 6. The differences and similarities of the three methods are compared and explained. Examples in applications are used to demonstrate these techniques.

In the third and fourth parts, modal analyses of passive and active guided-wave devices are presented. Passive guided-wave devices function mainly as power dividers, wavelength filters, resonators, and wavelength multiplexers. In each of these system functions, there are several different devices that could be used. Thus, devices that perform the same system function are discussed and analyzed together. Their performance is compared.

Active devices utilize electro-optical effects of the electrical signals to operate. Discussion of active guided-wave devices is complex because there are different physical mechanisms involved. How these mechanisms work is reviewed.The electrical performance, as well as the optical performance of these devices are analyzed.

In summary, when optics are presented as solutions of Maxwell's equations, the inter-relation between plane wave, diffraction, and modal analysis becomes clear. For example, the use of modal analysis is not limited to waveguides and fibers. There can be modes and modal expansion in plane wave analysis, as well as in diffraction optics. As we learn optics step by step in the order of the mathematical complexity and device configuration, we learn optical analysis from various perspectives.

1 Optical plane waves in an unbounded medium

Engineers involved in design and the use of optical and opto-electronic systems are often required to analyze theoretically the propagation and the interaction of optical waves using different methods. Sometimes it is diffraction analysis; on other occasions, modal analysis. They are all solutions of Maxwell's equations, yet they appear to be very different. All optical analyses should be presented as solutions of Maxwell's equations so that the inter-relations between different analytical techniques are clear. In order to avoid unnecessary mathematical complexity, the simplest analysis should be presented first. In this book, optics will be presented first by plane wave analysis, followed by diffraction and modal analyses, in increasing order of complexity.

Plane waves are the simplest form of optical waves that can be derived rigorously from Maxwell's equations. Plane wave analysis can be used to derive ray analysis, which is the basis of traditional optics. It can be applied directly to analyze many optical phenomena such as refraction, reflection, dispersion, etc. It can also be used to demonstrate sophisticated concepts such as superposition, interference, resonance, guided waves, and Fourier optics. Plane wave analyses will be the focus of discussion in Chapters 1 and 2.

However, plane wave analysis cannot be used to analyze diffraction, laser modes, optical signal processing, and propagation in small optical components such as fibers and waveguides, etc. These analyses will be the focus of discussion in subsequent chapters.

1.1 Introduction to optical plane waves

Plane wave analysis is presented here in full detail, so that the mathematical derivations and details can be fully exhibited and the physical significances of these analyses are fully explained.

1.1.1 Plane waves and Maxwell's equations

All optical waves are solutions of the Maxwell's equations (assuming there are no free carriers),

$$\nabla \times \underline{E} = \frac{-\partial \underline{B}}{\partial t}, \quad \nabla \times \underline{H} = \frac{\partial \underline{D}}{\partial t} \tag{1.1}$$

Here $\underline{E}$ is the electric field vector, $\underline{H}$ is the magnetic field vector, $\underline{D}$ is the displacement vector, and $\underline{B}$ is the magnetic induction vector. For isotropic media,

$$\underline{B} = \mu \underline{H}, \quad \underline{D} = \varepsilon \underline{E} \tag{1.2}$$

Let $\underline{i_x}$, $\underline{i_y}$, and $\underline{i_z}$, be unit vectors in the *x*, *y*, and *z* directions of an *x*-*y*-*z* rectangular coordinate system. Then $\underline{E}$, $\underline{H}$ and the position vector $\underline{r}$ can be written as

$$\underline{E} = E_x \underline{i_x} + E_y \underline{i_y} + E_z \underline{i_z} \qquad \underline{H} = H_x \underline{i_x} + H_y \underline{i_y} + H_z \underline{i_z} \tag{1.3a}$$

$$\underline{r} = x \underline{i_x} + y \underline{i_y} + z \underline{i_z} \tag{1.3b}$$

A special solution of Eqs. (1.1) and (1.2) is a plane wave that has no amplitude variation transverse to its direction of propagation. If we designate the *z* direction as the direction of propagation, this means that

$$\frac{\partial}{\partial x} = 0 \quad \text{and} \quad \frac{\partial}{\partial y} = 0 \tag{1.4}$$

Substituting $\partial/\partial x = 0$ and $\delta/\delta y = 0$ into the $\nabla \times \underline{E}$ and $\nabla \times \underline{H}$ equations leads to two distinct groups of equations:

$$\frac{\partial E_y}{\partial z} = \mu\, \partial H_x/\partial t, \qquad \frac{\partial H_x}{\partial z} = \varepsilon \partial E_y/\partial t; \quad or \quad \frac{\partial E_y{}^2}{\partial z^2} = \mu\varepsilon \frac{\partial^2}{\partial t^2} E_y \tag{1.5a}$$

and

$$\frac{\partial H_y}{\partial z} = -\varepsilon \partial E_x/\partial t, \qquad \frac{\partial E_x}{\partial z} = -\mu \partial H_y/\partial t; \quad or \quad \frac{\partial H_y{}^2}{\partial z^2} = \mu\varepsilon \frac{\partial^2}{\partial t^2} H_y \tag{1.5b}$$

Clearly, these are two separate independent sets of equations. E_y and H_x are related only to each other, and H_y and E_x are related only to each other. Solutions of Eq. (1.5a) are plane waves with *y* polarization of the electric field (or *x* polarization in magnetic field). Solutions of Eq. (1.5b) are plane waves with *x* polarization in the electric field $\underline{E}$ (or *y* polarization in magnetic field $\underline{H}$).

(a) The *y*-polarized plane wave

For a cw optical plane wave with a single angular frequency ω that has a time variation, $e^{j\omega t}$, and for lossless media (i.e. the medium has a real value of ε), there is a well-known solution of Eq. (1.5a) in the complex notation. It is

$$E_y = E_y^f e^{-j\beta z} e^{j\omega t}, \quad H_x = H_x^f e^{-j\beta z} e^{j\omega t}, \quad H_x^f = -\sqrt{\frac{\varepsilon}{\mu}} E_y^f, \tag{1.6a}$$

where $\beta = \omega\sqrt{\mu\varepsilon}$. The real time domain expression for the complex E_y shown in (1.6a) is $\left|E_y^f\right| \cos(\beta z - \omega t + \varphi)$ where φ is the phase of $\left|E_y^f\right|$ at $z = 0$ and $t = 0$. The angular frequency ω is related to the optical frequency *f* by $\omega = 2\pi f$. This wave is known as a *y*-polarized forward propagating wave in the +*z* direction. The phase of

E_y, i.e. $\beta z - \omega t = \beta(z - v_p t)$, is a constant when $z = v_p t$. Thus v_p is known as the phase velocity of the plane wave.

If the medium in which the plane wave propagates is free space, then $\varepsilon = \varepsilon_o$ and the free space phase velocity is $c_o = 1/\sqrt{\mu\varepsilon_o} \equiv 3 \times 10^8$ m s^{-1}. In free space, the optical wave length for a frequency f is λ_o, where $f\lambda_o = c_o$. If the medium is a lossless dielectric material with a permittivity ε, then its index of refraction is $n = \sqrt{\varepsilon/\varepsilon_o}$, $\beta = n\beta_o = n\omega\sqrt{\mu\varepsilon_o}$. If ε is a function of wavelength, the medium is said to be dispersive.

There is also a second solution for the same polarization of the electric field,

$$E_y = E_y^b e^{j\beta z} e^{j\omega t}, \quad H_x = H_x^b e^{j\beta z} e^{j\omega t}, \quad H_x^b = \sqrt{\frac{\varepsilon}{\mu}} E_y^b \tag{1.6b}$$

This solution is a backward propagating wave because the phase of E_y, i.e. $\beta z + \omega t = \beta(z + v_p t)$, at any time t is a constant when $z = -v_p t$ and $v_p = \omega/\beta$.

If the permittivity has a loss component, $\varepsilon = \varepsilon_r - j\varepsilon_\sigma$, then

$$\beta = \omega\sqrt{\mu(\varepsilon_r - j\varepsilon_\sigma)} = \beta_r - j\beta_\sigma \tag{1.7}$$

The phase velocity of light is now $v_p = c = \omega/\beta_r$. The amplitude of the plane wave decays as $e^{-\beta_\sigma z'}$ for forward waves and $e^{+j\beta_\sigma z'}$ for backward waves. In comparison with the phase velocity of free space, the ratio of the phase velocities, c_o/c, is the effective refractive index of the plane wave, $n = c_o\beta_r/\omega = c_o/c$. The wavelength in the medium is $\lambda = \lambda_o/n$. In addition to β, or phase velocity, the loss of optical waves in the medium is an important consideration in applications.

(b) The x-polarized plane wave

A similar solution exists for the x-polarized electric field and H_y. For the forward wave,

$$H_y = H_y^f e^{-j\beta z} e^{j\omega t}, \quad E_x = E_x^f e^{-j\beta z} e^{j\omega t}, \quad E_x^f = \sqrt{\frac{\mu}{\varepsilon}} H_y^f \tag{1.8a}$$

For the backward wave,

$$H_y = H_y^b e^{+j\beta z} e^{j\omega t}, \quad E_x = E_x^b e^{j\beta z} e^{j\omega t}, \quad E_x^b = -\sqrt{\frac{\mu}{\varepsilon}} H_y^b \tag{1.8b}$$

In summary, both equations (1.5a) and (1.5b) are second-order differential equations. Mathematically, each of them has two independent solutions, which are the forward and the backward propagating waves. However, Eqs. (1.5a) and (1.5b) are also two separate set of equations. The solution for Eq. (1.5a) describes a plane wave polarized in the y direction. The solution of Eq. (1.5b) describes a plane wave polarized in the x direction. Both waves have the same direction of propagation. $\underline{\beta}$ is usually designated as a propagation vector along the direction of propagation z that has magnitude β,

$$\underline{\beta} = \beta i_{\underline{z}}, \quad \underline{z} = z i_{\underline{z}}, \quad \beta z = \underline{\beta} \bullet \underline{z} \tag{1.9}$$

The forward wave has $+\underline{\beta}$, the backward wave has $-\underline{\beta}$.

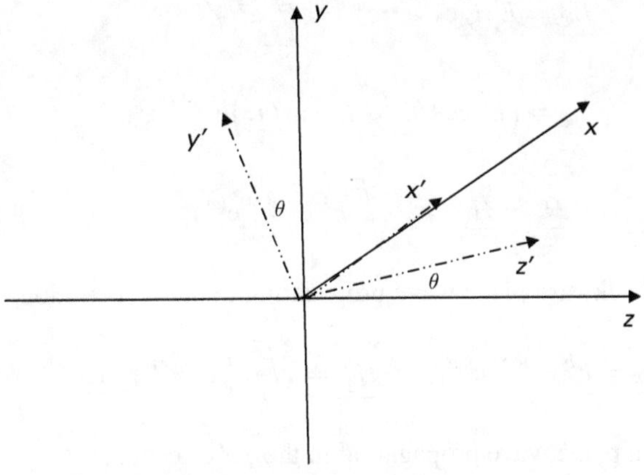

Figure 1.1 Illustration of x-y-z and x'-y'-z' coordinates.

It is important to note that, along any direction of propagation, there are always plane waves with two orthogonal polarizations. In each polarization, there are always two solutions, the forward wave and the backward wave. The propagation constant β and phase velocity will depend on the medium and the frequency.

1.1.2 Plane waves in an arbitrary direction

Frequently, plane waves in other directions of propagation need to be expressed mathematically for analysis. As an example, let there be another x'-y'-z' rectangular coordinate which is related to the x-y-z coordinate by

$$\underline{i_{x'}} = \underline{i_x}, \quad \underline{i_{y'}} = \cos\theta \underline{i_y} - \cos\left(\frac{\pi}{2} - \theta\right)\underline{i_z}, \quad \underline{i_{z'}} = \cos\left(\frac{\pi}{2} - \theta\right)\underline{i_y} + \cos\theta \underline{i_z} \qquad (1.10)$$

The x-y-z and the x'-y'-z' coordinates are illustrated in Figure 1.1. The x'-y'-z' coordinate is just the x-y-z coordinate rotated by angle θ about the x axis. The x and x' axes are the same.

Let there be a plane wave propagating along the z' direction. The solutions for the y' and x' polarized plane waves have already been given in Eqs. (1.6) and (1.8). However, these solutions could also be expressed in the x, y, and z coordinates, where

$$\beta z' = \underline{\beta} \bullet \underline{z'} = \beta \cos\theta z + \beta \cos\left(\frac{\pi}{2} - \theta\right) y \qquad (1.11)$$

$$\underline{\beta} = \beta \underline{i_{z'}} = \beta \cos\theta \underline{i_z} + \beta \cos\left(\frac{\pi}{2} - \theta\right) \underline{i_y} \qquad (1.12)$$

$$e^{\pm j\beta z'} = e^{\pm j\underline{\beta} \bullet \underline{z'}} = e^{\pm j\underline{\beta} \bullet \underline{r}} \qquad (1.13)$$

For the y' polarized plane wave propagating in the $+z'$ direction,

$$\underline{E_{y'}} = E^f_{y'}\underline{i_{y'}}e^{-j\underline{\beta}\cdot\underline{z'}}e^{j\omega t} = \underline{E^f_{y'}}e^{-j\underline{\beta}\cdot\underline{r}}e^{j\omega t}$$

$$= \left(E^f_{y'}\cos\theta\underline{i_y} - E^f_{y'}\sin\theta\underline{i_z}\right)e^{-j\underline{\beta}\cdot\underline{r}}e^{j\omega t} \tag{1.14}$$

$$\underline{H_x} = \underline{H_{x'}} = -\sqrt{\frac{\varepsilon}{\mu}}E^f_{y'}e^{-j\underline{\beta}\cdot\underline{r}}e^{j\omega t}\underline{i_x} \tag{1.15}$$

For the y' polarized backward plane wave propagating in the $-z'$ direction,

$$\underline{E_{y'}} = E^b_{y'}e^{+j\underline{\beta}\cdot\underline{r}}e^{j\omega t}\underline{i_{y'}}, \quad \underline{H^b_{x'}} = \sqrt{\frac{\varepsilon}{\mu}}E^b_{y'}e^{+j\underline{\beta}\cdot\underline{r}}e^{j\omega t}\underline{i_{x'}} \tag{1.16}$$

For the x' polarized plane wave propagating in the $+z'$ direction,

$$\underline{E_{x'}} = \underline{E_x} = E^f_{x'}e^{-j\underline{\beta}\cdot\underline{r}}e^{j\omega t}\underline{i_{x'}} \tag{1.17}$$

$$\underline{H_{y'}} = \sqrt{\frac{\varepsilon}{\mu}}E^f_{x'}e^{-j\underline{\beta}\cdot\underline{r}}e^{j\omega t}\underline{i_{y'}} = \sqrt{\frac{\varepsilon}{\mu}}E^f_{x'}\left(\cos\theta\underline{i_y} - \sin\theta\underline{i_z}\right)e^{-j\underline{\beta}\cdot\underline{r}}e^{-j\omega t} \tag{1.18}$$

For the x' polarized backward wave plane wave propagating in the $-z'$ direction,

$$\underline{E_{x'}} = \underline{E_x} = E^b_{x'}e^{+j\underline{\beta}\cdot\underline{r}}e^{j\omega t}\underline{i_{x'}} \tag{1.19}$$

$$\underline{H_{y'}} = -\sqrt{\frac{\varepsilon}{\mu}}E^b_{x'}e^{+j\underline{\beta}\cdot\underline{r}}e^{j\omega t}\underline{i_{y'}} \tag{1.20}$$

The preceding example can be generalized for any orientation of the x', y', and z' coordinates with respect to the x, y, and z coordinates. Any plane wave propagating in the z' direction can have two mutually perpendicular polarizations, $\underline{i_a}$ and $\underline{i_b}$. $\underline{i_{z'}}$, $\underline{i_a}$ and $\underline{i_b}$ are mutually perpendicular to each other, i.e. $\underline{i_a}\cdot\underline{i_b} = \underline{i_a}\cdot\underline{\beta} = \underline{i_b}\cdot\underline{\beta} = 0$.

$$\text{Let } \underline{i_a} = \underline{i_{x'}} \quad \text{and} \quad \underline{i_b} = \underline{i_{y'}} \tag{1.21}$$

Then the general solutions for the case of $\underline{i_a}$ polarization are:

$$\underline{E^f_a} = E^f_a e^{-j\underline{\beta}\cdot\underline{r'}}e^{j\omega t}\underline{i_{x'}} \quad \underline{H^f_a} = \sqrt{\frac{\varepsilon}{\mu}}E^f_a e^{-j\underline{\beta}\cdot\underline{r'}}e^{j\omega t}\underline{i_{y'}} \tag{1.22}$$

$$\underline{E^b_a} = E^b_a e^{+j\underline{\beta}\cdot\underline{r'}}e^{j\omega t}\underline{i_{x'}} \quad \underline{H^b_a} = -\sqrt{\frac{\varepsilon}{\mu}}E^b_a e^{+j\underline{\beta}\cdot\underline{r'}}e^{j\omega t}\underline{i_{y'}} \tag{1.23}$$

$$\underline{\beta} = \beta_{x'}\underline{i_{x'}} + \beta_{y'}\underline{i_{y'}} + \beta_{z'}\underline{i_{z'}} \qquad \beta^2 = \beta_{x'}{}^2 + \beta_{y'}{}^2 + \beta_{z'}{}^2 \tag{1.24}$$

Here, $\underline{\beta}$ makes angles $\theta_{x'}$, $\theta_{y'}$, and $\theta_{z'}$ with respect to the x', y', and z' axes, with $\beta_{x'}/\beta = \cos\theta_{x'}$, $\beta_{y'}/\beta = \cos\theta_{y'}$, and $\beta_{z'}/\beta = \cos\theta_{z'}$. The general solutions for the case of $\underline{i_b}$ polarization are:

$$\underline{E}_b^f = E_b^f e^{-j\underline{\beta}\cdot\underline{r}'} e^{j\omega t} \underline{i_{y'}} \quad \underline{H}_a^f = -\sqrt{\frac{\varepsilon}{\mu}} E_b^f e^{-j\underline{\beta}\cdot\underline{r}'} e^{j\omega t} \underline{i_{x'}} \tag{1.25}$$

$$\underline{E}_b^b = E_a^b e^{+j\underline{\beta}\cdot\underline{r}'} e^{j\omega t} \underline{i_{y'}} \quad \underline{H}_a^b = \sqrt{\frac{\varepsilon}{\mu}} E_a^b e^{+j\underline{\beta}\cdot\underline{r}'} e^{j\omega t} \underline{i_{x'}} \tag{1.26}$$

It is important to recognize that when there is a wave solution containing various terms, any term that has the form shown in Eqs. (1.17) to (1.26) represents a plane wave propagating in the direction of $\underline{\beta}$.

1.1.3 Evanescent plane waves

Eqs. (1.22) to (1.26) described propagating plane waves that have real $\beta_{x'}$, $\beta_{y'}$, and $\beta_{z'}$ values. The maximum real $\beta_{x'}$ and $\beta_{y'}$ values of propagating plane waves are limited to $\sqrt{\beta_{x'}{}^2 + \beta_{y'}{}^2} < \omega\sqrt{\mu\varepsilon}$, i.e. $0 < \theta_{x'}$, $\theta_{y'}$, and $\theta_{z'} < \pi/2$. Nevertheless, Maxwell's equation is still satisfied even if $\sqrt{\beta_x^2 + \beta_{y'}^2}$ is larger than β. In that case Eq. (1.24) can only be satisfied if $\beta_{z'}$ is imaginary. When $\beta_{z'}$ is imaginary, the z' variation is a real decaying or growing exponential function, $e^{\pm\sqrt{\beta_{x'}{}^2+\beta_{y'}{}^2-\beta^2}z'}$. In any passive medium, the plane wave cannot grow without energy input. Thus the solution must decay exponentially in the z' direction. Any solution with imaginary $\beta_{z'}$ is called an evanescent wave. Such solutions do not propagate in the z direction. They do not have a phase velocity. Evanescent waves are excited usually in the vicinity of a boundary with an incident wave applied across the boundary. It is only a near field, meaning that it is negligible at locations far away from the boundary.[1] It is interesting to note that when $\beta_{x'} = \beta$, $\beta_{y'} = \beta_{z'} = 0$, it is no longer a plane wave propagating in the z' direction. It is a plane wave propagating in the $+x'$ direction.

1.1.4 Intensity and power

In optics, only time-averaged power can be detected directly by means of detectors or by recording media such as film. The time-averaged power per unit area is known commonly as the intensity. In comparison with rf and microwaves, intensity analysis plays a much more important role in optics. From text books on electromagnetic theory, it is well known that the total time-averaged power in the direction of propagation is [1]

$$P_{av} = \frac{1}{2}\mathrm{Re}\int_S \underline{E} \times \underline{H}^* \cdot \underline{i_{z'}} \mathrm{d}s = \int_S \underline{I} \cdot \underline{i_{z'}} \mathrm{d}s, \quad \underline{I} = \frac{1}{2}Re[\underline{E} \times \underline{H}] \tag{1.27}$$

[1] It is important to note that although the mathematical solution of a plane wave exists for β_x or β_y values larger than $\omega\sqrt{\mu\varepsilon}$, such a solution is important only if those plane waves are excited in specific applications such as total internal reflection. Otherwise, the solutions have no practical significance.

The integration is carried out over the entire surface of the plane wave, S. The * designates the complex conjugate of the variable. Re designates the real part of the complex quantity. Therefore the time-averaged power per unit area in the direction of propagation z' in either polarization is

$$I_{z'} = \frac{1}{2}\mathrm{Re}\, E_a H_a^* = \frac{1}{2}\sqrt{\frac{\varepsilon}{\mu}} E_a E_a^* \quad \text{or} \quad I_{z'} = \frac{1}{2}\mathrm{Re}\, E_b H_b^* = \frac{1}{2}\sqrt{\frac{\varepsilon}{\mu}} E_b E_b^* \qquad (1.28)$$

Note that although the total I *is the sum of the* Is *in each polarization, the total* I *carries no information about polarization breakdown. Although the complex amplitude of the plane wave has a phase, its intensity* I *has no phase information. For plane waves in a lossless medium, i.e.* $\varepsilon_\sigma = 0$*, its intensity* I *is a constant. In media with loss, the decay of the time-averaged power is* $e^{-2\beta_\sigma z'}$ *for a forward wave and* $e^{+2\beta_\sigma z'}$ *for a backward wave.*

In microwaves, $\underline{I}$ is known as the Poynting vector. In x-y-z coordinates, the intensity along the z direction is ½ Re $E_x H_y^*$, the intensity along the y direction is ½ Re $E_z H_x^*$, and the intensity along the x direction is ½Re $E_y H_z^*$.

1.1.5 Superposition and plane wave modes

Plane waves in different direction of propagation (or plane wave modes) can be superimposed simultaneously. This is known as the superposition theory in linear media. Many interesting optical phenomena can be understood by superposition of plane waves. Three examples are presented here to illustrate the effects of superposition. They are important concepts in many applications.

(a) Plane waves with circular polarization

Let us consider superposition of two plane waves of equal magnitude, polarized in x and y, with a $\pi/2$ phase difference.

$$\underline{E} = E_o\left(\underline{i_x} + j\underline{i_y}\right) \qquad (1.29)$$

The real time domain form of this wave is

$$\underline{E} = E_o\left[\cos(\beta z - \omega t + \varphi)\underline{i_x} + \sin(\beta z - \omega t + \varphi)\underline{i_y}\right] \qquad (1.30)$$

So that, at any time t, the polarization rotates at different z positions. This type of wave is known as a circular polarized optical wave because the polarization of $\underline{E}$ rotates as it propagates. When these two waves have unequal amplitudes they give rise to an elliptical polarized plane wave.

(b) Interference of coherent plane waves

Let us consider two plane waves of equal amplitude at the same ω and y polarization. They propagate at different directions of propagation $\underline{\beta}$ in the x–z plane. Their $\underline{\beta}$s lie in the x–z plane and make angles, θ and ζ, with respect to the z axis. Mathematically, the waves are

$$E_o e^{-j\beta \sin\theta x} e^{-j\beta \cos\theta z} e^{j\omega t} \underline{i_y} + E_o e^{-j\beta \sin\zeta x} e^{-j\beta \cos\zeta z} e^{j\omega t} \underline{i_y} \tag{1.31}$$

According to Eq. (1.28), its time-averaged intensity in the z direction is

$$I_z = \sqrt{\frac{\varepsilon}{\mu}} |E_o|^2 [1 + \cos\beta((\sin\theta - \sin\zeta)x + (\cos\theta - \cos\zeta)z)] \tag{1.32}$$

Therefore, for $\theta \neq \zeta$, we would detect a sinusoidal intensity interference pattern of the two waves in the x direction. As z changes, the interference pattern in x will change. If we could record this intensity interference pattern, for example, by the transparency of a film, we could reproduce the plane waves by illuminating this film with another input plane wave. This is the very basic principle on which holography and phased array detection are based [2,3,4].

However, if the two waves do not have the same ω or a definite phase relation between them, then $I_z = \sqrt{\varepsilon/\mu}|E_o|^2$. In other words, there is no interference pattern unless the two waves are coherent.[2] The total intensity of incoherent waves is just the sum of the intensities of individual waves. It is also important to note that when the two coherent plane waves have cross polarizations, the total intensity is also just the sum of the two intensities without the interference effect.

(c) Representation by summation of plane waves

Let there be a linearly polarized TEM electric field propagating in the z direction with xy variation $g(x,y)$ at $z = 0$. It is well known that g can be represented by its Fourier transform. Let G be the Fourier transform of g.

$$G(f_x, f_y) = F_t(g) = \int_{-\infty}^{+\infty}\int_{-\infty}^{+\infty} g(x,y) e^{-j2\pi(f_x x + f_y y)} \mathrm{d}x\mathrm{d}y \tag{1.33}$$

$$g(x,y) = F_t^{-1}(G) = \int_{-\infty}^{+\infty}\int_{-\infty}^{+\infty} G(f_x, f_y) e^{j2\pi(f_x x + f_y y)} \mathrm{d}f_x \mathrm{d}f_y \tag{1.34}$$

G is the magnitude of the Fourier component at (f_x, f_y). When $g(x,y)$ contains only slow variations in x and y, G will have significant values only at low spatial frequencies f_x and f_y. In that case the integration in Eq. (1.34) could be approximated by just the integration of G within a limited range of f_x and f_y.

Let

$$2\pi f_x = \beta_{x'} = \beta \cos\theta_{x'}, \qquad 2\pi f_y = \beta_{y'} = \beta \cos\theta_{y'} \tag{1.35}$$

$$\mathrm{d}f_x = -\beta \sin\theta_{x'} d\theta_{x'} \qquad \mathrm{d}f_y = -\beta \sin\theta_{y'} d\theta_{y'} \tag{1.36}$$

[2] If the two waves have a randomly time-varying relative phase relation, for example from two independent lasers, the time-averaged detected intensity also will not have the interference pattern.

Eq. (1.33) gives the magnitude of the Fourier component that has an $x'y'$ variation of $e^{-j\beta\cos\theta_{x'}x'}e^{-j\beta\cos\theta_{y'}y'}$. If we also let $\beta = (2\pi c_o/\lambda_o)\sqrt{\mu\varepsilon_r} = 2\pi c/\lambda$ and consider only those Fourier components with $\sqrt{\beta_{x'}^2 + \beta_{y'}^2} < \beta$ (i.e. $0 < \theta_{x'}$ and $\theta_{y'} < \pi/2$), then Eq. (1.33) gives the magnitude of the Fourier component that has the $x'y'$ variation of a plane wave propagating in a direction $\underline{\beta}$, which has direction cosines $\theta_{x'}$, $\theta_{y'}$, and $\theta_{z'}$. $\theta_{z'}$ is related to $\theta_{x'}$ and $\theta_{y'}$ by Eq. (1.24). However, for propagating plane waves, $\theta_{x'}, \theta_{y'}$, and $\theta_{z'}$ must be real. The maximum and minimum values of f_x and f_y for real values of $\theta_{x'}$ and $\theta_{y'}$, are $-\beta/2\pi$ and $+\beta/2\pi$. Moreover, $\sqrt{\beta_{x'}^2 + \beta_{y'}^2} < \beta$.[3] It means that, the Fourier components that correspond to propagating waves with real values of $\theta_{x'}$ and $\theta_{y'}$, are only Fourier components that have low values of $f_{x'}$ and $f_{y'}$.

In summary, when g contains only variations in x and y very much slower than λ, $G(f_x,f_y)$ has significant values only at those f_x and f_y less than the limit $f_{\max}$. If $f_{\max} < \beta/2\pi$, it means the electric field $g(x,y)$ could be represented by superposition of just plane waves propagating in different directions $\underline{\beta}$. Under that condition, Eq. (1.34) can be approximated by

$$g(x,y) = F_t^{-1}(G) \cong \int_{-f_{\max}}^{+f_{\max}}\int_{-f_{\max}}^{+f_{\max}} G(f_x,f_y)e^{j2\pi(f_x x + f_y y)}df_x df_y \qquad (1.37)$$

There are three important concepts introduced here: (1) We have shown that plane waves can be used to represent an arbitrary field with slow x' *and* y' *variations. This is equivalent to the modal expansion concept used in microwaves. Here, plane waves are the modes of unbounded medium. (2) Fields with any* x'y' *variation can be represented by their Fourier components.*[4] *This means that many modern Fourier analysis techniques can be applied to optics. This is the basis of Fourier optics [2] and optical image processing [4]. (3) Knowing the plane wave composition at* z = 0*, we have determined the* xy *variation of each plane wave components at any distance* z *later. Thus it allows us to predict the electric field that propagates to* z *via plane wave analysis. Note that as component plane waves propagate the total optical radiation spreads or contracts. This phenomenon is also known as the diffraction of optical radiation. More details on diffraction will be presented in Chapter 3.*

It is interesting to note that if $G(f_x,f_y)$ contains frequency components with large f_x and f_y such that $f_x^2 + f_y^2 > \beta^2/4\pi^2$, then the z variation of the plane waves for those components will exponentially decay. This means that those components will contribute only to the near field and they will not propagate far in the z direction. Only the frequency components with $f_x^2 + f_y^2 < \beta^2/4\pi^2$ will propagate, so the fields at some z distance away will not be exactly the same as $g(x,y)$.

[3] Evanescent plane waves in the z direction could have $\beta_{x'}$ or $\beta_{y'}$ larger than β. However they are not propagating waves.

[4] Fields with rapid xy variation would yield Fourier components that are evanescent local waves in the z direction.

1.1.6 Representation of plane wave as optical rays

When an optical wave has a finite beam size, the diffraction effect will spread the beam as it propagates. Diffraction analysis allows us to analyze the fields at various positions that are not at the center of the beam. Diffraction will be discussed in Chapters 3 and 4. However, in many situations, we are only interested in analyzing the optical beam near the center; then diffraction is not important. In those situations, a local optical beam with finite beam size can be approximated by a plane wave, as long as the beam size is much larger than the optical wavelength and the variation of the beam within a distance of wavelengths is very small.[5] *The plane waves could then be considered simply as optical rays. Furthermore, in the analysis of natural light, which has many wavelengths or frequency components, with no specific phase relation among the different components, the phase interference effects of optical light are not important. Only the location, intensity, and direction of the propagated beam are detectable and important.*

Let the beam propagates in a y–z plane at an angle θ with respect to the z axis. The y–z plane is located at a constant x position. Let the beam originate at $z = 0$. In the traditional ray analysis used in the literature, its position at variance z distance from the z axis, i.e its y position at z, is given by $r(z)$ and its direction is given by $r'(z)$ which is $\mathrm{d}r(y)/\mathrm{d}z$. Note that, for a ray making an angle θ with respect to the z axis, $r' = \sin\theta$. Then the ray can be represented at a given z by its ray matrix

$$\begin{vmatrix} r(z) \\ r'(z) \end{vmatrix} \tag{1.38}$$

Note that for a ray making an angle θ with respect to the z axis, $r' = \sin\theta$. When the beam reaches a new y position later at z', where $d = z' - z$, its $r(z')$ and $r'(z')$ at z' are related to $r(z)$ and $r'(z)$ by

$$r'(z') = r'(z), \qquad r(z') = r(z) + r'(z)d \tag{1.39}$$

In other words, the relation can be expressed by a ray matrix,

$$\begin{vmatrix} r(z') \\ r'(z') \end{vmatrix} = \begin{vmatrix} 1 & d \\ 0 & 1 \end{vmatrix} \begin{vmatrix} r(z) \\ r'(z) \end{vmatrix} \tag{1.40}$$

A ray in an arbitary direction in the xyz coordinate could be considered as a ray in the y'–z plane of a new $x'y'z'$ coordinate. Similar to Section 1.1.2 the $x'y'z'$ coordinate is a rotation of the xyz coordinate. The expressions for r and r' in $x'y'z'$ have been given in Eqs. (1.39) and (1.40). They could be expressed in terms of xyz through coordinate transformation as we have done in Section 1.1.2.

The ray representation is only an approximation. It ignores the size of the optical beam and the size of the medium in which the beam propagates. It ignores diffraction effects. It does not give the intensity of the beam unless it is specified separately. When the

[5] For example, the free space wavelength of visible light ranges from 0.4 to 0.7 μm. A uniform visible light beam a fraction of a millimeter wide can be approximated by a plane wave near the center of the beam. The approximation is good within short distances of propagation, such as a few centimeters or more.

polarization of the ray is important in some applications, it must be specified in addition to the ray matrix for r and r'.

1.2 Mirror reflection of plane waves

Reflection properties of optical light can be analyzed very simply by plane waves.

1.2.1 Plane waves polarized perpendicular to the plane of incidence

Let there be a plane wave, polarized in the x direction and propagating in the $\underline{\beta}$ direction in the y–z plane that makes an angle θ_i with respect to the z axis,

$$\underline{\beta} = \beta \sin\theta_i \underline{i_y} + \beta \cos\theta_i \underline{i_z}, \quad \underline{E_{i\perp}} = E_o e^{-j\beta \sin\theta_i y} e^{-j\beta \cos\theta_i z} e^{j\omega t} \underline{i_x} \tag{1.41}$$

In this case, the electric field is perpendicular to the plane of incidence, which is the y–z plane. Thus the electric and magnetic fields are designated by $\underline{E_\perp}$ and $\underline{H_\perp}$. The incident wave generates a reflected electric field plane wave,

$$\underline{\beta_r} = \beta \sin\theta_r \underline{i_y} + \beta \cos\theta_r \underline{i_z}, \quad \underline{E_{r\perp}} = E_r e^{-j\beta \sin\theta_r y} e^{+j\beta \cos\theta_r (z-z')} e^{j\omega t} \underline{i_x} \tag{1.42}$$

When this wave is incident on a ideal planar mirror (or an ideal conductor with infinite conductivity) at $z = z'$, extending from $x = -\infty$ to $+\infty$ and from $y = -\infty$ to $+\infty$, the boundary condition at $z = z'$ is that the total electric field tangential to the boundary, i.e. $\underline{E_{i\perp}} + \underline{E_{r\perp}}$, be zero at $z = z'$. The boundary condition at $z = z'$ demands that

$$\theta_r = \pi - \theta_i \quad \text{and} \quad E_r = -E_o e^{-j\beta \cos\theta_i z'} \tag{1.43}$$

or

$$\underline{E_{r\perp}} = -E_o e^{-j\beta \sin\theta_i y} e^{+j\beta \cos\theta_i (z-z')} e^{-j\beta \cos\theta_i z'} e^{j\omega t} \underline{i_x} \tag{1.44}$$

From Eqs. (1.18) and (1.20) of Section 1.1, the magnetic field for the incident wave is

$$\underline{H_{i\perp}} = \sqrt{\frac{\varepsilon}{\mu}} E_o (\cos\theta_i \underline{i_y} - \sin\theta_i \underline{i_z}) e^{-j\beta \sin\theta_i y} e^{-j\beta \cos\theta_i z} e^{-j\omega t} \tag{1.45}$$

The magnetic field for the reflected wave is

$$\underline{H_{r\perp}} = \sqrt{\frac{\varepsilon}{\mu}} E_o (\cos\theta_i \underline{i_y} + \sin\theta_i \underline{i_z}) e^{-j\beta \sin\theta_i y} e^{+j\beta \cos\theta_i (z-z')} e^{-j\beta \cos\theta_i z'} e^{-j\omega t} \tag{1.46}$$

The relation given in Eq. (1.43) is commonly known as the law of reflection. The reflection changes the direction of propagation from $\underline{\beta}$ to $\underline{\beta_r}$ which is a mirror reflection of $\underline{\beta}$. The polarizations of the incident and reflected electric field are the same, but the orientations of the incident and reflected magnetic field are different. The magnetic field will induce surface current in the conductor at $z = z'$. For ideal mirrors, the ratio $|E_r|/|E_o|$, called the reflectivity R of the mirror, is one. For actual mirrors with reflectivity $R < 1$, $E_r = -RE_o e^{-j\beta \cos\theta_i z'}$.

1.2.2 Plane waves polarized in the plane of incidence

The second independent solution for plane wave propagating in the same $\underline{\beta}$ direction has $\underline{H}$ directed along the x direction and $\underline{E}$ polarized in the y–z plane, which is the plane of incidence. The incident wave is designated as $\underline{E_{//}}$ and $\underline{H_{//}}$.

$$\underline{E_{i//}} = E_o[-\cos\theta_i \underline{i_y} + \sin\theta_i \underline{i_z}]\mathrm{e}^{-j\beta\sin\theta_i y}\mathrm{e}^{-j\beta\cos\theta_i z}\mathrm{e}^{j\omega t} \tag{1.47}$$

$$\underline{H_{i//}} = E_o\sqrt{\frac{\varepsilon}{\mu}}\mathrm{e}^{-j\beta\sin\theta_i y}\mathrm{e}^{-j\beta\cos\theta_i z}\mathrm{e}^{j\omega t}\underline{i_x} \tag{1.48}$$

The reflected plane wave is

$$\underline{E_{r//}} = \Gamma E_o[+\cos\theta_i \underline{i_y} + \sin\theta_i \underline{i_z}]\mathrm{e}^{-j\beta\sin\theta_i y}\mathrm{e}^{+j\beta\cos\theta_i(z-z')}\mathrm{e}^{-j\beta\cos\theta_i z'}\mathrm{e}^{j\omega t} \tag{1.49}$$

$$\underline{H_{r//}} = \Gamma E_o\sqrt{\frac{\varepsilon}{\mu}}e^{-j\beta\sin\vartheta_i y}e^{+j\beta\cos\vartheta_i(z-z')}e^{-j\beta\cos\vartheta_i z'}e^{j\omega t}\underline{i_x} \tag{1.50}$$

Note that only the y component of the total electric field is zero at $z = z'$ for ideal mirrors with $\Gamma = 1$.

1.2.3 Plane waves with arbitrary polarization

For plane waves with an electric field polarized in any other direction, it can always be decomposed into the summation of two mutually perpendicular polarized electric field plane wave components, one polarized perpendicular to the plane of incidence and one polarized in the plane of incidence. There is a change in the reflected electric and magnetic field from that of the incident field at the reflection boundary. According to Sections 1.2.1 and 1.2.2, the polarization of the reflected beam will depend on the decomposition. Results obtained in Eqs. (1.41) to (1.50) could be applied to any plane waves in any direction of propagation in any polarization by a change of the x-y-z coordinates to new x'-y-z' coordinates. In the new coordinates the direction of the incident beam is in the y'-z' plane.

1.2.4 The intensity

According to Eq. (1.28) of Section 1.1, the intensities of the incident and reflected waves along their directions of propagation are

$$I_i = \frac{1}{2}\sqrt{\frac{\varepsilon}{\mu}}|E_o|^2, \quad I_r = \frac{1}{2}\sqrt{\frac{\varepsilon}{\mu}}|E_r|^2 \tag{1.51}$$

1.2.5 Ray representation of reflection

The reflection at z' of a light beam could again be described by the ray matrix representation discussed in Section 1.1.5. In that case, the r and the r' of the incident and the reflected beams at z' in the plane of incidence are related by

$$\begin{vmatrix} r_r(z') \\ r'_r(z') \end{vmatrix} = \begin{vmatrix} 1 & 0 \\ 0 & -1 \end{vmatrix} \begin{vmatrix} r_i(z') \\ r'_i(z') \end{vmatrix} \tag{1.52}$$

The phase of the reflected beam and the polarizations of the beams are not included in the ray representation.

1.2.6 Reflection from a spherical mirror

Let there be a mirror that is a section of a sphere with radius R. Here, in this section, R is not the reflectivity of the mirror as it is commonly used in the literature. Figure 1.2 shows the cross-sectional view of a spherical mirror in the y–z plane and an incident beam. The spherical mirror is centered at the origin of the x-y-z coordinate. It is much larger than the size of the incident beam. Consider a small incident beam, uniform within a lateral area that is much larger than the optical wavelength. It is sufficiently wide so that it can be approximated by a plane wave near the center of the beam. Let the incident beam be represented approximately by a plane wave polarized in the x direction at an angle θ_i with respect to the z axis in the y–z plane.

$$\underline{\beta} = \beta \sin\theta_i \underline{i_y} + \beta \cos\theta_i \underline{i_z}, \quad \underline{E_{i\perp}} = E_o e^{+j\beta \sin\theta_i y} e^{-j\beta\cos\theta_i z} e^{j\omega t} \underline{i_x} \tag{1.53}$$

$$\underline{H_{i\perp}} = \sqrt{\frac{\varepsilon}{\mu}} E_o (\cos\theta_i \underline{i_y} - \sin\theta_i \underline{i_z}) e^{+j\beta\sin\theta_i y} e^{-j\beta\cos\theta_i z} e^{-j\omega t} \tag{1.54}$$

In Figure 1.2, θ_i is shown as a negative angle. The slope of the incident beam is $\tan\theta_i$. When the beam is incident on the mirror at z', the mirror at that location can be approximated by a planar mirror tangential to the sphere. This flat tangential mirror makes an angle φ with respect to the y axis. According to Sections 1.2.1 and 1.2.5, the reflected beam will make an angle $-\pi + 2\varphi + \theta_i$ with respect to the z axis. The slope of

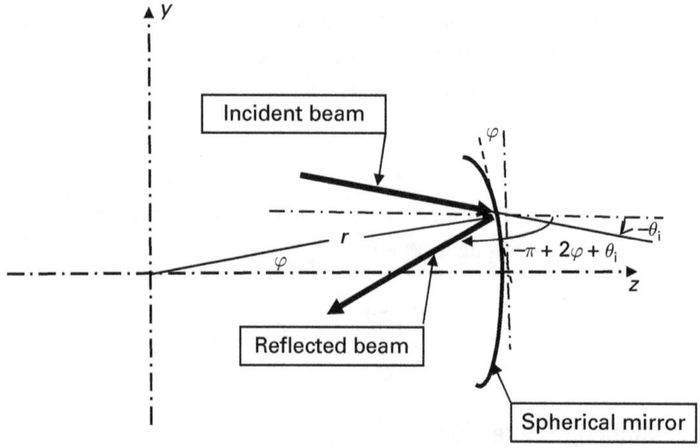

Figure 1.2 The cross-sectional view in the y–z plane for an optical beam reflected by a spherical mirror. The local incident beam at the incident angle $-\theta$ is reflected by the curved mirror. The plane tangential to the spherical mirror at the incident location makes an angle $-\varphi$ with respect to the vertical axis. The reflected beam makes an angle $-\pi + 2\varphi + \theta_i$ with respect to the $+z$ axis.

the reflected beam is $\tan(-\pi + 2\varphi + \vartheta_i) = \dfrac{\tan 2\varphi + \tan\theta_i}{1 + \tan 2\varphi \tan\theta_i} \cong -\dfrac{2r(z')}{R} + \tan\theta_i$. The same conclusion is obtained when the electric field is polarized in the y–z plane.[6] Therefore, the reflection from a spherical mirror in the ray representation is

$$\begin{vmatrix} r_r(z') \\ r'_r(z') \end{vmatrix} = \begin{vmatrix} 1 & 0 \\ -\dfrac{2}{R} & 1 \end{vmatrix} \begin{vmatrix} r_i(z') \\ r'_i(z') \end{vmatrix} \tag{1.55}$$

For incident beams parallel to the z axis, the reflected beams will be focused at $z = R/2$. Therefore, this location is called the focus of the spherical mirror.[7]

1.3 Refraction of plane waves

Refraction properties of optical radiation could also be derived directly by plane wave analysis.

The law of refraction is concerned with the reflection and the change of direction of propagation of optical light incident obliquely onto a planar boundary of two materials that have different dielectric permittivities, ε_1 and ε_2, or indices of refraction, n_1 and n_2. For the sake of simplicity, the media are assumed to be lossless in Sections 1.3.1 to 1.3.4.[8] Refraction is used in designing optical components ranging from eye glasses and cameras, to telescopes. Refraction and reflection of plane waves will be discussed first, followed by ray optical analysis and analysis of components such as prisms, lenses, and gratings.

1.3.1 Plane waves polarized perpendicular to the plane of incidence

Let there be an incident plane wave polarized in the x direction, perpendicular to the plane of incidence, and propagating in a direction in the y–z plane with an angle θ_i with respect to the z axis in media 1.

$$\underline{\beta} = \beta_1 \sin\theta_i \underline{i_y} + \beta_1 \cos\theta_i \underline{i_z}, \quad \underline{E_{i\perp}} = E_o e^{-j\beta_1 \sin\theta_i x} e^{-j\beta_1 \cos\theta_i z} e^{j\omega t} \underline{i_x} \tag{1.56}$$

$$\underline{H_{i\perp}} = \sqrt{\frac{\varepsilon_1}{\mu}} E_o (\cos\theta_i \underline{i_y} - \sin\theta_i \underline{i_z}) e^{-j\beta_1 \sin\theta_i y} e^{-j\beta_1 \cos\theta_i z} e^{-j\omega t} \tag{1.57}$$

Let there be a plane boundary at z', extending from $x = -\infty$ to $+\infty$ and from $y = -\infty$ to $+\infty$, with medium #1 at $z < z'$ and medium #2 at $z > z'$. The boundary separates medium #1 from medium #2. In addition to the transmitted wave in medium #2, there is a reflected wave in medium #1. The boundary condition at $z = z'$ is that the electric field $\underline{E}$ and the magnetic field $\underline{H}$ tangential to the boundary, i.e. E_x, E_y, H_x, and H_y, must be continuous across the boundary at z'. Since the incident wave is polarized in the x

[6] Since the mirror is curved, the locally reflected beam is no longer strictly a plane wave. The use of plane wave for local analysis is an approximation.

[7] The analysis presented here does not include rays at angles of incidence oblique to meridian planes. It is presented here only to demonstrate the very basic concept.

[8] In media with losses, β_1 and β_2 will be complex. Waves will be attenuated as they propagate. The matching of attenuated waves at the boundary becomes much more complex than the simple relationship presented here.

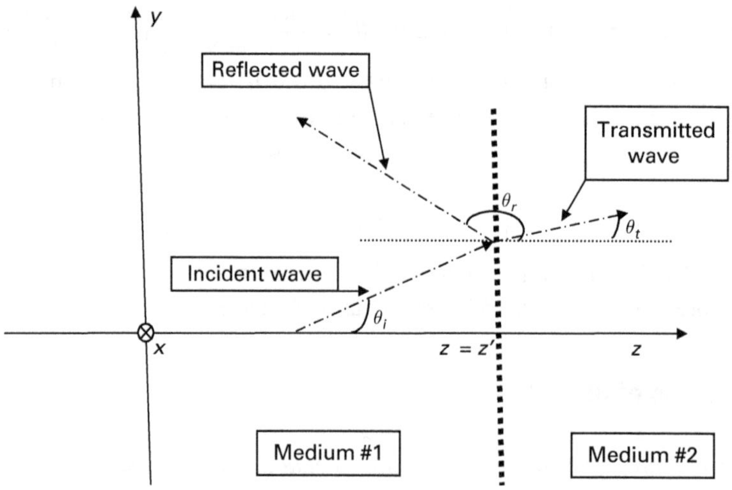

Figure 1.3 Reflection and transmission at a planar dielectric interface. The incident beam makes an angle θ with respect to the +z axis. The transmitted beam refracted from the vertical interface makes an angle θ_t. The reflected beam makes an angle θ_r.

direction, the reflected and transmitted wave must also be polarized in the x direction. The reflected wave in medium #1 is[9]

$$\underline{\beta_r} = \beta_1 \sin\theta_r \underline{i_y} + \beta_1 \cos\theta_r \underline{i_z}, \quad \underline{E_{r\perp}} = \Gamma_{\perp 12} E_o e^{-j\beta_1 \sin\theta_r y} e^{-j\beta_1 \cos\theta_r z} e^{j\omega t} \underline{i_x} \tag{1.58}$$

$$\underline{H_{r\perp}} = -\sqrt{\frac{\varepsilon_1}{\mu}} \Gamma_{\perp 12} E_o (\cos\theta_r \underline{i_y} + \sin\theta_r \underline{i_z}) e^{-j\beta_1 \sin\theta_r y} e^{-j\beta_1 \cos\theta_r z} e^{-j\omega t} \tag{1.59}$$

The transmitted plane wave in media #2 is

$$\underline{\beta_t} = \beta_2 \sin\theta_t \underline{i_y} + \beta_2 \cos\theta_t \underline{i_z}, \quad \underline{E_{t\perp}} = T_{\perp 12} E_o e^{-j\beta_2 \sin\theta_t y} e^{-j\beta_2 \cos\theta_t z} e^{j\omega t} \underline{i_x} \tag{1.60}$$

$$\underline{H_{t\perp}} = \sqrt{\frac{\varepsilon_2}{\mu}} T_{\perp 12} E_o \left(\cos\theta_t \underline{i_y} - \sin\theta_t \underline{i_z} \right) e^{-j\beta_2 \sin\theta_t y} e^{-j\beta_2 \cos\theta_t z} e^{j\omega t} \tag{1.61}$$

Figure 1.3 illustrates the incident wave, the reflected wave, and the transmitted wave in media #1 and #2, plus the boundary at $z = z'$. The continuity conditions of tangential $\underline{E}$ and $\underline{H}$ at $z = z'$ at all time t demand that

$$\theta_r = \pi - \theta_i, \quad \beta_2 \sin\theta_t = \beta_1 \sin\theta_i \quad \text{or} \quad n_2 \sin\theta_t = n_1 \sin\theta_i \tag{1.62}$$

[9] Note the notations. $\Gamma_{\perp 12}$ and $T_{\perp 12}$ stand for reflection and transmission coefficients of the electric field perpendicular to the plane of incidence from medium #1 to medium #2. The coefficients may be different when the polarization is changed or the direction of propagation is reversed.

$$T_{\perp 12} = 1 + \Gamma_{\perp 12}, \quad T_{\perp 12} = (1 - \Gamma_{\perp 12})\frac{n_1 \cos\theta_i}{n_2 \cos\theta_t} \tag{1.63}$$

or

$$\Gamma_{\perp 12} = \frac{n_1 \cos\theta_i - n_2 \cos\theta_t}{n_1 \cos\theta_i + n_2 \cos\theta_t} = -\frac{\sin(\theta_i - \theta_t)}{\sin(\theta_i + \theta_t)}, \quad T_{\perp 12} = \frac{2n_1 \cos\theta_i}{n_1 \cos\theta_i + n_2 \cos\theta_t} \tag{1.64}$$

In this case, the intensities of the incident, reflected, and transmitted waves in the z direction are:

$$I_{\perp i} = \sqrt{\frac{\varepsilon_1}{\mu}}|E_o|^2 \cos\theta_i, \quad I_{\perp r} = \sqrt{\frac{\varepsilon_1}{\mu}}|\Gamma_{\perp 12}E_o|^2 \cos\theta_r, \quad I_{\perp t} = \sqrt{\frac{\varepsilon_2}{\mu}}|T_{\perp 12}E_o|^2 \cos\theta_t \tag{1.65}$$

The intensities are conserved in the z direction, i.e.

$$I_{\perp i} = I_{\perp t} + I_{\perp r}$$

1.3.2 Plane waves polarized in the plane of incidence

The second independent solution of the plane wave propagating in the same $\underline{\beta}$ direction has the electric field polarized in the plane of incidence. Its $\underline{H}$ is directed along the x direction, and its $\underline{E}$ is polarized in the y–z plane.

$$\underline{E_{i//}} = E_o[-\cos\theta_i \underline{i_y} + \sin\theta_i \underline{i_z}]\mathrm{e}^{-j\beta_1 \sin\theta_{i1} y}\mathrm{e}^{-j\beta_1 \cos\theta_i z}\mathrm{e}^{j\omega t} \tag{1.66}$$

$$\underline{H_{i//}} = E_o\sqrt{\frac{\varepsilon_1}{\mu}}\mathrm{e}^{-j\beta_1 \sin\theta_i y}\mathrm{e}^{-j\beta_1 \cos\theta_i z}\mathrm{e}^{j\omega t}\underline{i_x} \tag{1.67}$$

The reflected wave in medium #1 and the transmitted wave in medium #2 are:

$$\underline{E_{r//}} = \Gamma_{//12}E_o[+\cos\theta_i \underline{i_y} + \sin\theta_i \underline{i_z}]\mathrm{e}^{-j\beta_1 \sin\theta_i y}\mathrm{e}^{+j\beta_1 \cos\theta_i (z-z')}\mathrm{e}^{-j\beta_1 \cos\theta_i z'}\mathrm{e}^{j\omega t} \tag{1.68}$$

$$\underline{H_{r//}} = +\Gamma_{//12}E_o\sqrt{\frac{\varepsilon_1}{\mu}}\mathrm{e}^{-j\beta_1 \sin\theta_{i1} y}\mathrm{e}^{+j\beta_1 \cos\theta_i (z-z')}\mathrm{e}^{-j\beta_1 \cos\theta_i z'}\mathrm{e}^{j\omega t}\underline{i_x} \tag{1.69}$$

$$\underline{E_{t//}} = T_{//12}E_o[-\cos\theta_t \underline{i_y} + \sin\theta_t \underline{i_z}]\mathrm{e}^{-j\beta_2 \sin\theta_t y}\mathrm{e}^{-j\beta_2 \cos\theta_t z}\mathrm{e}^{j\omega t} \tag{1.70}$$

$$\underline{H_{t//}} = T_{//12}E_o\sqrt{\frac{\varepsilon_2}{\mu}}\mathrm{e}^{-j\beta_2 \sin\theta_t y}\mathrm{e}^{-j\beta_2 \cos\theta_t z}\mathrm{e}^{j\omega t}\underline{i_x} \tag{1.71}$$

The boundary conditions at $z = z'$ requires:

$$\beta_2 \sin\theta_t = \beta_1 \sin\theta_i \quad \text{or} \quad n_2 \sin\theta_t = n_1 \sin\theta_i \tag{1.72}$$

$$(\Gamma_{//12} - 1)\cos\theta_i = -T_{//12}\cos\theta_t, \qquad (1 + \Gamma_{//12})n_1 = T_{//12}n_2 \tag{1.73}$$

In other words,

$$T_{//12} = \frac{2n_1 \cos\theta_i}{n_2 \cos\theta_i + n_1 \cos\theta_t}, \quad \Gamma_{//12} = \frac{n_2 \cos\theta_i - n_1 \cos\theta_t}{n_2 \cos\theta_i + n_1 \cos\theta_t} = \frac{\tan(\theta_i - \theta_t)}{\tan(\theta_i + \theta_t)} \quad (1.74)$$

The intensities in the z direction for the incident, transmitted, and reflected plane waves are:

$$I_{//i} = \sqrt{\frac{\varepsilon_1}{\mu}}|E_o|^2 \cos\theta_i, \quad I_{//r} = \sqrt{\frac{\varepsilon_1}{\mu}}|\Gamma_{//12}E_o|^2 \cos\theta_i, \quad I_{//t} = \sqrt{\frac{\varepsilon_2}{\mu}}|T_{//12}E_o|^2 \cos\theta_t \quad (1.75)$$

Again, the intensities in the z direction are conserved, i.e. $I_{//i} = I_{//r} + I_{//t}$.

It is important to note that all optical reflection, refraction, and diffraction effects are calculated based on meeting the boundary conditions by waves that satisfy Maxwell's equations. The energy in the waves is conserved. Note that the transmission and reflection coefficients of optical waves are dependent on the polarization of the electric field, while the intensity of the wave is not affected.

1.3.3 Properties of refracted and transmitted waves

(a) Transmission and reflection at different incident angles

It is interesting to note that, at normal incidence, $\theta_i = \theta_t = 0$ and $\theta_r = \pi$. T and Γ are the same for polarizations either perpendicular to the plane of incidence or in the plane of incidence. The direction of propagation of the transmitted wave is the same as the incident wave, while the reflected wave has a reverse direction of propagation. There is no change of polarization of the reflected and transmitted waves from the incident wave.

$$T_{12} = \frac{2n_1}{n_1 + n_2}, \quad \Gamma_{12} = \frac{n_2 - n_1}{n_2 + n_1} \quad (1.76)$$

$$I_i = I_t + I_r \quad (1.77)$$

It is important to realize the relative importance of this result in practical applications. At an interface of free space with $n_1 = 1$ and glass with $n_2 = 1.5$, $T_{12} = 0.8$ and $\Gamma_{12} = 0.2$, which is small. Therefore in many applications of glass components, such as imaging through a lens, the reflection may not be analyzed. The situation is very different when medium #2 has a large index of refraction such as a III–V semiconductor. If $n_2 = 3.5$, then $T_{12} = 0.56$ and $\Gamma_{12} = 0.44$ at normal incidence.

At other angles of incidence, Γ and T will vary dependent on the angle of incidence and the polarization. The magnitude of reflection increases at large θ_i. It is interesting to note that when $\theta_i + \theta_t = \pi/2$, $\Gamma_{//12} = 0$ in Eq. (1.74). The θ_i that satisfies this condition is traditionally known as the Brewster angle. At this angle the incident and the reflected plane waves are polarized perpendicular to each other in the plane of incidence. The Brewster angle has many practical applications because at this angle the reflection is zero without any anti-reflection coating.

(b) Total internal reflection

When $n_1 > n_2$, at the angle of incidence θ_i, such that $n_1 \sin\theta_i = n_2$, $\theta_t = \pi/2$, $\Gamma = 1$ and $I_t = 0$. This means that, for a plane wave with any polarization there is no energy transmitted in the z direction. For $\theta_i > \sin^{-1} n_2/n_1$ and E_i polarized perpendicular to the plane of incidence, the boundary condition in Eq. (1.62) demands that $\cos\theta_t = j\sqrt{(n_1^2 \sin^2\theta_1/n_2^2) - 1}$. Therefore, we have an evanescent wave in medium 2 in which the propagation constant in the z direction of the transmitted wave shown in Eq. (1.60) is imaginary. The reflection coefficient is $\Gamma_{12} = |\Gamma_{12}|e^{j\phi_{12}}$. From Eq. (1.64), we obtain $|\Gamma_{\perp 12}| = 1$ and

$$\varphi_{\perp 12} = \tan^{-1}\frac{-2n_1\cos\theta_i\sqrt{n_1{}^2\sin^2\theta_i - n_2{}^2}}{n_1{}^2\cos^2\theta_i - (n_1{}^2\sin^2\theta_i - n_2{}^2)}$$

From Eq. (1.74), a similar conclusion can be reached for plane waves polarized in the plane of incidence. Again, $|\Gamma_{//12}| = 1$. However, the phase angle is different from $\phi_{\perp 12}$.

$$\varphi_{//12} = \tan^{-1}\frac{-2n_2\cos\theta_i\sqrt{n_2{}^2\sin^2\theta_i - n_1{}^2}}{n_2{}^2\cos^2\theta_i - (n_2{}^2\sin^2\theta_i - n_1{}^2)}$$

In summary, the incident plane wave with any polarization is said to be totally internally reflected at the boundary for $\theta_i > \sin^{-1} n_2/n_1$, but the phase angle is dependent on polarization. Total internally reflected waves have only an evanescent tail in the lower index medium. Total internal reflection is utilized extensively in optical fibers and waveguides to minimize the loss due to the surroundings, by using a cladding layer that has a lower index of refraction so that losses in the surrounding media at distances further away from the interface than the length of the evanescent tail do not cause much propagation loss to the totally internal reflected optical wave.

(c) Refraction and reflection of arbitrary polarized waves

For plane waves with arbitrary polarization, results derived in Eqs. (1.56) to (1.75) are applicable when the electric field is first decomposed into two components, one polarized perpendicular to the plane of incidence and the second polarized in the plane of incidence. Although these two components have the same direction of propagation of reflected and transmitted waves (see Eqs. (1.62) and (1.72)), their polarization, transmission coefficient T, and reflection coefficient Γ are different.

(d) Ray representation of refraction

It was shown in Sections 1.1.6 and 1.2.5 that natural light with finite beam width and location can be represented by its ray matrix. There is also a matrix representation of the refracted (i.e. transmitted) beam as follows

$$\begin{vmatrix} r(z') \\ r'(z') \end{vmatrix} = \begin{vmatrix} 1 & 0 \\ 0 & \dfrac{n_1}{n_2} \end{vmatrix} \begin{vmatrix} r_{in}(z') \\ r_{in}'(z') \end{vmatrix} = \begin{vmatrix} 1 & 0 \\ 0 & \dfrac{\sin\theta_2}{\sin\theta_1} \end{vmatrix} \begin{vmatrix} r_{in}(z') \\ r_{in}'(z') \end{vmatrix} \quad (1.78)$$

Note that the ray matrix representation is independent of polarization. It does not tell us the magnitude, the size, or the polarization of the refracted beam. The reflected beam is not included in Eq. (1.78). The ray matrix representation of the reflected beam for a mirror is given in Eq. (1.52) in Section 1.2.5.

1.3.4 Refraction and dispersion in prisms

Prisms are optical components used to redirect the direction of propagation of an optical beam. Note that the permittivity ε and the index of refraction n_2 of materials are usually wavelength (i.e. ω or λ) dependent. This means that the transmitted plane wave at the boundary will have different directions of propagation at different wavelengths. This is known commonly as the dispersion. In a prism spectrometer, the incoming optical beam may have many wavelength components. The collimated incident beam passes through a prism. At the exit of the prism, different wavelength components propagate in different directions due to the dispersion effect. The exit beams in different directions are focused by a lens to different positions. An exit slit located on the focal plane of the lens selects the radiation in a specific wavelength range to be detected. The spectral width of the detected radiation is determined by the width of the slit. As the prism rotates, the detected radiation displays the spectral component of the incident radiation as a function of the prism angle.

(a) Plane wave analysis of prisms

A prism is usually a dielectric cylinder with a triangular cross-section made from material with a refractive index n_2. This index n_2 is larger than the index of the surrounding medium, which has index n_1. Usually the surrounding medium is free space with $n_1 = 1$. The triangular cross section of a prism in the y–z plane is shown in Figure 1.4. The prism is uniform in the x direction. It has a vertex angle, $A + B$, and a base angle, $(\pi/2 - A)$, for the front surface, and base angle, $(\pi/2 - B)$, for the back surface. The dimensions of the surfaces of the triangle are larger than the width of the optical beam, which is much larger than the optical wavelength itself.

Let there be an optical incident beam propagating in a direction θ_i from the z axis in the y–z plane. For uniform beams that have a beam width much larger than the optical wavelength, the beams can be represented by plane waves near the center of the beam. The incident beam, the refracted beam in the prism, and the transmitted beam of the prism are also illustrated in Figure 1.4. The analysis of the wave propagation in prisms is simply a detailed analysis of the directions of the refracted beams at each dielectric interface, as follows.

In order to analyze the beam propagation, let us designate x'-y'-z' coordinates and x''-y''-z'' coordinates, as shown in Figure 1.4. The y' axis is parallel to the front prism surface

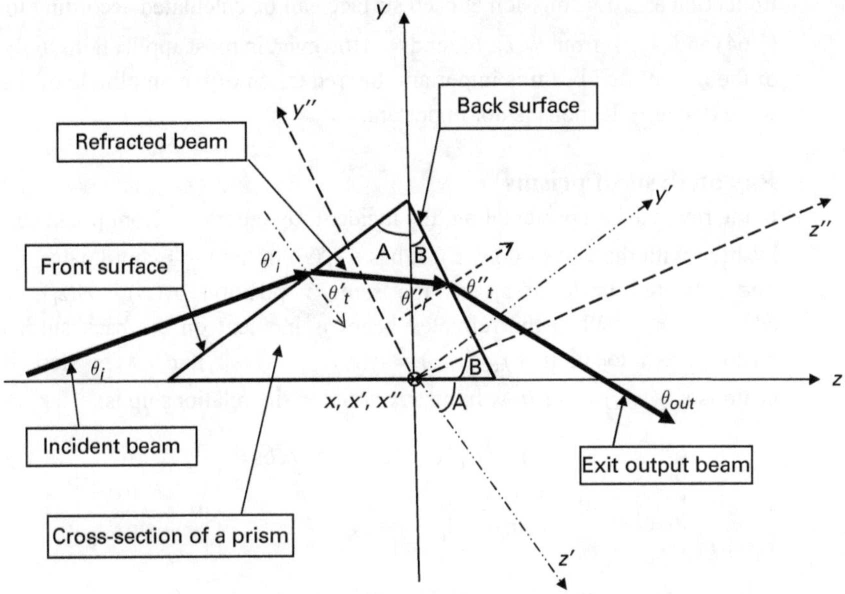

Figure 1.4 Incident, refracted, and transmitted wave in a prism. The prism has a vertex angle $A + B$. The incident beam angle is θ_i. It is refracted by the front prism surface. The transmitted beam from the front surface makes an angle θ'_t with respect to the vertical axis of the front prism surface. It is refracted again by the back prism surface. The output beam angle is θ_{out}.

and the z' axis is perpendicular to the front surface. The y'' axis is parallel to the back prism surface and the z'' axis is perpendicular to the back surface. The x'-y'-z' and x''-y''-z'' coordinates are related to the x-y-z coordinates by:

$$\underline{i}_{y'} = \cos A\underline{i}_y + \sin A\underline{i}_z, \quad \underline{i}_{z'} = -\sin A\underline{i}_y + \cos A\underline{i}_z \tag{1.79}$$

$$\underline{i}_{y''} = \cos B\underline{i}_y - \sin B\underline{i}_{z'} \quad \underline{i}_{z''} = +\sin B\underline{i}_y + \cos B\underline{i}_z \tag{1.80}$$

In Figure 1.4, the incident beam in material #1 is directed in the θ_i' direction in the y'–z' plane, where $\theta_i' = \theta_i + A$. The refracted beam from the front surface is directed in the $\underline{\theta'_t}$ direction in the y'–z' plane. According to Eqs. (1.62) and (1.72), $\theta_t' = \sin^{-1}(n_1 \sin\theta_i'/n_2)$. The angle θ_t that this beam makes with respect to the z axis in the x–y plane is $\theta_t = A - \theta_t'$. In the x''-y''-z'' coordinates, the refracted beam at the angle θ_t in the x-y-z coordinates makes an angle θ_i'' with respect to the z'' axis, where $\theta_i'' = \theta_t + B = A - \theta_t' + B$. Its exit beam in medium #1 makes an angle θ_t'' with respect to the z'' axis, where $\theta_t'' = \sin^{-1}(n_2 \sin\theta''/n_1)$. In the x-y-z coordinates, this exit beam makes an angle θ_{out} with respect to the z axis, $\theta_{\text{out}} = B - \theta_t''$. This analysis of beam direction is independent of polarization.

There are also reflected beams at each surface. Reflections need to be considered whenever the difference of refractive indices at the interface of the prism is large.

Reflection and transmission at each surface can be calculated according to Eqs. (1.63), (1.64) and (1.74) from θ_i', θ_t', θ_i'', and θ_t''. However, in most applications, only the analysis of the transmitted beam is important, the reduction of the amplitude of the transmitted wave due to reflections is not important.

(b) Ray analysis of prisms

In the ray matrix representation, the incident ray enters the front prism surface at the z_i location with the y position $r_i(z_i)$. It has a $r_i'(z_i)$ in the x-y-z coordinates, $r_i'(z_i) = \sin\theta_i$. The refracted beam at z_i has the same y position, $r_t(z_i) = r_i(z_i)$, and a slope, $r_t'(z_i) = \sin\theta_t$. When the refracted beam is incident on the back surface at the z_{out} position, its y location is $r_{\text{out}}(z_{\text{out}}) = r_i(z_i) + (z_{\text{out}} - z_i)\tan\theta_t$. The slope of the output beam is $r_{\text{out}}'(y_{\text{out}}) = \sin\theta_{\text{out}}$. In matrix notation the relationship is:

$$\begin{vmatrix} r_{\text{out}} \\ r_{\text{out}}' \end{vmatrix} = \begin{vmatrix} 1 & 0 \\ 0 & \dfrac{\sin\theta_{\text{out}}}{\sin\theta_t} \end{vmatrix} \begin{vmatrix} 1 & (z_{\text{out}} - z_i)/\cos\theta_t \\ 0 & 1 \end{vmatrix} \begin{vmatrix} 1 & 0 \\ 0 & \dfrac{\sin\theta_t}{\sin\theta_i} \end{vmatrix} \begin{vmatrix} r_i \\ r_i' \end{vmatrix}. \tag{1.81}$$

In the case of thin prisms, $A+B$ is small, and $z_{\text{out}} - z_i \approx 0$.

$$\begin{vmatrix} r_{\text{out}} \\ r_{\text{out}}' \end{vmatrix} = \begin{vmatrix} 1 & 0 \\ 0 & \dfrac{\sin\theta_{\text{out}}}{\sin\theta_i} \end{vmatrix} \begin{vmatrix} r_i \\ r_i' \end{vmatrix} \tag{1.82}$$

In other words, a thin prism does not change the position of the beam. It only changes its direction. Furthermore, for a small incident angles θ_i,

$$\theta_{\text{out}} = \theta_i - \frac{n_2 - n_1}{n_1}(A + B). \tag{1.83}$$

Note that the reflected beams are not included in the ray representation above. The magnitude and polarization of the beams are also not included in the ray representation. These quantities may not be important in applications that use natural light in components that have low value of n_2. *For applications, such as image formation, the ray representation is a simple method for analyzing the direction, position, and propagation distance of the beam that are most important.*

(c) Thin prism represented as a transparent layer with a varying index

It is interesting to view this result from another viewpoint. In a thin prism we could also consider the prism as a dielectric layer that has an n_2 layer with thickness τ embedded in a medium with index n_1. The thickness τ varies at different position y. From Figure 1.4, we obtain $\tau = (y_{\text{vert}} - y)(\tan A + \tan B) \cong (y_{\text{vert}} - y)(A + B)$. Here, y_{vert} is the vertex of the prism. Let there be a plane wave propagating in the z direction in a medium that has index n_1. The beam is centered at y_i. After transmitting through this composite dielectric layer, the electric field for this beam is

$$E_{out} = E_o e^{-jn_1\beta_o z} e^{-j(n_2-n_1)\beta_o \tau} e^{j\omega t} = E_o e^{-jn_1\beta_o z} e^{-j(n_2-n_1)\beta_o y_{vert}} e^{-jn_1\beta_o \sin\phi_{out} y} e^{j\omega t}, \quad (1.84)$$

where $n_1 \sin\varphi_{out} = -(n_2 - n_1)(A + B)$.

Any plane wave that has an $\exp(-jn_1\beta_o \sin\varphi_{out} y)$ variation in y is a plane wave propagating at an angle φ_{out} in the y–z plane. This φ_{out} agrees with the θ_{out} given in Eq. (1.83) above. In other words, we have just introduced an important new concept. Transmission through a thin prism could also be represented by transmission of a plane wave through a medium with a phase transmission coefficient that is a linear function of y,

$$t = t_o e^{-j(n_2-n_1)\beta_o (y_{vert}-y)(\tan A + \tan B)} \quad (1.85)$$

$E_{out} = tE_{in}$. Note that the results in Eqs. (1.84) and (1.85) are independent of polarization.

In other words, when a plane wave is transmitted through a refractive medium with variable refractive index given in Eq. (1.85), it produces an output beam in a different direction of propagation. The conclusion is also valid for a small incidence angle θ_i. Conversely, any transmission medium with a phase transmission coefficient that has a linear y variation will tilt the incident beam to a new direction of propagation like a prism.

1.3.5 Refraction in a lens

A lens is probably the most commonly used optical component. It is used principally for imaging and instrumentation. Ray analysis is the principle tool used for lens design. The design of a compound lens is very complex. A detailed discussion on ray analysis of lens design is beyond the scope of this book. However, an analysis of a simple spherical lens for meridian rays will be beneficial to illustrate the basic principle of a lens.[10] It will be presented first by ray analysis, then as a transparent medium with a quadratic varying phase in transmission.

(a) Ray analysis of a thin lens

Let us consider a simple spherical lens whose geometrical configuration is shown in Figure 1.5. The right surface of the lens is described by

$$x''^2 + y''^2 + z''^2 = r_1^2 \quad (1.86)$$

The left surface of the lens is described by

$$x'^2 + y'^2 + (z' - z_1)^2 = r_2^2 \quad (1.87)$$

[10] Like prism analysis, reflection exists at any dielectric interface. There are reductions of the amplitude of the transmitted wave as it propagates through the lens. Reflections in lenses are analyzed when it is necessary. Thus only ray analysis of the transmitted beam will be presented here.

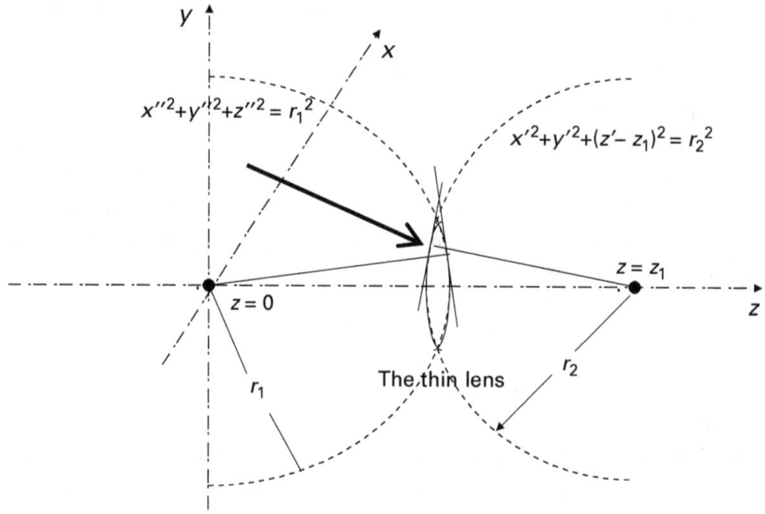

Figure 1.5 The geometrical configuration of a spherical lens. Two spherical surfaces centered about $z = z_1$ and $z = 0$ are shown. The front (left) and back (right) surfaces are made from the interception section of these two spherical surfaces.

The origin of the spherical surfaces are at $z = 0$ and $z = z_1$. The refractive index of the lens is n_2. It is placed in a medium that has refractive index n_1. In free space, $n_1 = 1$. The thin lens and the dotted spheres shown in Figure 1.5 are the cross-sectional view of the lens and the spheres in the y–z plane.

In order to demonstrate the properties of a lens with simple ray analysis, let us consider an optical beam incident on the lens in the y–z plane at an angle $-\theta_i$ with respect to the z axis. This beam is incident on the lens at the z' and y' positions. The refracted beam is transmitted through the lens and excites an output beam. In the thin lens approximation, the y' positions of the beam at the front and the back surfaces of the lens are the same. At the y' position of the front surface, the curved spherical lens surface can be approximated locally by a plane tangential to the sphere centered at z_1. This plane makes an angle A with respect to the Y axis. Similarly, at the back surface of the lens, the curved surface can be approximated by a plane tangential to the sphere centered at $z = 0$. This plane makes an angle B with respect to the y axis. Thus the change of direction of the beam going through the lens at this location is approximately the same as a beam going through a prism with the vertex angle, $A+B$. From Section 1.3.4, we obtain

$$\theta_{\text{out}} = -\theta_i - \frac{n_2 - n_1}{n_1}(A + B) = -\theta_i - \frac{n_2 - n_1}{n_1}\left(\frac{1}{r_1} + \frac{1}{r_2}\right)y \tag{1.88}$$

If we designate

$$\frac{1}{f} = \frac{n_2 - n_1}{n_1}\left(\frac{1}{r_1} + \frac{1}{r_2}\right) \tag{1.89}$$

then the refraction of a beam through a thin spherical lens placed at z' can be expressed in a ray optical representation as

$$\begin{vmatrix} r_{\text{out}}(z') \\ r'_{\text{out}}(z') \end{vmatrix} = \begin{vmatrix} 1 & 0 \\ -\dfrac{1}{f} & 1 \end{vmatrix} \begin{vmatrix} r_i(z') \\ r'_i(z') \end{vmatrix} \tag{1.90}$$

When the incident beam is parallel to the z axis, $\theta_i = 0$. Parallel beams at different y positions would all focus on the z axis at the position $z' + f$. Therefore $z = f$ is commonly known as the focal length of the lens. For parallel beams incident at small θ_i, θ_{out} will still be related to θ_i by Eq. (1.88). Thus they will be focused to a point at $z = z' + f$ and $y = \theta_i f$. The plane at $z = z' + f$ is known as the focal plane of the lens. The preceding analysis has only been carried out for optical beams incident in the y–z plane. However, in a cylindrically symmetric configuration, the x and y axes can be rotated about the z axis. Thus the results can be generalized to three dimensions for any beam incident in the meridian plane.

The objective of the analysis presented here is only to demonstrate simple lens properties by plane wave analysis. The analysis of practical lenses is much more complex than the preceding discussion. It involves oblique rays, skewed rays, astigmatism, etc. and is beyond the scope of this book [5].

(b) Thin lens represented as a transparency with varying index

Similar to the discussion in Section 1.3.4 (c) for the prism, it is instructive to represent a thin lens as a transparent planar medium with a varying phase change. Consider an incident plane wave that has a beam size small compared to the size of the lens. It propagates in the direction of z axis.

$$E_i = E_o e^{-j\beta_1 z} e^{j\omega t}$$

Let us consider this small beam in the y–z plane near $x = 0$. At the transverse position $(x = 0, y)$, it passes through the lens beginning at $z = z_1 - r_2$ and ending at $z = r_1$. Its phase at the output will depend on y because the ray goes through a higher index region with thickness, $z'' - z'$, at $y = y' = y''$. The change in its phase, in comparison to a beam in free space without the lens, is:

$$\begin{aligned} \Delta\phi &= -\beta_1(n_2 - n_1)(z'' - z') \\ &= -\beta_1(n_2 - n_1)\left(r_1 \left\{ 1 - \frac{y^2}{r_1^{\,2}} \right\}^{1/2} - z_1 + r_2 \left\{ 1 - \frac{y^2}{r_2^{\,2}} \right\}^{1/2} \right) \end{aligned} \tag{1.91}$$

Here, $z'' > z'$ and x and $y << r_1$ and r_2 inside the lens. Binomial expansion can be used again for the terms in the curly brackets. When the first-order approximation is used for a thin lens, we obtain

$$\Delta\phi = -\beta(n_2 - n_1)\left[r_1 + r_2 - z_1 - \frac{y^2}{r_1} - \frac{y^2}{r_2} \right]. \tag{1.92}$$

The focal length of a thin spherical lens is given in Eq. (1.89) as $1/f = (n_2 - n_1)(1/r_1 + 1/r_2)$. Thus, for any wave passing through a thin lens near $x = 0$, we can now multiply the incident wave on the lens by a phase function,

$$t_l = \mathrm{e}^{-j\beta_1(n_2-n_1)(r_1+r_2-z_1)}\mathrm{e}^{j\frac{\beta_1}{2f}(y^2)} \tag{1.93}$$

to obtain the wave that has passed through the lens. The preceding result can be extended to incident rays in any meridian plane by rotating the x and y axes with respect to the z axis. Therefore the general result in the x-y-z coordinates for a plane wave incident on the lens in the z direction is

$$t = \mathrm{e}^{j\frac{\beta_1}{2f}(x^2+y^2)} \tag{1.94}$$

$$E_{\text{out}} = tE_o\mathrm{e}^{-j\beta_1 r_1}\mathrm{e}^{j\omega t} \tag{1.95}$$

The output electric field at the back side of the lens is

$$E_{\text{out}} = E_o\mathrm{e}^{-j\beta_1(r_1+f)}\left[\mathrm{e}^{j\beta_1 f}\mathrm{e}^{j\frac{\beta_1}{2f}(x^2+y^2)}\mathrm{e}^{j\omega t}\right] \tag{1.96}$$

The quantity in the brackets represents a spherical wave, $\mathrm{e}^{j\beta_1\sqrt{f^2+x^2+y^2}}\mathrm{e}^{j\omega t}$, in the form of the first term of a binomial expansion at $z = r_1$. The spherical wave is focused on to the location $z = r_1 + f$.

This is a very simple result that can be applied to any incident wave passing perpendicularly through a lens. It should be emphasized that this is a thin lens approximation. Only an ideal lens can be represented by Eq. (1.94). A practical lens will have other higher-order phase shifts, which are considered as distortions from an ideal lens. Note that the output after leaving the lens is no longer a plane wave. Fourier analysis discussed in Section 1.1.5 (c) must be used to find its plane wave components.

The importance of this representation is to recognize that whenever a medium has a quadratic phase variation in transmission, it functions as a lens.

1.4 Geometrical relations in image formation

Image formation is one of the most important applications in optics. It has been presented extensively in traditional optical literatures. It is also a very specialized topic. The geometrical relation between an object and its image is presented here only to demonstrate the basic relation between ray analysis and image formation.

Consider a point optical source placed at $x = 0$ and $z = -p$ at the position $y = h_{\text{ob}}$, a thin lens with focal length f is placed at $z = 0$, centered at $x = 0$ and $y = 0$ and perpendicular to the z axis. Figure 1.6 illustrates the configuration. From the discussion in Section 1.1.5 (c), we can consider that the point source yields a summation of plane wave component beams in different directions. Let us consider two incident component rays. (a) A component ray that propagates parallel to the z axis. According to discussion in the previous section, this

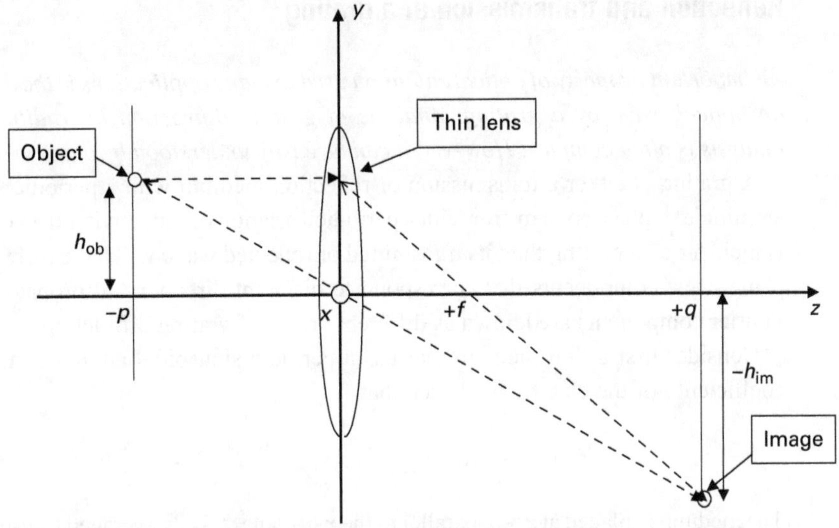

Figure 1.6 Illustration of the geometrical relations in imaging. The object h_{ob} long is placed at $-p$. The lens with focal length f is placed at 0. The image h_{im} long appears at q.

ray will be redirected after the lens. It passes through the focus of the lens at $z = f$. (b) A component ray that propagates toward the center of the lens at $y = 0$. This ray is not redirected in its direction of propagation because it passes locally through two parallel dielectric interfaces with negligible separation at $y = 0$. The two rays meet at the image point. If ray analysis is carried out for rays in other directions of propagation, they will also meet at the same image point. In other words, the optical light from the object point source is refocused by the lens to the image point. The relation between the positions of the object and the image is determined geometrically.

From Figure 1.6, it is clear that

$$\frac{h_{ob}}{f} = \frac{h_{im}}{q - f} \text{ for ray } (a), \qquad \frac{h_{ob}}{p} = \frac{h_{im}}{q} \text{ for ray } (b) \tag{1.97}$$

or,

$$\frac{1}{p} + \frac{1}{q} = \frac{1}{f}, \qquad \frac{h_{im}}{h_{ob}} = \frac{q}{p} \tag{1.98}$$

Any extended object at $z = -p$ can be represented by the summation of point objects at different h. Therefore the magnification ratio of the image to the extended object is q/p. When $p = \infty$, $q = f$, and $h_{im} = 0$. Thus the object is focused by the lens to $z = f$. Conversely, when $p = f$, $q = \infty$. A point source is collimated by the lens to a parallel beam.

Eqs. (1.97) and (1.98) represent the geometrical relations between an object and its image.

1.5 Reflection and transmission at a grating

An important property of optical radiation used in many applications is the diffraction of an optical wave by a grating. Analysis of grating diffraction in traditional optical analysis is often complex. However, it can be easily understood by plane wave analysis.

A grating is either a transmission or reflection medium with a periodic variation of amplitude or phase, or a mirror with a periodic variation of reflectivity. If an optical wave is incident on a grating then its transmitted or reflected wave will have different Fourier plane wave components that correspond to different directions of propagation. These Fourier components are known as different orders of grating diffraction.

Consider first a thin medium that has a periodic sinusoidal amplitude transmission coefficient t of the electric field such that

$$t = t_o(1 + \Delta t \cos 2\pi f_g y) = t_o + t_o \frac{\Delta t}{2} e^{j2\pi f_g y} + t_o \frac{\Delta t}{2} e^{-j2\pi f_g y} \tag{1.99}$$

The medium is placed at $z = 0$, parallel to the x–y plane. t_o is its averaged transmission, and Δt is the magnitude of the periodic variation in the y direction. $\Delta t \leq 1$, so t is always positive. The minimum transmission is $t_o - \Delta t$; the maximum transmission is $t_o + \Delta t$. The periodic variation has a grating period T_g per unit length in the y direction where $T_g = 1/f_g$

Let there be an incident plane wave polarized in the x direction and propagating at an angle θ with respect to the z axis.

$$\underline{E_{ix}} = E_x \underline{i_x} = E_o e^{-j\beta \cos\theta z} e^{-j\beta \sin\theta y} e^{j\omega t} \underline{i_x}, \quad \underline{H_{iy}} = H_y \underline{i_y} = \sqrt{\frac{\varepsilon}{\mu}} E_o e^{-j\beta \cos\theta z} e^{-j\beta \sin\theta y} e^{-j\omega t} \underline{i_y} \tag{1.100}$$

The output plane wave at $z > 0$ after the grating is

$$\underline{E_{ox}} = E_{ox} \underline{i_x}, E_{ox} = E_o t_o \left[e^{-j\beta \sin\theta y} + \frac{\Delta t}{2} e^{-j\beta(\sin\theta - 2\pi f_g/\beta) y} + \frac{\Delta t}{2} e^{-j\beta(\sin\theta + 2\pi f_g/\beta) y} \right] e^{-j\beta \cos\theta z} e^{j\omega t} \tag{1.101}$$

$$\underline{H_{oy}} = H_{oy} \underline{i_y}, \quad H_{oy} = \sqrt{\frac{\varepsilon}{\mu}} E_{ox} \tag{1.102}$$

The output wave has three components, a plane wave propagating in the incident direction, a plane wave propagating at an angle θ^{+1}, called the +1 order diffracted wave where $\theta^{+1} = \sin^{-1}(\sin\theta + 2\pi f_g/\beta)$, and a plane wave propagating at an angle θ^{-1}, called the –1 order diffracted wave, where $\theta^{-1} = \sin^{-1}(\sin\theta - 2\pi f_g/\beta)$. Note that θ^{-1} and θ^{+1} of any propagating diffracted wave must be less than $\pm\pi/2$, otherwise that order of the diffracted wave is an evanescent wave. When the diffracted wave for a specific order is evanescent, we say that the grating is cut off for that order. A similar result is obtained for a y polarized incident wave. If t depends on polarization, the magnitude of the diffracted wave will be polarization dependent. However, the diffraction angles will not be polarization dependent.

The sinusoidal grating transmission function in Eq. (1.99) is used here because it is simple to analyze. When the periodic transmission function t has a non-sinusoidal periodic variation, it can be expressed as a Fourier series with periodicity T_g. For example, for a grating with an on–off periodic variation of Δt, Δt can be expressed as a repetition of individual on–off sections.

$$\Delta t(y) = \Delta t_o \sum_m \text{rect}\left(\frac{mT_g - y}{\delta/2}\right) = \sum_n \Delta t_n \cos(2\pi n f_g y) \tag{1.103}$$

rect(τ) is defined as

$$\text{rect}(\tau) = 1 \quad \text{for} \quad |\tau| \leq 1 \quad \text{and} \quad \text{rect}(\tau) = 0 \quad \text{for} \quad |\tau| > 1 \tag{1.104}$$

Here, δ is the width of individual on-section, $\delta < T_g$. $T_g - \delta$ is the width of individual off-section. $\Delta t(y)$ can also be expressed by its Fourier series. Each Fourier component has a magnitude Δt_k. For the kth Fourier component,

$$\Delta t_k = 2\int_0^{T_g} \Delta t(y)\cos(2\pi k f_g y)\text{d}y \tag{1.105}$$

Each $\pm n$ Fourier component has an angle of diffraction, $\theta^{\pm n} = \sin^{-1}(\sin\theta \pm 2n\pi f_g/\beta)$ They are the $\pm n$th order of diffracted waves. Only those with $|\theta^{\pm n}| < \pi/2$ are propagating waves.

Plane wave analysis provides a simple way to understand grating diffraction. Note that the direction of the nth-*order diffracted wave is dependent on* β, *which is proportional to the optical frequency* ω, *or wavelength. This is known as the dispersion of the grating diffraction. Different orders of diffraction will have different angles for the diffracted beam. Some orders of diffraction may be cut off.*

1.6 Pulse propagation of plane waves

When the amplitude of the plane wave is time dependent, the wave is a pulse. Let there be a plane wave pulse in the z direction, polarized in the x direction,

$$\underline{E_x} = E_x \underline{i_x} = A(t)e^{-j\beta(\omega_o)z}e^{j\omega_o t}\underline{i_x} \tag{1.106}$$

Here, in order to emphasize the dispersion effect, we have written β as $\beta(\omega)$. At $z = 0$,

$$E_x = A(t)e^{j2\pi f_o t} \tag{1.107}$$

A(t) can be represented by its Fourier transform pairs,

$$F_A(f) = \int_{-\infty}^{+\infty} A(t)e^{-j2\pi ft}\text{d}t, \quad A(t) = \int_{-\infty}^{+\infty} F_A(f)e^{+j2\pi ft}\text{d}f \tag{1.108}$$

Here F_A is the component of A at frequency f. Therefore, at $z = 0$,

$$E_x(z=0) = \int_{-\infty}^{+\infty} F_A(f) e^{+j2\pi(f_o - f)t} df \quad (1.109)$$

It is a sum of plane waves at different frequencies $f_o - f$. Each component at $f_o - f$ propagates with a different β to the position z. Therefore, at z

$$E_x(z) = \int_{-\infty}^{+\infty} F_A(f) e^{-j\beta(f_{oo} - f)z} e^{-j2\pi(f_o - f)t} df \quad (1.110)$$

Usually, $f_o - f << f_o$. Therefore, $\beta(f)$ can be represented by its Taylor's series,

$$\beta(f) = \beta(f_o) + \left[\frac{\partial \beta}{\partial f}(f_o)\right](-f) + \left[\frac{1}{2}\frac{\partial^2 \beta}{\partial f^2}(f_o)\right](-f)^2 + \cdots \quad (1.111)$$

When the second- and higher-order terms can be neglected, we obtain

$$E_x = \left[\int_{-\infty}^{+\infty} F_A(f) e^{j2\pi\left(\frac{\partial \beta}{\partial \omega}\right)\big|_{f_o} z} e^{+j2\pi f t} df\right] e^{-j\beta(f_o)z} e^{j\pi f_o t} \quad (1.112)$$

where

$$v_g = \partial \omega / \partial \beta|_{f_o} \quad (1.113)$$

is known as the group velocity.

In a realistic situation, pulse distortion is important if the distance of propagation is very long and the pulse duration of $A(t)$ is short. If the group velocity is independent of f, the quantity $e^{j2\pi\left(\frac{\partial \beta}{\partial \omega}\right)|_{f_o} z}$ can be factored out of the integral. The pulse is then propagated to z without distortion, i.e. $A(t)$ is unchanged. The only change is a change of the phase of E_x from $e^{-j\beta(f_o)z}$ which equals $2\pi(\partial\beta/\partial\omega)|_{f_o} z$. Otherwise, there will be distortion, or change of $A(t)$. Clearly, when higher-order terms in Eq. (1.111) cannot be neglected, there will be additional distortion.

Chapter summary

Basic plane wave analysis is presented. A plane wave is the simplest rigorous solution of Maxwell's equations. Yet it can be used to illustrate many basic concepts in optics. Under appropriate circumstances, an optical ray could be represented locally approximately by a plane wave. Optical properties such as reflection, refraction, and focusing can also be derived from plane wave analysis. The plane wave presented here shows how the traditional analysis is related to Maxwell's equations. However, much more complex analyses are required for optical components design and image transfer [5]. Traditional optics is better suited for these applications. On the other hand, plane wave analysis shows optical properties that are not emphasized in traditional optics. These include the dependence of refraction on polarization, the differentiation between the amplitude (including phase) and intensity of the wave, the importance of change in polarization, the phase interference effects, etc.

Sophisticated engineering analytical techniques can be illustrated by plane wave analysis. Concepts such as evanescent waves are introduced. Thin refractive components are representable by a transparent medium with phase variation. Grating diffraction is presented as another example of how phase variation can be used to understand simply a complex phenomenon.

Note that, an arbitrary optical field can be represented by summation of plane waves in the form of Fourier transformation, which is the basis of optical signal processing. Representation of an arbitrary radiation pattern by superposition of plane waves is also probably the simplest form of modal analysis in which the modes are just the plane waves.

Plane wave analysis is also an important vehicle to learn the basic mathematics of wave solutions. For example, there are always two independent solutions, the forward and the backward waves, and two mutually perpendicular polarizations for each direction of propagation. Optical interactions in all components are analyzed by matching the boundary conditions at the interfaces.

References

1. David M. Pozar, *Microwave Engineering*, John Wiley & Sons, 2005.
2. Joseph W. Goodman, *Introduction to Fourier Optics*, McGraw-Hill, 1968.
3. P. Hariharan, *Optical Holography Principles, Techniques, and Applications*, Cambridge University Press, 1996.
4. W. Thomas Cathey, *Optical Information Processing and Holography*, John Wiley & Sons, 1974.
5. M. Born and E. Wolf, *Principles of Optics*, Pergamon Press, 1959.

2 Superposition of plane waves and applications

The basics of many applications such as anti-reflection and reflection coatings, beam splitters, interferometers, resonators, holography, and planar waveguides, etc. can be analyzed by superposition and multiple reflections of plane waves. These analyses demonstrate the usefulness of simple plane wave analysis in another dimension. This is the focus of Chapter 2.

However, there are many shortcomings of plane wave analysis. It does not provide a full characterization of many applications because it ignores lateral variation of beams that occur in real components. For example, the consequence of the finite size of the beam is not included in plane wave analysis. Other analytical tools such as Fourier transform and convolution theory also cannot be presented by plane wave analysis. Laser cavity modes and Gaussian beams are not plane waves. These analyses are presented in Chapters 3 and 4.

2.1 Reflection and anti-reflection coatings

Reflection and transmission of plane waves at a dielectric interface can often be increased or reduced by coating the surface with transparent dielectric layers that have appropriate refractive indices. It is an anti-reflection coating when it is designed for maximum transmission and a reflection coating when it is designed for maximum reflection. It is a beam splitter when a specific ratio of reflected and transmitted intensities is required for some applications.

Consider an x-polarized plane wave propagating in the $+z$ direction in a medium with refractive index n_1. If this wave is incident perpendicularly onto another unbounded medium that has a refractive index n_2 at $z > d$, the reflection Γ_{12} and transmission T_{12} of this plane wave at the boundary is given by Eq. (1.64) in Section 1.3.1 as:

$$\Gamma_{12} = \frac{1 - \dfrac{n_2}{n_1}}{1 + \dfrac{n_2}{n_1}}, \qquad T_{12} = \frac{2}{1 + \dfrac{n_2}{n_1}} \tag{2.1}$$

The magnitude of the transmitted and reflected waves in Eq. (2.1) can be changed by adding layers of transparent materials with appropriate refractive indices in front of medium #2.

Let us consider a single transition layer of a transparent dielectric material that has a refractive index n_t and thickness d. It is placed from $z = 0$ to $z = d$ in front of the medium

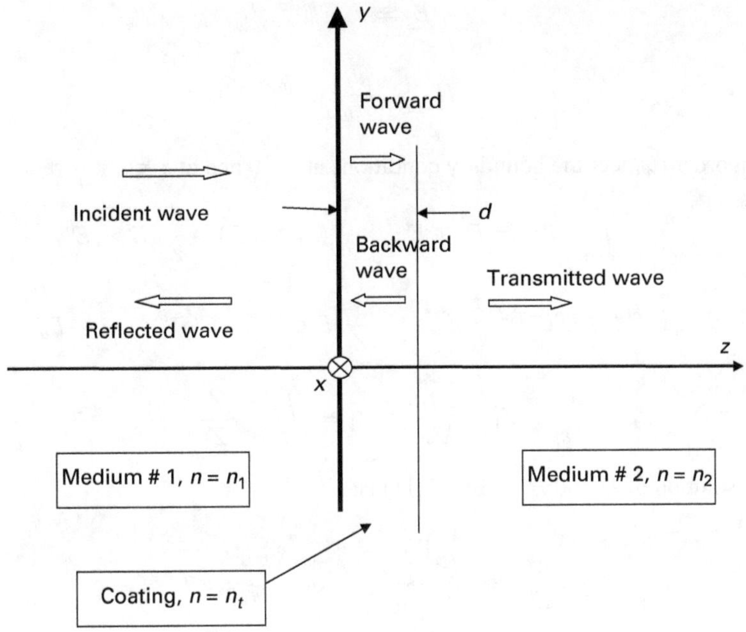

Figure 2.1 Anti-reflection and reflection coatings. A coating with index n_t is placed between $z = 0$ and $z = d$. The incident and reflected waves in medium #1, the forward and backward waves in the coating, and the transmitted wave in medium #2 are shown.

with n_2, as shown in Figure 2.1. Let there be an x-polarized wave for $z < 0$, incident on the interface along the z direction. Its x-y-z variation is:

$$\underline{E_i} = E_i e^{-j\beta_1 z} \underline{i_x} \tag{2.2}$$

$$\underline{H_i} = \sqrt{\frac{\varepsilon_1}{\mu}} E_i e^{-j\beta_1 z} \underline{i_y} \tag{2.3}$$

For simplicity, the time variation of $e^{j\omega t}$ is not shown here explicitly.

There will also be a reflected wave in $z < 0$,

$$\underline{E_r} = R E_i e^{+j\beta_1 z} \underline{i_x} \tag{2.4}$$

$$\underline{H_r} = -\sqrt{\frac{\varepsilon_1}{\mu}} R E_i e^{+j\beta_1 z} \underline{i_y} \tag{2.5}$$

There are two plane waves in the transition layer in $0 < z < d$, a forward wave and a backward wave.

$$\underline{E_t} = (E^f e^{-j\beta_t z} + E^b e^{+j\beta_t z}) \underline{i_x} \tag{2.6}$$

$$\underline{H_t} = \sqrt{\frac{\varepsilon_t}{\mu}} (E^f e^{-j\beta_t z} - E^b e^{+j\beta_t z}) \underline{i_y} \tag{2.7}$$

There is a transmitted wave in the unbounded medium with refractive index n_2 at $z > d$.

$$\underline{E_o} = E_o \mathrm{e}^{-j\beta_2(z-d)} \underline{i_x} \tag{2.8}$$

$$\underline{H_o} = \sqrt{\frac{\varepsilon_2}{\mu}} E_o \mathrm{e}^{-j\beta_2(z-d)} \underline{i_y} \tag{2.9}$$

In order to meet the boundary conditions at $z = 0$ and at $z = d$, it is required that

$$E_i + RE_i = E^f + E^b, \quad \frac{1}{Z_1}(E_i - RE_i) = \frac{1}{Z_t}(E^f - E^b) \tag{2.10}$$

$$E^f \mathrm{e}^{-j\beta_t d} + E^b \mathrm{e}^{+j\beta_t d} = E_o, \quad \frac{1}{Z_t}(E^f \mathrm{e}^{-j\beta_t d} - E^b \mathrm{e}^{+j\beta_t d}) = \frac{1}{Z_2} E_o \tag{2.11}$$

$$Z_t = \sqrt{\frac{\mu}{\varepsilon_t}}, \quad Z_1 = \sqrt{\frac{\mu}{\varepsilon_1}}, \quad Z_2 = \sqrt{\frac{\mu}{\varepsilon_2}}, \quad \frac{Z_t}{Z_1} = \frac{n_1}{n_t}, \quad \frac{Z_2}{Z_t} = \frac{n_t}{n_2} \tag{2.12}$$

The solution of E^f and E^b in Eq. (2.11) is:

$$\frac{E^b}{E^f} \mathrm{e}^{j2\beta_t d} = \frac{1 - \frac{n_2}{n_t}}{1 + \frac{n_2}{n_t}}, \quad \text{or} \quad \frac{E^b}{E^f} = \mathrm{e}^{-j2\beta_t d} \Gamma_{t2} \tag{2.13}$$

$$\Gamma_{t2} = \frac{1 - \frac{n_2}{n_t}}{1 + \frac{n_2}{n_t}} = -\Gamma_{2t}, \quad T_{t2} = \frac{2}{1 + \frac{n_2}{n_t}}, \quad \Gamma_{t2} + T_{t2} = 1 \tag{2.14}$$

From Eq. (2.10),

$$R = \frac{\Gamma_{1t} + \Gamma_{t2} \mathrm{e}^{-j2\beta_t d}}{1 + \Gamma_{1t}\Gamma_{t2} \mathrm{e}^{-j2\beta_t d}} \tag{2.15}$$

$$\Gamma_{1t} = \frac{1 - \frac{n_t}{n_1}}{1 + \frac{n_t}{n_1}} = -\Gamma_{t1}, \quad T_{1t} = \frac{2}{1 + \frac{n_t}{n_1}}, \quad \Gamma_{1t} + T_{1t} = 1 \tag{2.16}$$

$$\frac{E_o}{E_i} = \frac{T_{1t} T_{t2}}{1 + \Gamma_{1t}\Gamma_{t2} \mathrm{e}^{-j2\beta_t d}} \mathrm{e}^{-j\beta_t d} \tag{2.17}$$

If we choose n_t and d such that

$$n_t = \sqrt{n_1 n_2} \quad \text{and} \quad e^{-j2\beta_t d} = -1 \tag{2.18}$$

Then

$$\frac{n_t}{n_1} = \frac{n_2}{n_t} = \sqrt{\frac{n_2}{n_1}}, \quad R = 0, \quad E_o = \pm j E_i \tag{2.19}$$

In this manner, we have obtained an anti-reflection coating that has no reflection in medium #1 and 100% transmission into medium #2 at a specific wavelength, at which d

is equal to ¼ of the wavelength. Identical results are obtained when the electric field is polarized in the y–z plane. Note that for a given d the anti-reflection effect is wavelength sensitive. As the wavelength deviates, the reflection will increase and the transmission will decrease. Thus there is an effective wavelength range of the anti-reflection coating.

In reality, there may not be a coating material that has exactly the required n_t. The wavelength range within which anti-reflection, reflection, or beam splitting is required for different applications may also need to be decreased or increased. Therefore, multiple layer coatings are used in most commercial devices. However, the basic principle is demonstrated by the above example. In a similar manner, coatings can be applied to enhance reflection or to split the incident beam into desired ratios of reflected and transmitted beams. Beam splitters can also be designed for beams incident at specific incident angles.

Note that the analysis presented here is similar to impedance transformation analysis of microwave transmission lines. The differential equation for E and H is identical to that for V and I of microwave transmission lines [1]. In microwaves, anti-reflection is called impedance matching.

Many of the analytical techniques developed for microwaves are also very useful for optical analysis, especially when we need to analyze multi-layer transitions. It is important for optics engineers to understand transmission line methods. However, a detailed discussion of that is beyond the scope of this book.

2.2 Fabry–Perot resonance

2.2.1 Multiple reflections and Fabry–Perot resonance

Although plane waves propagating between two boundaries have already been analyzed in the previous section by matching the total fields at the boundaries, an alternate way to analyze it is to consider an incident plane wave multiply reflected and transmitted at the two boundaries. Much more physical insight on resonance could be gained by presenting this alternate approach.

Let us first consider a plane wave incident on the first boundary at $z = 0$, without considering the second boundary at $z = d$. This incident wave E_i would excite a reflected backward wave $\underline{E_{r1}}$ in medium #1 and a transmitted forward wave $E^f{}_1$ in the transition medium. Let the boundary at $z = 0$ have a reflection coefficient Γ_{1t} and transmission coefficient T_{1t} for the incident wave. The x-y-z variations without showing the time variation $e^{j\omega t}$ are:

$$\underline{E_1^f} = T_{1t}E_i e^{-j\beta_t z}\underline{i_x} \quad \text{for } d > z > 0, \qquad \underline{E_{r1}} = \Gamma_{1t}E_i e^{+j\beta_1 z}\underline{i_x} \quad \text{for } z < 0 \tag{2.20}$$

This boundary will have reflection coefficient Γ_{t1} and transmission coefficient T_{t1} for any plane wave incident on it in the reverse direction from the transition medium.

As $E^f{}_1$ propagates to $z = d$, it excites a reflected wave $\underline{E^b{}_1}$ in the transition medium and a transmitted wave $\underline{E_{o1}}$ in medium #2. Let the boundary at $z = d$ have reflection coefficient Γ_{t2} and transmission coefficient T_{t2} for any forward wave propagating in the transition medium. Then we obtain:

$$\underline{E_{o1}} = T_{t2}(T_{1t}E_i e^{-j\beta_t d})e^{-j\beta_2(z-d)}\underline{i_x} \quad \text{for} \quad z > d \tag{2.21}$$

$$\underline{E_1^b} = \Gamma_{t2}(T_{1t}E_i e^{-j\beta_t d})e^{+j\beta_t(z-d)}\underline{i_x} \quad \text{for} \quad 0 < z < d \tag{2.22}$$

The reflected wave $\underline{E^b_1}$ propagates back to $z = 0$ and excites another transmitted backward wave $\underline{E_{r2}}$ in medium #1 and a reflected forward wave $\underline{E^f_2}$ in the transition medium.

$$\underline{E_2^f} = \Gamma_{t1}\Gamma_{t2}(T_{1t}E_i e^{-j\beta_t d})e^{-j\beta_t d}e^{-j\beta_t z}\underline{i_x} \quad \text{for} \quad 0 < z < d \tag{2.23}$$

$$\underline{E_2^b} = \Gamma_{t1}\Gamma_{t2}^2(T_{1t}E_i e^{-j\beta_t d})e^{-2j\beta_t d}e^{+j\beta_t(z-d)}\underline{i_x} \quad \text{for} \quad 0 < z < d \tag{2.24}$$

$$\underline{E_{r2}} = T_{t1}\Gamma_{t2}(T_{1t}E_i e^{-j\beta_t d})e^{-j\beta_t d}e^{j\beta_1 z}\underline{i_x} \tag{2.25}$$

As $\underline{E^f_2}$ reaches $z = d$, it excites a forward $\underline{E_{o2}}$ in the transition medium and an $\underline{E^b_2}$ in medium #2.

$$\underline{E_{o2}} = T_{t2}(\Gamma_{t1}\Gamma_{t2}T_{1t}E_i e^{-j3\beta_t d})e^{-j\beta_2(z-d)}\underline{i_x} \quad \text{for} \quad z > d \tag{2.26}$$

As $\underline{E^b_2}$ reaches $z = 0$, it excites a forward $\underline{E^f_3}$ in the transition region and an $\underline{E_{r2}}$ in medium #1.

$$\underline{E_{r3}} = T_{t1}\Gamma_{t1}\Gamma_{t2}^2(T_{1t}E_i e^{-j\beta_t d})e^{-3j\beta_t d}e^{+j\beta_1 z}\underline{i_x} \quad \text{for} \quad 0 < z \tag{2.27}$$

Consequentially, the forward and the backward waves in the transition medium continue to generate backward reflected waves at $z < 0$ and transmitted output waves at $z > d$. The amplitudes of the total forward and backward propagating waves are related to the incident wave by:

$$\frac{E_o}{E_i} = T_{1t}T_{t2}e^{-j\beta_t d}[1 + \Gamma_{t1}\Gamma_{t2}e^{-j2\beta_t d} + \Gamma_{t1}^2\Gamma_{t2}^2e^{-j4\beta_t d} + \ldots] = \frac{T_{1t}T_{t2}e^{-j\beta_t d}}{1 - \Gamma_{t1}\Gamma_{t2}e^{-j2\beta_t d}} \tag{2.28}$$

for $z > d$,

$$\frac{E_r}{E_i} = \Gamma_{1t} + \Gamma_{t2}T_{1t}T_{t1}e^{-j2\beta_t d}\frac{1}{1 - \Gamma_{t1}\Gamma_{t2}e^{-j2\beta_t d}} = \frac{\Gamma_{1t} + \Gamma_{t2}e^{-j2\beta_t d}}{1 - \Gamma_{t1}\Gamma_{t2}e^{-j2\beta_t d}} \tag{2.29}$$

for $z < 0$, and

$$\frac{E^f}{E_i} = T_{1t}\frac{1}{1 - \Gamma_{t1}\Gamma_{t2}e^{-j2\beta_t d}}, \quad \frac{E^b}{E_i} = T_{1t}\Gamma_{t2}\frac{e^{-j2\beta_t d}}{1 - \Gamma_{t1}\Gamma_{t2}e^{-j2\beta_t d}} \tag{2.30}$$

for $0 < z < d$.

When the Γs and *T*s in Eqs. (2.14) and (2.16) in the previous section are used in Eqs. (2.28) and (2.29) the solutions of E_o and E_r become identical to the results in Eqs. (2.15) and (2.17) of the previous section. However, the above results are more general. They also apply if the boundaries are partially reflecting mirrors. Note that, the reflection coefficients, Γ, and transmission coefficients, Ts, at the two mirrors are related by

$$\Gamma_{1t} + T_{1t} = 1, \quad \Gamma_{t2} + T_{t2} = 1, \quad \Gamma_{1t} = -\Gamma_{t1}, \quad \Gamma_{2t} = -\Gamma_{t2} \tag{2.31}$$

When the mirrors have high reflection coefficients, Eq. (2.30) shows that the forward and backward plane waves in the transmission medium can be very large whenever $e^{-j2\beta_t d} \cong 1$. *At the wavelengths and separation of the mirrors, d, that satisfy this condition, the round-trip phase shift δ of the plane wave is 2qπ. This means that the multiply reflected forward and backward waves in the transmission medium reinforce each other. The stored energy of the plane waves becomes very high for a small incident* E_i. *The optical component is said to be in resonance at these wavelengths.*

2.2.2 Properties of Fabry–Perot resonance

So far, only a formal analysis has been presented. In order to understand the physical significance of E_o and E_r, let us consider a case in which medium #1 and medium #2 are identical, the mirrors are symmetrical, and Γs and Ts are real.

$$\Gamma_{1t} = \Gamma_{2t} = \Gamma, \quad T_{1t} = T_{2t} = T, \quad R_m = \Gamma_{1t}^{\ 2} = \Gamma_{2t}^{\ 2}, \quad T_m = T_{1t}^{\ 2} = T_{2t}^{\ 2} \tag{2.32}$$

R_m and T_m are the optical power reflection and transmission coefficients of the mirrors.

The time-averaged incident and reflected powers in medium #1 and the transmitted power in medium #2 are:

$$P_i = \frac{1}{2} n_1 \sqrt{\frac{\varepsilon_o}{\mu}} E_i E_i * \tag{2.33}$$

$$P_r = \frac{1}{2} n_1 \sqrt{\frac{\varepsilon_o}{\mu}} \frac{4R_m \sin^2(\beta_t d)}{(1 - R_m)^2 + 4R_m \sin^2(\beta_t d)} E_i E_i * \tag{2.34}$$

$$P_o = \frac{1}{2} n_1 \sqrt{\frac{\varepsilon_o}{\mu}} \frac{(1 - R_m)^2}{(1 - R_m)^2 + 4R_m \sin^2(\beta_t d)} E_i E_i * \tag{2.35}$$

Since the sine function is nonlinear, P_o remains very small at wavelengths other than the resonance wavelengths. When R_m is close to 1, P_o/P_i is close to 1 within a narrow wavelength range from the resonance wavelengths at which $\delta = 2\beta_t d \cong 2q\pi$ (q = any integer).

For a given d, resonance occurs at frequencies f_q such that

$$f_q = \frac{q}{2d\sqrt{\varepsilon_t \mu}} = \frac{qc_o}{2n_t d} \tag{2.36}$$

The separation of adjacent resonance frequencies, known as the free spectral range (FSR) of the Fabry–Perot resonance, is

$$\Delta f_q = f_{q+1} - f_q = \frac{1}{2d\sqrt{\varepsilon_t \mu}}. \tag{2.37}$$

If we let $\Delta\delta = \delta - q\pi$, then T can be expressed as

$$T = \frac{P_o}{P_i} = \frac{(1-R_m)^2}{(1-R_m)^2 + 4R_m\left(\sin\frac{\Delta\delta}{2}\right)^2}. \tag{2.38}$$

T is at its maximum when $\Delta\delta = 2q\pi$; it drops to ½ when

$$\Delta\delta^2 = (\delta - q\pi)^2 \cong \frac{(1-R_m)^2}{R_m} \tag{2.39}$$

The reflected power P_r is

$$\frac{P_r}{P_i} = \frac{4R_m \sin\left(\frac{\Delta\delta}{2}\right)^2}{(1-R_m)^2 + 4R_m \sin^2\left(\frac{\Delta\delta}{2}\right)}. \tag{2.40}$$

If we let ω_o be the center of the resonance frequency, $\omega_o = q\pi/2\sqrt{\mu\varepsilon_t}d$, and q = any integer, then the half linewidth $\Delta\omega$ in which T drops to ½ is

$$\Delta\omega = \frac{(1-R_m)c_o}{\sqrt{R_m}n_t d} \tag{2.41}$$

Figure 2.2 shows the ratio of P_o/P_i in Eqs. (2.33) and (2.35) for a typical Fabry–Perot resonator as a function of $\beta_t d$ for several R_m. Clearly the resonance can serve as a very narrow band filter when the reflectivity of the mirrors is high.

There are two ways to measure the quality of the resonance. A measure commonly used to gauge the resonance in the optics literature is the finesse, F, which is the FSR divided by the full line width. Using the half linewidth $\Delta\omega$ in Eq. (2.41), we obtain

$$F = \frac{\sqrt{R_m}\pi}{(1-R_m)} \tag{2.42}$$

In engineering applications, a measure commonly used to gauge any resonator characteristics is the Q factor. It is defined for any resonator without any excitation as

$$Q = \omega_o \frac{\text{energy stored}}{\text{power dissipated}} \tag{2.43}$$

The bandwidth is related to Q by $\Delta\omega = \omega_o/Q$.

For the Fabry–Perot resonator under consideration, its homogeneous solution consists of plane waves inside the resonator reflected back and forth between mirrors, and partially transmitted at each reflection. For a plane wave with electric field amplitude

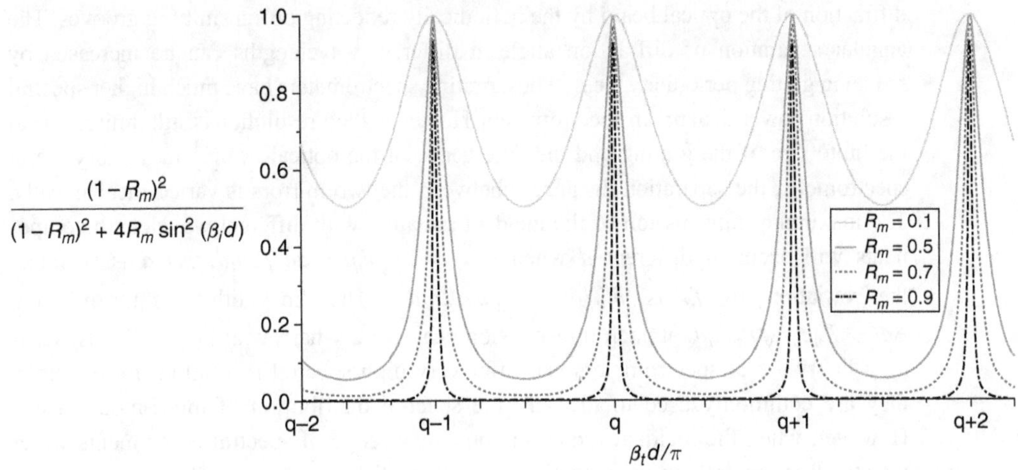

Figure 2.2 Ratio of transmitted to incident power in a typical Fabry–Perot resonator with different reflectors.

E, its time-averaged stored energy per unit area is $(\varepsilon_t/2)EE^*d$. When it reaches the mirror, the time-averaged transmitted power per unit area is $1/2T_m\sqrt{(\varepsilon_t/\mu)}EE^*$. The ratio of energy stored/power loss is $n_t d/T_m c_o$. As the plane wave is reflected in each trip at the two mirrors, this ratio is repeated. Therefore the total ratio of energy stored/power loss is also $n_t d/T_m c_o$. The Q and the bandwidth of the resonator at resonance are

$$Q = \omega_o \frac{n_t d}{T_m c_o} \qquad \Delta\omega = \frac{T_m c_o}{n_t d} \tag{2.44}$$

Comparing the bandwidth shown in Eq. (2.41) and Eq. (2.44), they agree with each other when the reflectivity R_m is high.

Fabry–Perot resonators are important in many applications, such as scanning interferometry and wavelength filtering. Fabry–Perot resonance properties that are commonly used in optics are line width, finesse, and free spectral range. The quantities commonly used in engineering are free spectral range, line width, and Q factor. These properties depend only on the reflectivity of the mirrors and the separation distance d in the lossless plane wave approximation. In reality, they may also be affected by the propagation loss of the medium and the diffraction losses.

2.2.3 Applications of the Fabry–Perot resonance

Fabry–Perot resonance has many applications, such as spectrometry, wavelength filtering, loss measurement, and time delay. However, different applications utilize different features of the resonance.

(a) The Fabry–Perot scanning interferometer

The operation of prism spectrometers discussed in Section 1.3.4 depends on the dispersion of the refractive index of the material; they have low spectral resolution. The operation of grating spectrometers discussed in Section 1.5 depends on the dispersion created by diffraction of the optical beam by the periodically reflecting or transmitting grooves. The angular separation of diffraction angle at different wavelengths can be increased by reducing grating periodicity, i.e f_g. Thus, grating spectrometers have much higher spectral resolution power than prism spectrometers. However, their resolution is still limited due to the finite size of the grating and the divergence of the optical beam.[1] In a Fabry–Perot spectrometer, the separation distance d between the two mirrors is varied mechanically. The maximum transmission of the incident radiation with different wavelength components will occur at different d when $f_q = q/2d\sqrt{\varepsilon_t \mu} = qc_o/2n_t d$. As d is scanned, the variation of f_q is $\delta f_q/\delta d = -qc_o/2n_t d^2$. The linewidth of transmission, $\Delta\omega = T_m c_o/\sqrt{R_m} n_t d$, at each transmission peak is very narrow at large d. Thus, scanning Fabry–Perot spectrometers can offer very high spectral resolution, for example, they are commonly used to measure the spectral distribution of multi-mode lasers. However, when the incident radiation contains a range of spectral components wider than the free spectral range, transmission peaks of different order q will occur within the same range of d.

Note also that at incident angles θ different than the normal incidence, the Fabry–Perot resonance peak will occur at $f_q = q/2d\cos\theta\sqrt{\varepsilon_t \mu} = qc_o/2n_t d\cos\theta$. Thus, for a divergent incident beam and for a given d, the transmission peak of the same frequency component will appear at different θ, caused by the different orders q. If a lens is used to focus the output beam, the output will appear as concentric circles.

In retrospect, while the prism spectrometer is the simpler tool to fabricate, it has a low resolution. Grating spectrometers can be designed to provide very high resolution. Both of them are used commonly to measure inputs with wide spectral content. Scanning Fabry–Perot spectrometers are useful for resolving closely spaced spectral components such as those emitted from a multi-mode laser. However, overlapping transmission of different orders of q within the scanning range needs to be resolved for inputs that have spectral contents wider than the free spectral range.

(b) Measurement of refractive properties of materials

From Eq. (2.37), it is clear that the refractive index n_t of the transmission medium between mirrors is related to d and the free spectral range of Fabry–Perot resonances by $n_t = c_o/2\Delta f_q d$. For a given d, if the frequencies of adjacent resonance transmission peaks can be measured accurately, one can obtain an accurate evaluation of the refractive index n_t of the medium between mirrors by Δf_q. Note that the accuracy of this measurement is independent of the reflectivity R_m. Therefore, it can be used to measure the refractive index of any material, even if the reflectivity between medium 1 and the transition medium is low or moderate.

When the transition medium has loss, $\beta_t = \beta_{tr} - j\beta_{t\sigma}$. Eq. (2.28) becomes

[1] The divergence of an optical beam that has a limited beam size will be discussed in the next chapter.

$$\frac{E_o}{E_i} = T_{1t}T_{t2}\mathrm{e}^{-j\beta_t d}[1 + \Gamma_{t1}\Gamma_{t2}\mathrm{e}^{-j2\beta_t d} + \Gamma_{t1}{}^2\Gamma_{t2}{}^2\mathrm{e}^{-j4\beta_t d} + \ldots] = \frac{T_{1t}T_{t2}\mathrm{e}^{-\beta_{t\sigma}d}\mathrm{e}^{-j\beta_{tr}d}}{1 - \Gamma_{t1}\Gamma_{t2}\mathrm{e}^{-2\beta_{t\sigma}d}\mathrm{e}^{-j2\beta_{tr}d}} \quad (2.45)$$

When R_m and T_m are real as shown in Eq. (2.32), the P_o is

$$P_o = \frac{1}{2}n_1\sqrt{\frac{\varepsilon_o}{\mu}}\frac{(T_m)^2\mathrm{e}^{-2\beta_{t\sigma}d}}{(1 - R_m\mathrm{e}^{-2\beta_{t\sigma}d})^2 + 4R_m\mathrm{e}^{-2\beta_{t\sigma}d}\sin^2(\beta_{tr}d)}E_iE_i^* \quad (2.46)$$

P_o is a maximum at $\beta_{tr}\, d = q\pi$, and a minimum at $\beta_{tr}\, d = (q + 1/2)\pi$. The ratio of P_o at its maximum to P_o at its minimum, called the contrast ratio, is

$$\frac{P_{o,\max}}{P_{o,\min}} = \frac{1 + R_m\mathrm{e}^{-2\beta_{t\sigma}d}}{1 - R_m\mathrm{e}^{-2\beta_{t\sigma}d}} \quad (2.47)$$

Therefore, for a known R_m, the contrast ratio can be used to determine $\beta_{t\sigma}$, i.e. the loss of the refractive material.

(c) Resonators for filtering and time delay of signals

From Eq. (2.41), it is clear that if d is large and T_m is small, the bandwidth of a Fabry–Perot resonator can be made very small. Therefore it can be used as a filter. Although plane wave Fabry–Perot resonators are too bulky to use for many applications, waveguide resonators have been used effectively as filters in other configurations.

It takes time for the stored energy in a resonator to decay, caused by dissipation. Thus, the decay time of energy in the resonator, τ, is

$$\frac{\mathrm{d(stored\ energy)}}{\mathrm{d}t} = \frac{1}{\tau}(\text{stored energy}) = \text{dissipated energy} \qquad \tau = \frac{Q}{\omega_o} = \frac{d/T_m}{c_o/n_t} \quad (2.48)$$

In many applications, a time delay is used for signal processing. For example, a long optical fiber is often used to delay a pulsed signal. The time delay that can be achieved by propagating in a medium such as a waveguide (or fiber) that has refractive index n_t and length L is n_tL/c_o. In a Fabry–Perot resonator, an input signal pulse will be reflected back and forth between mirrors. Therefore the output pulse, which is emitted later, is a delayed signal. The output pulsed signals will decay in time. It is common to regard τ as the useful time period of signals. Therefore, the last useful output pulse would have increased the delay time by $1/T_m$.

2.3 Reconstruction of propagating waves

When the intensity of the interference pattern between an object wave and a reference wave is recorded as the index or transmittance variation in a recording medium, the transmitted waves of another incident beam in the form of the original reference beam through the recorded medium will then reproduce the original object wave and its conjugate.[2] This is the basic principle of holography. It can be illustrated very simply by plane wave analysis.

[2] The principle is also applicable to recordings in reflection.

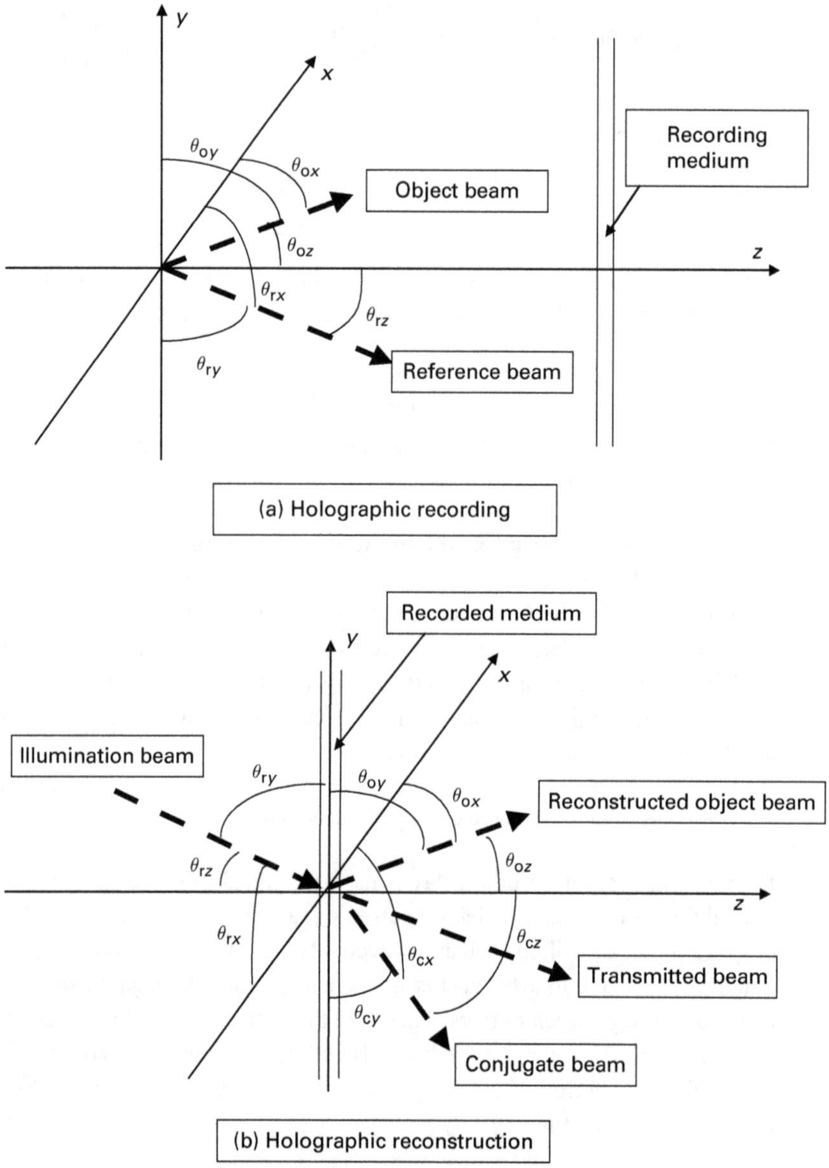

Figure 2.3 Holographic recording and reconstruction of a plane wave. (a) The interference pattern of the intensity of the object and the reference beam is recorded by the recording medium. (b) Upon the illumination of a beam identical to the reference beam used in the recording process in (a), the reconstructed object beam, the un-diffracted beam and the conjugate beam are created.

Consider an object beam, $A_o e^{-j\underline{\beta_o}\cdot\underline{r}} e^{j\omega t}$, and a reference beam, $A_r e^{-j\underline{\beta_r}\cdot\underline{r}} e^{j\omega t}$. Both are polarized in the x direction. They are incident on a recording medium at $z = 0$, as illustrated in Figure 2.3(a). The total electric field and the time-averaged intensity of the incident and the reference beams are:

$$E_t = (A_o e^{-j\beta\cos\theta_{ox}x} e^{-j\beta\cos\theta_{oy}y} e^{-j\beta\cos\theta_{oz}z} + A_r e^{-j\beta\cos\theta_{rx}x} e^{-j\beta\cos\theta_{ry}y} e^{-j\beta\cos\theta_{rz}z}) e^{j\omega t} \quad (2.49)$$

$$\begin{aligned} I &= \frac{1}{2}\sqrt{\frac{\varepsilon}{\mu}} E_t E_t^* \\ &= \frac{1}{2}\sqrt{\frac{\varepsilon}{\mu}}\{A_o^2 + A_r^2 + A_o A_r [e^{-j\beta(\cos\theta_{ox} - \cos\theta_{rx})x} e^{-j\beta(\cos\theta_{oy} - \cos\theta_{ry})y} \\ &\quad + e^{+j\beta(\cos\theta_{ox} - \cos\theta_{rx})x} e^{+j\beta(\cos\theta_{oy} - \cos\theta_{ry})y}]\} \end{aligned} \quad (2.50)$$

Here, $\cos\theta_{ox}, \cos\theta_{oy},$ and $\cos\theta_{oz}$ are direction cosines of the object beam with respect to the *x, y*, and *z* axes. Similarly, $\cos\theta_{rx}$, $\cos\theta_{ry}$, and $\cos\theta_{rz}$ are direction cosines of the reference beam. A recording medium is placed at $z = 0$ to record I. Let the transparency of the recorded medium, t, be proportional to I, i.e. $t(x,y) = T_o I$. If an illumination plane wave $E_{in} = A_3 e^{-j\underline{\beta_{ox}}\cdot\underline{r}} e^{j\omega t}$ is incident on the recording medium as shown in Figure 2.3 (b), the transmitted wave tE_{in} is

$$\begin{aligned} tE_{in} &= T_o \frac{1}{2}\sqrt{\frac{\varepsilon}{\mu}} A_{in} e^{-j\beta\cos\theta_{rx}x} e^{-j\beta\cos\theta_{ry}y} e^{-j\beta\cos\theta_{rz}} \\ &\{A_o^2 + A_r^2 + A_o A_r [e^{-j\beta(\cos\theta_{ox} - \cos\theta_{rx})x} e^{-j\beta(\cos\theta_{oy} - \cos\theta_{ry})y} \\ &\quad + e^{+j\beta(\cos\theta_{ox} - \cos\theta_{rx})x} e^{+j\beta(\cos\theta_{oy} - \cos\theta_{ry})y}] e^{j\omega t}\} \end{aligned} \quad (2.51)$$

There are three output terms. The first term, which represents a transmitted illumination beam, is

$$T_o \frac{1}{2}\sqrt{\frac{\varepsilon}{\mu}} A_{in}(A_o^2 + A_r^2) e^{-j\beta\cos\theta_{rx}x} e^{-j\beta\cos\theta_{ry}y} e^{-j\beta\cos\theta_{rz}z} e^{j\omega t} \quad (2.52)$$

The second term, which represents a beam identical to the object wave, called a reconstructed wave, is

$$T_o \frac{1}{2}\sqrt{\frac{\varepsilon}{\mu}} A_{in} A_o A_r e^{-j\beta\cos\theta_{ox}x} e^{-j\beta\cos\theta_{oy}y} e^{-j\beta\cos\theta_{oz}z} e^{j\omega t} \quad (2.53)$$

The third term, which represents a beam tilted into a new direction with respect to the *x* and *y* axes, called a conjugate beam, is

$$T_o \frac{1}{2}\sqrt{\frac{\varepsilon}{\mu}} A_{in} A_o A_r e^{-j\beta(\cos\theta_{ox} - 2\cos\theta_{rx})x} e^{-j\beta(\cos\theta_{oy} - 2\cos\theta_{ry}y)} e^{-j\beta\cos\theta_{cz}z} e^{j\omega t} \quad (2.54)$$

The conjugate beam has direction cosines $\cos\theta_{cx}$, $\cos\theta_{cy}$, and $\cos\theta_{cz}$, where $\cos\theta_{cx} = \cos\theta_{ox} - 2\cos\theta_{rx}$, $\cos\theta_{cy} = \cos\theta_{oy} - 2\cos\theta_{ry}$, and $\cos\theta_{cz} = \sqrt{1 - \cos^2\theta_{cx} - \cos^2\theta_{cy}}$

In short, if the recording medium with transmittance, t(x,y), is illuminated by E_{in}, it recreates the object beam and its conjugate. The relative magnitude of the transmitted beams can be adjusted by the magnitude of A_r, A_o, and A_{in}. If there is more than one object wave, then all the object waves will be recreated by E_{in}. For example, the object waves can

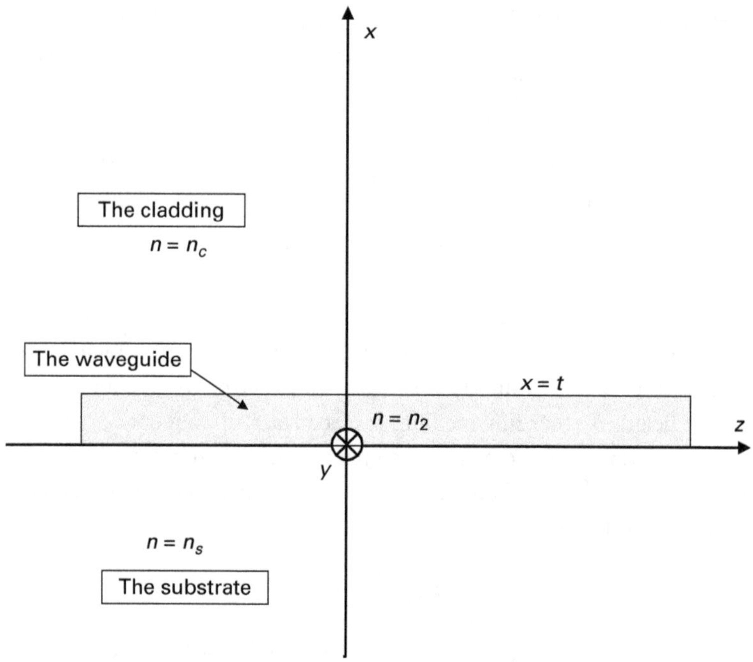

Figure 2.4 Illustration of a planar optical waveguide. The core with index n_2 has a planar surface parallel to the y–z plane, placed between $x = 0$ and $x = t$. The substrate is below $x = 0$. The cladding is above $x = t$.

be the Fourier components of a complex object. The same Fourier components are created in the object and conjugate beams. This is the basic principle of holography.

2.4 Planar waveguide modes viewed as internal reflected plane waves

Optical planar waveguides can also be understood from the analysis of plane wave propagation in multi-layered media. A typical optical planar waveguide is illustrated in Figure 2.4. It has a high-index layer, n_2, surrounded by a cladding with index n_c and a substrate with index n_s. The width of the middle waveguide layer, the cladding and the substrate, extends to both $y = \pm\infty$ and $z = \pm\infty$. The thickness of the substrate and cladding also extends to infinity in the x direction. If we analyze the optical plane waves, propagating in a multi-layered media such as that shown in Figure 2.4, we find that there are three typical cases.

2.4.1 Plane waves incident from the cladding

Consider a plane wave that is incident obliquely on the layered structure shown in Figure 2.4 from $x > t$. Let us assume the plane wave is polarized in the y direction.[3] It propagates in the x–z plane in a direction that makes an angle θ_j with respect to the x axis

[3] For an electric field polarized in the plane of incidence, there will be a similar set of equations. However, the boundary conditions at $z = 0$ and $z = t$ will lead to a different set of solutions than those shown in this section. In this case, the magnetic field will be in the y direction. In Chapter 6, we will show that these solutions are the TM modes.

in various layers. The angle, θ_j, will be different in different layers where j designates the layer with index n_j. For example, the incident plane wave in the cladding with index n_c will have a functional form,

$$E_i \underline{i_y} = A e^{-jn_c k \sin\theta_c z} e^{+jn_c k \cos\theta_c x} \underline{i_y} \tag{2.55}$$

k is the free space propagation constant, $k = 2\pi/\lambda_o$. There will be a reflected wave in the cladding, excited by the incident wave,

$$E_{rc1} \underline{i_y} = A e^{jn_c k \cos\theta_c t} \Gamma_{cw} e^{-jn_c k \sin\theta_c z} e^{-jn_c k \cos\theta_c (x-t)} \underline{i_y} \tag{2.56}$$

There will be a forward transmitted wave in the waveguide layer excited by the incident wave,

$$E_{fw1} \underline{i_y} = A e^{jn_c k \cos\theta_c t} T_{cw} e^{-jn_2 k \sin\theta_2 z} e^{-jn_2 k \cos\theta_2 (x-t)} \underline{i_y} \tag{2.57}$$

The continuity of the tangential electric field demands that $n_1 k \sin\theta_1 = n_2 k \sin\theta_2$ at the boundary $x = t$. The amplitudes A_{arc} and A_{fw}, including their phase, will be given by A and the reflection and transmission coefficients at $x = t$. When the transmitted wave E_{fw1} reaches $x = 0$, it excites a transmitted wave in the substrate E_{ts} and a reflected wave E_{rw} in the waveguide.

$$E_{ts1} \underline{i_y} = A T_{cw} T_{ws} e^{jn_c k \cos\theta_c t} e^{-jn_2 k \cos\theta_2 t} e^{-jn_s k \sin\theta_s z} e^{+jn_s k \cos\theta_s x} \underline{i_y} \tag{2.58}$$

$$E_{rw1} \underline{i_y} = A_{rw} e^{-jn_2 k \sin\theta_2 z} e^{-jn_2 k \cos\theta_2 x} \underline{i_y} \tag{2.59}$$

The boundary condition at $x = 0$ is $n_s k \sin\theta_s = n_2 k \sin\theta_2$. The amplitudes A_{ts} and A_{tw} are given by A_{fw} and the transmission and the reflection coefficients at $x = 0$. All the reflection and transmission coefficients at the $x = t$ and $x = 0$ boundaries are given in Section 1.3.1. Similar to the discussions presented in Section 2.2.1, for waveguide structures with $n_1 < n_s < n_2$, there will be multiple reflected and transmitted waves. In addition to the multiple forward and backward waves in the waveguide, the total reflected waves in the cladding and the total transmitted waves in the substrate are also the sum of the waves after each reflection and transmission.

Note that at the maximum incidence angle $\theta_c = \pi/2$, the maximum θ_2 is $\theta_{2,\max} = \sin^{-1}(n_c/n_2)$. In a typical waveguide, $n_2 > n_s > n_c$. Thus, for any plane wave incident from the cladding, all the plane waves in the waveguide layer have angle $\theta_2 < \theta_{2,\max}$. The transmitted wave in the substrate will have θ_s limited to $\theta_s = \sin^{-1}(n_c/n_s)$.

When

$$2n_2 k \cos\theta_2 t + \varphi_{wc} + \varphi_{ws} = 2q\pi, \tag{2.60}$$

where q is an integer, and the multiply reflected and transmitted waves in the waveguide layer will be in phase with each other. Here φ_{wc} and φ_{ws} are the phase angles of Γ_{wc} and Γ_{ws} for this polarization of the electric field at the waveguide-to-cladding and the waveguide-to-substrate interfaces. At these specific angles of incidence, similar to the Fabry–Perot resonance the sum of the waves in the waveguide layer could have very large amplitude.

2.4.2 Plane waves incident from the substrate

Similarly when there is a plane wave incident on the waveguide from the substrate side, there will be a transmitted and a reflected beam in the waveguide and a reflected beam in the substrate at the $x = 0$ interface. The θ_s can vary within the range $0 < \theta_s < \pi/2$. However, there are two different cases.

(a) Incident plane waves with $\sin^{-1}(n_c/n_s) < \theta_s < \pi/2$

Let the incident wave have θ_s such that $\sin^{-1}(n_c/n_s) < \theta_s < \pi/2$. Since in most waveguides, $n_c < n_s < n_2$, the plane wave in the cladding is an evanescent wave. At the θ_s limit, the plane waves in the waveguide have $\theta_2 = \pm\sin^{-1}(n_s/n_2)$. Moreover, at the discrete incidence angles in which the condition in Eq. (2.60) is satisfied, there will be resonance in the waveguide, as discussed in the previous section.

(b) Incident plane waves with $0 < \theta_s < \sin^{-1}(n_c/n_s)$

If the incident angle θ_s is small enough such that $0 < \theta_s < \sin^{-1}(n_c/n_s)$, there will also be transmitted waves in the cladding. The transmitted plane waves in the cladding have $n_c\sin\theta_c = n_s\sin\theta_s$. The solutions will be similar to those shown in Eqs. (2.55) to (2.59) with the subscript c replaced by the subscript s and vice versa. The resonance condition for the waves in the waveguide will be the same as that given in Eq. (2.60). Other properties of the reflected and transmitted waves will be similar to those discussed in Section 2.4.1.

Note that, at the same θ_c, θ_s, and θ_2, the plane wave solution for incident wave in the substrate with $0 < \theta_s < \sin^{-1}(n_c/n_s)$ discussed here and the plane wave solution for incident wave in the cladding with $0 < \theta_c < \pi/2$ discussed in Section 2.4.1 constitute two equivalent but independent solutions[4] of Maxwell's equations. The solutions could have the same θ values; they are degenerate. Any linear combination of these degenerate solutions is also a plane wave solution at the same angles. This feature is utilized in Chapter 6 to give the air modes.

2.4.3 Plane waves incident within the waveguide: the planar waveguide modes

When $n_c < n_s < n_2$, there is no solution at $\theta_2 > \sin^{-1}(n_s/n_2)$ for plane waves incident from either the substrate or the cladding. However, if there is a plane wave with $\theta_2 > \sin^{-1}(n_s/n_2)$ already excited in the waveguide there will be multiply totally internally reflected plane waves at both the $x = 0$ and $x = t$ boundaries. The sum of the plane waves, reflecting back and forth between the interfaces at $x = 0$ and $x = t$ are the sum of the solutions of the Maxwell's equation that satisfies all the boundary conditions.

[4] The equivalent solutions have the same angles θ_c, θ_s, and θ_2, and the same resonance condition.

However, the sum of all the multiply reflected waves is zero because of the cancellation of the phases $e^{-jmkt2n_2\cos\theta_2}$. There is only one exception. When the condition in Eq. (2.60) is satisfied, all the waves then add to each other. Therefore there is a solution of the Maxwell's equation only when the resonance condition is satisfied. Each of these solutions is a discrete planar waveguide solution of the structure. There are only finite and discrete values of θ_2 that satisfy Eq. (2.60). Let us designate the mth discrete θ_2 at resonance to be θ_m, $m = 0, 1, 2, 3 \ldots$

$$2n_2 k\cos\theta_m t + \varphi_{wc} + \varphi_{ws} = 2(m+1)\pi \tag{2.61}$$

The mth reflected plane waves in the waveguide will be[5]:

$$E_m = A\left(e^{-jm(2n_2k\cos\theta_m t+\varphi_{wc}+\varphi_{ws})}e^{+jkn_2k\cos\theta_m 2x} - e^{-jm(2n_2k\cos\theta_m t+\varphi_{wc}+\varphi_{ws})}e^{-jn_2k\cos\theta_m x}\right) \times e^{-jkn_2\sin\theta_m z} \tag{2.62}$$

Here $|\Gamma|$ for total internal reflection is 1. From Section 1.3.3(b), we got two separate φ answers for the electric field polarized perpendicular to the plane of incidence and for the electric field polarized in the plane of incidence. The φ for these two cases are

$$\varphi_{\perp wc} = \tan^{-1}\frac{-2n_2\cos\theta_m\sqrt{n_2{}^2\sin^2\theta_m - n_c{}^2}}{n_2{}^2\cos^2\theta_m - (n_2{}^2\sin^2\theta_m - n_c{}^2)} \tag{2.63}$$

$$\varphi_{\perp ws} = \tan^{-1}\frac{-2n_2\cos\theta_m\sqrt{n_2{}^2\sin^2\theta_m - n_s{}^2}}{n_2{}^2\cos^2\theta_m - (n_2{}^2\sin^2\theta_m - n_s{}^2)} \tag{2.64}$$

$$\varphi_{//wc} = \tan^{-1}\frac{-2n_c\cos\theta_m\sqrt{n_c{}^2\sin^2\theta_m - n_2{}^2}}{n_c{}^2\cos^2\theta_m - (n_c{}^2\sin^2\theta_m - n_2{}^2)} \tag{2.65}$$

$$\varphi_{//w2} = \tan^{-1}\frac{-2n_s\cos\theta_m\sqrt{n_s{}^2\sin^2\theta_m - n_2{}^2}}{n_s{}^2\cos^2\theta_m - (n_s{}^2\sin^2\theta_m - n_2{}^2)} \tag{2.66}$$

Therefore, radiation in two different polarizations will have two different sets of solutions.

In short, planar waveguide modes are modes excited within the waveguide layer at resonance values of θ_2 where $\theta > \sin^{-1}(n_s/n_2)$. Note that the direction of propagation of the planar waveguide mode is z. There is no energy directed in the y direction. There are two separate cases. For an electric field polarized along the x axis, there is a set of modes satisfying Eqs. (2.61), (2.63) and (2.64). Although the total electric field is perpendicular to z, the total magnetic field has components in both the y and the z

[5] The waves in the waveguide at these discrete θ_m values of θ_2 cannot be excited by an incident propagating wave from either the cladding or the substrate.

directions; therefore these modes are called TE modes. For an electric field polarized in the plane of incidence, there is a set of modes satisfying Eqs. (2.61), (2.65) and (2.66). These modes have a magnetic field in the x direction, perpendicular to the propagation direction z, but the electric field has components in both the y and z directions; thus they are called the TM modes.

All planar waveguide modes have evanescent fields in the cladding and substrate. Therefore, even if there is scattering or absorption in the cladding or the substrate, they still have low propagation losses in the z direction. Planar waveguide modes cannot be excited by a propagating plane wave in the cladding or the substrate.[6]

2.4.4 The hollow dielectric waveguide mode

Different than the planar waveguide, the hollow dielectric waveguide has the same material geometry as the planar waveguide shown in Figure 2.4, except n_c and n_s are both considerably larger than n_2. The solutions presented in Sections 2.4.1 and 2.4.2(b) are applicable. However, the hollow dielectric waveguide refers to the situation where the excitation takes place inside the waveguide layer. When the excitation of the waves takes place inside the waveguide, there will be multiple reflected plane waves back and forth between boundaries inside the waveguide, and there will also be corresponding radiated waves in the cladding and in the substrate that are propagating away from the boundaries. No matter what the indices of the layers, whenever $2n_2k\cos\theta_m t + \varphi_{wc} + \varphi_{ws} = 2m\pi$ the multiply reflected E_{rw} and transmitted E_{fw} in the waveguide layer will be in phase with each other. Thus we again have resonance of the multiply reflected and transmitted waves in the waveguide.

The reflection coefficients from the waveguide to the cladding and substrate at the $x = t$ and the $x = 0$ boundaries will depend on the polarization of the electric field. They are

$$\Gamma_{\perp wc} = \frac{n_2\cos\theta_2 - n_c\cos\theta_1}{n_2\cos\theta_2 + n_c\cos\theta_1} \qquad \Gamma_{\perp ws} = \frac{n_2\cos\theta_2 - n_s\cos\theta_s}{n_2\cos\theta_2 + n_s\cos\theta_s} \tag{2.67}$$

$$\Gamma_{//ws} = \frac{n_s\cos\theta_2 - n_2\cos\theta_s}{n_s\cos\theta_2 + n_2\cos\theta_s} \qquad \Gamma_{//wc} = \frac{n_c\cos\theta_2 - n_2\cos\theta_c}{n_c\cos\theta_2 + n_2\cos\theta_c} \tag{2.68}$$

The resonance condition is

$$2n_2k\cos\theta_m t + \varphi_{wc} + \varphi_{ws} = 2m\pi \tag{2.69}$$

Note that the transmission coefficients to the propagating wave in the cladding and the substrate are related to the reflection coefficients by

[6] Planar waveguide modes can be excited from plane waves in a prism that has an index $> n_2$.

$$T_{\perp wc} = 1 + \Gamma_{\perp wc}, \quad T_{\perp ws} = 1 + \Gamma_{\perp ws}, \quad T_{//wc} = \frac{n_c}{n_2}(1 + \Gamma_{\perp wc}),$$
$$T_{//ws} = \frac{n_s}{n_2}(1 + \Gamma_{//ws}) \tag{2.70}$$

If n_c and n_s are significantly larger than n_2 then the reflection coefficients in Eqs. (2.67) and (2.68) can be large, and the transmission coefficients will be small. Similar to the resonance effect discussed in Sections 2.4.1 and 2.4.2, it means that when the reflection coefficients Γ_{wc} and Γ_{ws} are large, the total field in the waveguide is much larger than the fields in the cladding and the substrate at resonance. The excited field at resonance will propagate in the z direction with just some radiation loss. Since the lowest n_2 is the free space, such a waveguide structure used at resonance is called a hollow dielectric waveguide.

In summary, the planar waveguide structure shown in Figure 2.4 could be analyzed simply by plane wave analysis. How such a structure functions depends on the excitation of modes. (1) For incident beams in the cladding or the substrate, there are the usual reflected and transmitted beams. As we have shown in Section 2.1, one can use it to control reflection or transmission. There are also the resonances at specific θ_2 *angles. (2) When* $n_c < n_s < n_2$ *and* $\theta_m > \sin^{-1}(n_s/n_2)$, *there is a non-zero solution only if the resonance condition is satisfied. The mth solution satisfying the resonance condition is the planar waveguide mode. It can only be excited from inside the waveguide. (3) If* $n_2 << n_s$ *and* n_c, *waves excited within the waveguide layer at the resonance angle propagate as hollow dielectric modes.*

Although the solutions of guided-wave modes propagating in a planar waveguide are obtained here by plane wave analysis, the total field of the guided-wave mode is no longer a TEM wave propagating in the z direction. For the electric field polarized in the y direction, we have the TE modes. The magnetic field has components in both x and z directions. For the magnetic field in the y direction, we have the TM modes. In other words, even though the plane wave analysis provided us with a solution, the properties of the waves can best be described in terms of modes. In Chapter 5, these modes and their interactions will be analyzed again by modal analysis. The modal analysis is used in Chapters 6, 7 and 8 to analyze devices based on the mutual interactions of modes. These tasks cannot be accomplished by plane wave analysis.

Chapter summary

The interference effects caused by superposition and multiple reflections of plane waves could be used very effectively to analyze and understand the gist of many applications. They include anti-reflection coatings, beam splitting, reflection coatings, the Fabry–Perot resonance, modes in planar waveguides, holography, various applications of the Fabry–Perot resonance, etc.

The analyses demonstrated the importance of the concept of superposition of waves. Pedagogically, it is interesting to note that the plane wave analysis has already yielded the modes in a planar waveguide. Yet, many properties and application of the waveguide devices can only be discussed effectively by modal analysis, not by plane wave analysis.

If we combine the discussions in Chapters 1 and 2, we see that plane wave analysis could be used as a first approximation to analyze many applications. It is important to note that plane wave analysis can be applied only to material structures with a planar boundary. The full analysis of these applications in the non-planar configurations of realistic components requires the use of diffraction or modal analysis that will be presented subsequently in this book.

Reference

[1] David M. Pozar, *Microwave Engineering*, John Wiley & Sons, 2005, Chapter 4.

3 Scalar wave equation and diffraction of optical radiation

For analysis of optical radiation propagating in realistic components that have finite boundaries and an optical radiation beam that has lateral variation, plane wave analysis cannot be used. Maxwell's equations with appropriate boundary conditions should be used. However, rigorous analysis using vector Maxwell's equations plus boundary conditions is very complex and tedious. Even if we find the solutions they might contain fine features (such as the fringe fields near the aperture) that are often of little or no significance for practical applications. In many cases, we need only a simple solution that can give us the main features (i.e. the amplitude and phase variations) of the dominant electromagnetic field at a distance moderately far away from the input aperture. We do not need to know the near field close to the aperture.

When one deals with radiation fields that have slow transverse variations and that interact with devices that have overall dimensions much larger than the optical wavelength λ, the fields are often transverse electric and magnetic (TEM). In TEM waves, both the dominant electric field and the dominant magnetic field polarization lie approximately in the plane perpendicular to the direction of propagation. The dominant electric and magnetic fields are also perpendicular to each other. The polarization does not change rapidly while the radiation propagates in an isotropic medium within moderate distances.[1] In this case, we usually need only to solve the scalar wave equations to obtain the amplitude and the phase of the dominant electric field along its polarization direction. The dominant magnetic field can be calculated directly from the dominant electric field. Conversely, we can also first solve the scalar equation of the dominant magnetic field, and the electric field can be calculated from the magnet field.

The condition under which the scalar wave equation is applicable will be discussed in Section 3.1. To find the solution of the scalar wave equation, commonly known as the Kirchoff's integral, involves a lot of mathematical details. A discussion of its derivation will divert our attention from the application of Kirchoff's integral. Therefore it is presented separately in the Appendix. Kirchoff's integral is presented in Section 3.2 without derivation. In the rest of the sections in this chapter we will focus on the applications of Kirchoff's integral. These applications lead directly to the traditional Fresnel and Fraunhofer diffraction patterns that determine the resolution of telescopes and microscopes, as well as to laser cavity modes and Gaussian beams described in Chapter 4. In addition, various mathematical techniques can be applied to the

[1] In birefringent media such as crystal, the polarization of electric field rotates.

Kirchhoff's integral. For example, under certain circumstances, the incident and diffracted fields are related by Fourier transform. Therefore, analytical techniques based on Fourier analysis, such as convolution, are applicable.

3.1 The scalar wave equation

The simplest way to understand why we can use a scalar wave equation is to consider Maxwell's vector wave equation in a homogeneous medium without free charge carriers. It can be written in terms of the rectangular coordinates as:

$$\nabla^2 \underline{E} - \frac{1}{c^2}\frac{\partial^2 \underline{E}}{\partial t^2} = 0\,, \quad \underline{E} = E_x \underline{i_x} + E_y \underline{i_y} + E_z \underline{i_z} \tag{3.1}$$

If $\underline{E}$ has only one dominant component, for example $E_x\ \underline{i_x}$, then E_y and E_z may be neglected. The unit vector $\underline{i_x}$ does not have to be displayed explicitly. In this approximation, the resultant equation is a scalar wave equation for E_x. Alternatively, the dominant component may be $\underline{E_y}$, and E_x and E_z can be dropped from Eq. (3.1).

In short, for TEM waves with constant polarization, we usually describe the dominant electromagnetic (EM) field by a scalar function U that is proportional to the dominant electric field without specifying explicitly its polarization.[2] In a homogeneous medium without free carriers, U satisfies the scalar wave equation,

$$\nabla^2 U - \frac{1}{c^2}\frac{\partial^2}{\partial t^2} U = 0 \tag{3.2}$$

In our presentation here, U is the instantaneous complex amplitude of the transverse electric field in its direction of polarization. U varies slowly in the transverse direction within a distance comparable to the wavelength. The dominant magnetic field can be calculated directly from the dominant electric field. From a different perspective, when we use the scalar wave equation, we have implicitly assumed that the curl equations in the Maxwell's equations do not yield significant magnitudes of electric field in other directions. There are other views of what constitutes U. In books such as that of Born and Wolf, *Principles of Optics*, it is shown that U can also be considered as a scalar potential for the optical field. In that case, electric and magnetic fields can be derived from the scalar potential [1].

Both the scalar wave equation (3.2) and the boundary conditions have been obtained from Maxwell's equations. If U represents the dominant electric field, the continuity of electric field is equivalent to the continuity of U across the boundary. The continuity of the magnetic field across the boundary is equivalent to the continuity of the normal derivative of U. In other words, the boundary conditions in vector Maxwell's equations are replaced by boundary conditions of U (i.e. the continuity of U and the normal derivative of U) across the boundary.

[2] All detectors convert the optical power into electrical current. In electromagnetic field theory, we learned that $I = 1/2|E|^2/\sqrt{\mu_o/\varepsilon}$, where E is the transverse electric field. In optics, U is usually normalized (i.e. $|U|$ is just proportional to the magnitude of the transverse electric field) such that UU^* is the intensity.

Note that U is only a solution for a given polarization of TEM wave. There are always two independent solutions representing fields in orthogonal polarizations. For each polarization, there are forward and backward waves that are two independent solutions.

For wave propagation in a complex medium, Eq. (3.2) can be considered as the equation for propagation of TEM waves in the local region. In order to obtain a global analysis of wave propagation in a complex medium, solutions obtained for adjacent local regions are then matched in both spatial and temporal variations at the boundaries of local regions.

For cw single-frequency radiation with a harmonic time variation, we usually write:

$$U(x,y,z;t) = U(x,y,z)e^{j\omega t} \tag{3.3}$$

Here $U(x, y, z)$ is complex, i.e. $U(x,y,z)$ has both amplitude and phase. Consequently U satisfies the Helmholtz equation,

$$\nabla^2 U + k^2 U = 0 \tag{3.4}$$

Here, $k = \omega/c = 2\pi/\lambda$; c = velocity of light $=1/\sqrt{\varepsilon\mu}$. The boundary conditions are the continuity of U and the normal derivative of U across the discontinuity.

3.2 The solution of the scalar wave equation: Kirchhoff's diffraction integral

Let us consider a radiation U_{in} incident on an opaque flat screen Σ at $z = 0$ that has a limited open aperture Ω. The screen extends to infinity in both x and y directions. The volume of space beyond the screen and the aperture is enclosed by a spherical boundary at $z >> 0$ with a very large radius R. Figure 3.1 illustrates the configuration. When the U within this enclosed volume does not have any optical radiation source, U satisfies the radiation condition at the spherical boundary [2].

$$\lim_{R\to\infty} R\left(\frac{\partial U}{\partial n} + jkU\right) = 0 \tag{3.5}$$

Here, $\partial/\partial n$ means the derivative normal to the boundary.

In most cases, we know the input U_{in} at $z = 0$ within the aperture Ω. We like to calculate U for an observer located at a position $\underline{r_o}$, some distance away from the Ω. In the appendix, U at $\underline{r_o}$ is shown to be related to U_{in} as[3]:

$$U(\underline{r_o}) = \frac{j}{\lambda}\iint_{\Omega} U_{in}\frac{e^{-jkr_{o1}}}{r_{o1}}\cos\alpha\, dx_1 dy_1 \tag{3.6}$$

Here $\underline{r_1}$ is any position x_1 and y_1 in the aperture at $z = 0$, and $r_{o1} = \sqrt{(x_o - x_1)^2 + (y_o - y_1)^2 + z_o{}^2}$. The integration is carried over the entire aperture Ω. α is the angle of $\underline{r_o}$ with respect to the z axis.

[3] This result has also been derived from Huygens' principle in classical optics.

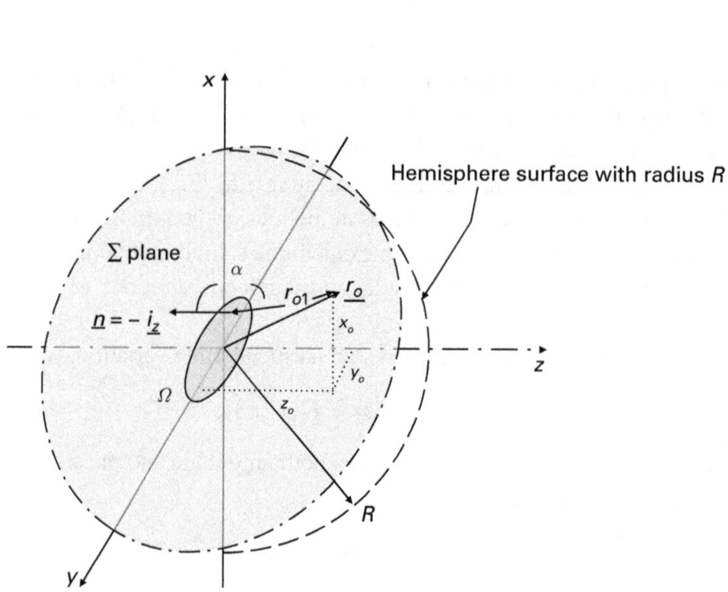

Figure 3.1 Geometrical configuration of the aperture and the semispherical volume for the Kirchoff's integral. The radiation is incident on Σ, which has an aperture Ω. The very large hemisphere with radius R is connected to Σ. The coordinates for the observation point r_o are x_o, y_o, and z_o.

In a paraxial approximation, the observer position (x_o,y_o,z_o) is in a direction not far away from the center direction of propagation and the observer is located at a distance reasonably far from the aperture, i.e. $\alpha \approx 180^\circ$ and $|\underline{r_{o1}}| \approx |z_o| \cong \rho$. Then, under the condition of paraxial approximation, α is now approximately a constant in the integrand of Eq. (3.6) over the entire aperture Ω, while the change of ρ in the denominator of the integrand also varies very slowly over the entire Ω. Thus, U can be simplified further to yield:

$$U(z \cong \rho) = \frac{-j}{\lambda \rho} \iint_{\Omega} U_{\text{in}} \mathrm{e}^{-jkr_{o1}} \mathrm{d}x_1 \mathrm{d}y_1 \tag{3.7}$$

Note that $k = 2\pi/\lambda$ is a very large quantity. A small change in r_{o1} in the exponential can significantly affect the value of the integral. Thus r_{o1} in the $\mathrm{e}^{-jkr_{o1}}$ factor in the integrand cannot be simplified, while the ρ factor in the denominator of the integrand can be considered as a constant in the paraxial approximation.

Both Eqs. (3.6) and (3.7) are known as Kirchhoff's diffraction formula. In the case of paraxial approximation, limited aperture and large r_{o1}, Eq. (3.7) yields the same result as Eq. (3.6). However, Eq. (3.7) is more commonly used in engineering literature.

Note that, in order to calculate U at $\underline{r_o}$ we need to know $U_{in}(x,y,0)$ in the aperture. Strictly speaking, when U is incident on the aperture, it creates a $U_{in}(x,y,0)$ that includes the incident U plus the fringe fields created by the induced currents on the screen. For example, the screen could be made of metal. There are induced currents at the rim of the aperture. However, the fringe fields are weak for large apertures. They are near-fields *that decay rapidly.*[4] *Therefore, we assume* $U(x, y, 0) \cong U_{\text{in}}(x, y, 0)$ *in many optical applications.*[5]

3.2.1 Kirchhoff's integral and the unit impulse response

Eq. (3.7) is sometimes presented in a different format for engineers. Let

$$\frac{-j}{\lambda \rho} \mathrm{e}^{-jkr_{o1}} = h[(x - x_o), (y - y_o), (z - z_o)]. \tag{3.8}$$

Then, we can write U in a different format,

$$U(\underline{r_o}) = \iint_{\Omega} U(x, y, 0) h[(x - x_o), (y - y_o), z_o] \mathrm{d}x \mathrm{d}y. \tag{3.9}$$

In this format, U at $\underline{r_o}$ and U at $(x,y,z = 0)$ are related by a transform relation through h. If $U(x, y, 0)$ is a unit impulse $\delta(x,y)$ (i.e. a point source), then the $U(\underline{r_o})$ obtained from the integration is approximately h for large apertures Ω. Thus, the h function is known as the unit impulse response function.

The expression $h(x - x_o, y - y_o, z - z_o)$ *has the same format as the electrical impulse response in system analysis.* $U(x, y, 0)$ *is just the source excitation at the* $z = 0$ *plane. "h" determines completely* $U(x_o, y_o, z_o)$ *from any input* $U(x, y, 0)$*. Eq. (3.9) is the foundation of many pattern recognition, filtering, and optical signal processing techniques.*

Note that unit impulse techniques used in system analyses usually use integrals within $-\infty$ *and* $+\infty$ *limits of integration, while the limits of integration in Eq. (3.9) are determined by the aperture size. Nevertheless, much can be learned from those techniques, especially when the aperture is large. Furthermore, as the integrand in Eq. (3.9) can also be written as a product of "*$h(x, x_o; y, y_o) U(x, y, 0)$*" and an unit step function of x and y representing* Ω*, the limits of integration can be extended to* $\pm\infty$*.*

3.2.2 Fresnel and Fraunhofer diffractions

In Eq. (3.7) or Eq. (3.9), we note that binomial expansion may be applied to ρ as follows:

$$\begin{aligned} \rho &= (z_o - z)\sqrt{1 + \frac{(x_o - x)^2 + (y_o - y)^2}{(z_o - z)^2}} \\ &= d\left[1 + \frac{1}{2d^2}(x_o^2 + y_o^2 - 2xx_o - 2yy_o + x^2 + y^2) + \text{higher-order terms}\right]. \end{aligned} \tag{3.10}$$

Here, $d = z_o - z$, and in paraxial approximation, $d >> |x_o - x|$ and $|y_o - y|$.

If d is sufficiently large so that we can drop the higher-order terms, we obtain from Eq. (3.10):

[4] See Section 1.1.5(c).

[5] This is the major difference between microwaves and optics. The induced fields are often important in microwaves, because of the much larger ratio of wavelength relative to the aperture size in microwaves.

$$U(\underline{r_o}) = \frac{-j}{\lambda\, d} \mathrm{e}^{-jkd} \mathrm{e}^{-jk\frac{x_o{}^2+y_o{}^2}{2d}} \iint_{\Omega} \left[U(z=0) \mathrm{e}^{-j2\pi\frac{x^2+y^2}{2\lambda d}} \right] \mathrm{e}^{+j2\pi\frac{xx_o}{\lambda d}} \mathrm{e}^{+j2\pi\frac{yy_o}{\lambda d}} \mathrm{d}x \mathrm{d}y\,. \tag{3.11}$$

This is known as the Fresnel diffraction integral, which describes diffraction effects.

If d is so large that the term involving $(x^2 + y^2)$ can also be neglected, then we obtain even a simpler diffraction integral,

$$U(\underline{r_o}) = \frac{-j}{\lambda\, d} \mathrm{e}^{-jkd} \mathrm{e}^{-jk\frac{x_o{}^2+y_o{}^2}{2d}} \iint_{\Omega} U(z=0) \mathrm{e}^{+j2\pi\frac{xx_o}{\lambda d}} \mathrm{e}^{+j2\pi\frac{yy_o}{\lambda d}} \mathrm{d}x \mathrm{d}y\,. \tag{3.12}$$

This is known as the Fraunhofer diffraction integral of the far radiation field. Note that U as a function of x_o and y_o is approximately a Fourier transform of U as a function of x and y.

It is important to note the implications and the differences of the results shown in Eqs. (3.10) to (3.12). (1) The major physical difference in Fresnel and Fraunhofer diffraction formula is the distance from the aperture. When the condition for Fraunhofer diffraction is met, $U(\underline{r_o})$ and $U(z=0)$ are related by a Fourier transform integral with finite limits of integration. Therefore, mathematical techniques in Fourier analysis could be used to analyze $U(\underline{r_o})$. (2) The preceding result is valid only if the radiation condition and the condition for paraxial approximation could be satisfied. (3) When U_{in} is incident on the aperture, part of the incident beam U_{in} is blocked by the screen Σ, which has the aperture Ω. Therefore the power carried by the diffracted beam is reduced from the power of the incident beam by the screen.

3.2.3 Applications of diffraction integrals

There are many applications that could be analyzed by Fraunhofer and Fresnel diffraction integrals. There are numerous examples given in existing books. Only a few applications are presented below to demonstrate the use and the significance of diffraction integrals.

(a) Far field diffraction pattern of an aperture

Far field diffraction from a uniform U_{in} incident on a rectangular aperture is the simplest example to illustrate the significance of Eq. (3.12).

Let the radiation U_{in} be a plane wave in the z direction that has amplitude A. It is normally incident on an opaque screen at $z = 0$ that has a rectangular open aperture with dimensions, $2l_x$ and $2l_y$, in the x and y directions, i.e.

$$U(x, y, z=0) = A\, \mathrm{rect}\left(\frac{x}{l_x}\right) \mathrm{rect}\left(\frac{y}{l_y}\right) \tag{3.13}$$

where

$$\begin{aligned} \mathrm{rect}(\chi) &= 0 \quad \text{for} \quad |\chi| \rangle\, 1, \\ \mathrm{rect}(\chi) &= 1 \quad \text{for} \quad |\chi| \le 1 \end{aligned} \tag{3.14}$$

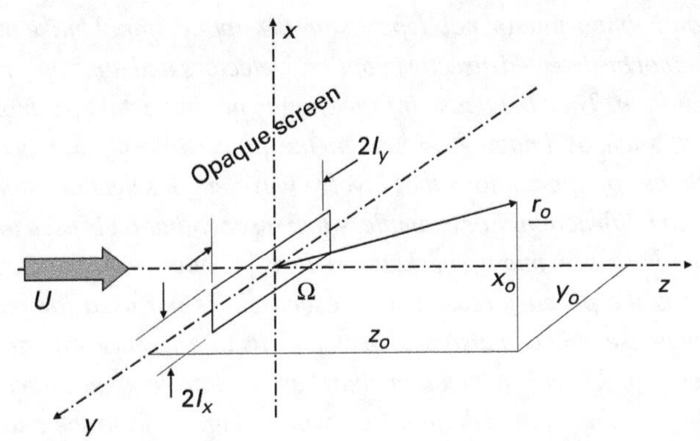

Figure 3.2 Geometrical configuration of a rectangular aperture. The radiation U is incident on a rectangular aperture Ω on an opaque screen, which is the x–y plane. For a far field, r_o is far away with large z_o coordinates. In the paraxial approximation, $|z_o| >> |x_o|$ and $|y_o|$.

Figure 3.2 illustrates the geometric configuration. Substituting $U(x,y,z=0)$ into Eq. (3.12), we obtain:

$$U(x_o, y_o, d) = \frac{-j\mathrm{e}^{-jkd}\mathrm{e}^{-j\frac{k}{2d}(x_o{}^2+y_o{}^2)}}{\lambda d} \iint_{\Omega} \left[A\mathrm{e}^{-jk\frac{x^2+y^2}{2d}}\right] \mathrm{e}^{j2\pi\left(\frac{x_o}{\lambda d}\right)x} \mathrm{e}^{j2\pi\left(\frac{y_o}{\lambda d}\right)y} \mathrm{d}x\mathrm{d}y. \tag{3.15}$$

When d is very much larger than the aperture, so that

$$k\frac{l_x{}^2 + l_y{}^2}{2d} \langle\langle 1 \tag{3.16}$$

then $\mathrm{e}^{-jk\frac{x^2+y^2}{2d}} \cong 1$. Since

$$\int_{-l_x}^{l_x} \mathrm{e}^{j2\pi\left(\frac{x_o}{\lambda d}\right)x}\mathrm{d}x = \frac{\mathrm{e}^{j2\pi\left(\frac{x_o}{\lambda d}\right)l_x} - \mathrm{e}^{-j2\pi\left(\frac{x_o}{\lambda d}\right)l_x}}{j2\pi\left(\frac{x_o}{\lambda d}\right)} = \frac{j2\sin\left[2\pi\left(\frac{x_o}{\lambda d}\right)l_x\right]}{j2\pi\left(\frac{x_o}{\lambda d}\right)}. \tag{3.17}$$

we obtain the far field U from Eq. (3.17) as,

$$U(x_o, y_o, d) = \frac{4j\mathrm{e}^{-jkd}\mathrm{e}^{-j\frac{k}{2d}(x_o{}^2+y_o{}^2)}}{\lambda d} A\, l_x l_y \mathrm{sinc}\left(\frac{2l_x x_o}{\lambda d}\right)\mathrm{sinc}\left(\frac{2l_y y_o}{\lambda d}\right),$$

where

$$\mathrm{sinc}(x) = \frac{\sin \pi x}{\pi x} \tag{3.18}$$

U is the classical Fraunhofer diffraction pattern of the rectangular aperture for a plane wave normally incident on the aperture. There are four comments. (1) The Fraunhofer diffraction pattern is ignored in geometric or ray optics because the transverse amplitude and phase variations are not important in those applications. The ray optics approximation corresponds to the situation where one is interested only in U as x_o/d *and* $y_o/d \to 0$ *in Eq. (3.18). (2) U in the far field has a spherical phase front centered about* $z = 0$*. Whether* *this phase variation is important or not depends on the application. Unlike microwaves, the electric field cannot be detected directly in optics. Detectors and films only measure the intensity of the radiation. Thus the phase information is not important for most classical optics applications, such as imaging. However, the phase information becomes very important for a number of applications that involve wavelength selection, signal processing, interference, and diffraction. For example, when laser radiation is used to illuminate an image pattern, there are many speckles created by interference effects of small irregularities. This is the primary reason why laser light is not used for photography. (3) The effect of the phase of U can also be detected by its interference with another U' or by diffraction effects of U with a medium that has an interference pattern, such as holography. (4) Besides the main radiation lobe within angle α from the z axis that has* $\alpha_x \cong x_o/d < \lambda/2l_x$ and $\alpha_y \cong y_o/d < \lambda/2l_y$, *there are side lobes with secondary peaks at* $2l_x x_o/\lambda d = n + 1/2$ and $2l_y y_o/\lambda d = n + 1/2$ *where n = 1, 2, 3,... In some applications the side lobes are very important.*

When the input is a plane wave incident on the screen in the y–z plane at an angle θ with respect to the z axis,

$$U(x,y,z=0) = Ae^{-jk\sin\theta x}\mathrm{rect}\left(\frac{x}{l_x}\right)\mathrm{rect}\left(\frac{y}{l_y}\right) \tag{3.19}$$

The far field is

$$U(x_o,y_o,d) = \frac{4je^{-jkd}e^{-j\frac{k}{2d}(x_o{}^2+y_o{}^2)}}{\lambda d} A\, l_x l_y \mathrm{sinc}\left(\frac{2l_x}{\lambda}\left(\sin\theta - \frac{x_o}{d}\right)\right)\mathrm{sinc}\left(\frac{2l_y y_o}{\lambda d}\right) \tag{3.20}$$

Therefore the diffracted wave is a wave with its main lobe in the direction $x_o/d = \sin\theta$.

(b) Far field radiation intensity pattern of a lens

If a point source is placed at the focus of a lens with infinite aperture, it creates a plane wave at the output of the lens. When the lens has a finite size, the output is equivalent to having this plane wave to pass through an additional aperture, as discussed in (a).

The intensity I at x_o and y_o for a point source placed at the focus of a lens with a rectangular aperture is:

$$I(x_o,y_o) = UU^* = \left[\frac{4Al_x l_y}{\lambda d}\mathrm{sinc}\left(\frac{2l_x x_o}{\lambda d}\right)\mathrm{sinc}\left(\frac{2l_y y_o}{\lambda d}\right)\right]^2. \tag{3.21}$$

Figure 3.3 illustrates I as a function of x_o when $y_o = 0$. Clearly I is inversely proportional to d^2, as we would expect for a divergent wave. The intensity I has a major radiation loop directed along the direction of propagation of the incident beam. I also has minor radiation loops in x directions when $x_o/d = (3/2)\lambda/l_x$, $(5/2)\lambda/l_x$, etc., and in y directions when $y_o/d = (3/2)\lambda/l_y$, $(5/2)\lambda/l_y$, etc.

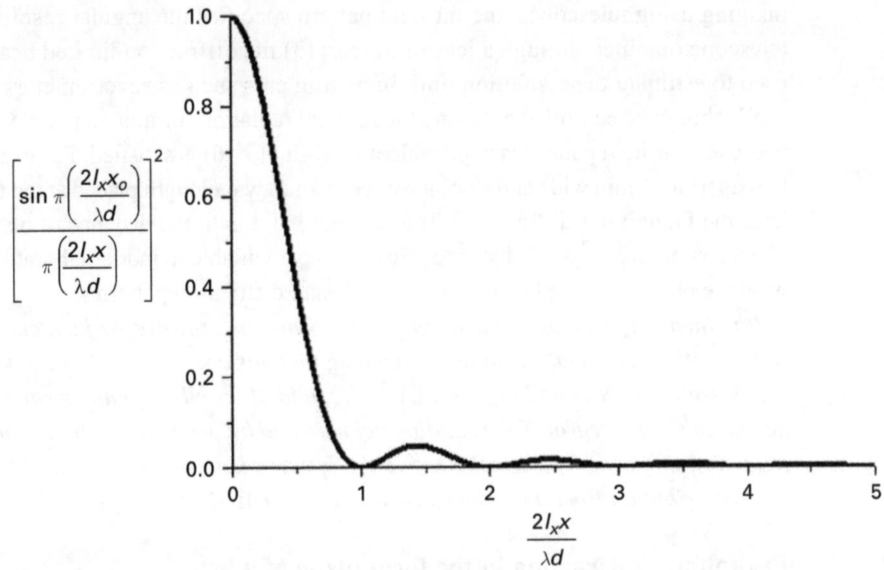

Figure 3.3 Diffraction pattern of the plane wave passing through the rectangular aperture Ω in Figure 3.1.

In optics, the minimum diffraction beam width of the major loop is defined as the angle θ between the direction of propagation and the first zero of I. Thus, for a rectangular aperture, the beam widths, θ_x and θ_y, are:

$$\theta_x = \frac{\lambda}{2l_x} \quad \text{and} \quad \theta_y = \frac{\lambda}{2l_y} \tag{3.22}$$

The preceding discussion in the rectangular coordinates demonstrated clearly the characteristics of the diffracted far field without complex mathematics. In practice, the apertures are round. For circular apertures with radius r', similar results have been described in classical optics books using cylindrical coordinates and Bessel functions.[6] In that case the beam width of the main radiation loop is given by [1]:

$$\theta_c = 0.62\lambda/r' \tag{3.23}$$

As a result, Eq. (3.23) is commonly used to specify the angular resolution of a lens. It is also applicable to mirrors. The difference in the results derived from circular and rectangular apertures is minor.

The diffraction beam width at the far field is often used to characterize the output radiation from many instruments without a detailed discussion of the beam pattern. For example: (1) the output from a laser is frequently described in trade brochures by its far field radiation beam width; (2) for communication among distant stations or

[6] The mathematics is much more complex for the calculation. It does not lead to any new insight. Therefore it is not presented in detail in this book.

imaging using telescopes, the far field pattern specifies the angular resolution of the telescope obtained through a lens or mirror; (3) the diffraction limited beam width is used to estimate the resolution limit in instruments such as spectrometers.

We should be careful about using the far field radiation formula in practical situations because a far field pattern is applicable only if Eq. (3.16) is satisfied. For example, for a lens aperture 1 mm wide and a point source at 1 μm wavelength placed at the focus of the lens, the Fraunhofer diffraction formula is not valid until the distance of observation is 30 meters or larger.[7] Such distances are often not available in indoor laboratories. Often, what we observe in the laboratory is the Fresnel diffraction pattern.

It is interesting to note that when a plane wave (microwave) is incident on a metal screen with a very small rectangular opening that has a size comparable to wavelength, the emerging wave is no longer a TEM wave, and Maxwell's equations need to be used, not Kirchhoff's integral. The radiation field created by the induced current on the edges of the opening needs to be included. As the opening is increased, the far field radiation field contribution from the induced current is decreased.

(c) Fraunhofer diffraction in the focal plane of a lens

A lens is a very common optical component. In imaging, the diffraction effect yields a finite spot size. The Fraunhofer diffraction in the focal plane of a lens is presented below. In addition to the diffraction limit of the spot size, there is a Fourier transform relationship between the incident field and the field at the focal plane. It is used in many signal processing functions, for example spatial filtering [3].

Consider a plane wave incident normally on a lens at $z = 0$. A plane aperture is placed immediately after the lens. Let the focal length of the lens be z_2. Then, from Section 1.3.5 (b), the transmitted $U(\underline{r_o})$ for a pane wave normally incident on the lens placed at $z = 0$, without any limitation on the aperture size, will be focused onto a spot at $\underline{r_2} = z_2\underline{i_z}$. The output from the lens, before the aperture, is a convergent wave. The lens, the focus, and the aperture are illustrated in Figure 3.4.

The spherical wave emerging from the lens can be expressed for $0 < z_o < z_2$ as

$$U = A\frac{e^{+jkr_{2o}}}{r_{2o}}, \qquad r_{2o} = |\underline{r_2} - \underline{r_o}| = \sqrt{(x_2 - x_o)^2 + (y_2 - y_o)^2 + (z_2 - z_o)^2} \tag{3.24}$$

Note that the + sign in the exponential combined with the $e^{+j\omega t}$ time variation represents a convergent wave. U immediately after the lens is given by Eq. (3.24) where $x_o = x$, $y_o = y$, and $z_o = 0$. When an aperture Ω is placed after the lens at $z = 0$, U at (x_o, y_o, z_o) for $z_2 > z_o > 0$ is given by Eq. (3.7).

$$U(x_o, y_o, z_o) = \frac{-jA}{\lambda z_2 r_{o1}} \iint_{\Omega} e^{-jk(r_{01} - r_{21})} dxdy \tag{3.25}$$

[7] It will be shown in Chapter 4 that the far field condition for the Gaussian modes of a laser is much less stringent than the condition in Eq. (3.16).

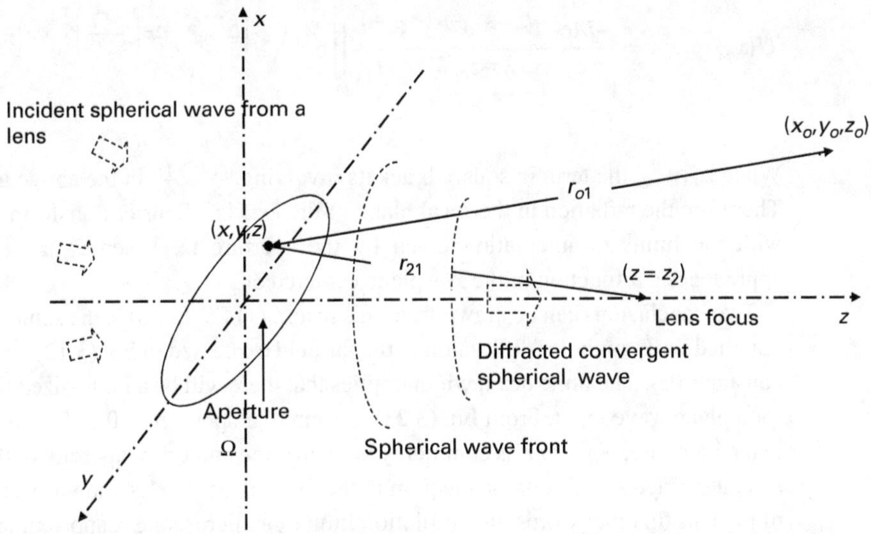

Figure 3.4 Illustration of a spherical wave incident on a plane aperture Ω. The incident wave is a converging spherical wave focused at (x_2, y_2, z_2). It passes through an opening aperture Ω of an opaque screen, which is the x–y plane. r_o is the observation point.

Here, the coordinates in the aperture (x_1, y_1, z) in Eq. (3.7) are replaced by (x, y, z) in Eq. (3.25). Using paraxial approximation and binomial expansion, and noting that the aperture is in the $(z = 0)$ plane, we obtain:

$$k(r_{o1} - r_{21}) = k[z_o - z_2] + \frac{k}{2}\left[\frac{x_o^2 + y_o^2}{z_o} - \frac{x_2^2 + y_2^2}{z_2}\right] - k\left[\frac{x_o x + y_o y}{z_o} - \frac{x_2 x + y_2 y}{z_2}\right] + \frac{k}{2}\left[\begin{array}{l}\dfrac{x^2 + y^2}{z_o} - \dfrac{x^2 + y^2}{z_2} + \left\{-\dfrac{1}{4}\dfrac{\left[(x_o - x)^2 + (y_o - y)^2\right]^2}{z_o^3} + \dfrac{1}{4}\dfrac{\left[(x_2 - x)^2 + (y_2 - y)^2\right]^2}{z_2^3}\right\} \\ + \text{ other higher-order terms}\end{array}\right] \tag{3.26}$$

For sufficiently large z_o and z_2,

$$\left|\frac{k}{8}\frac{\left[(x_o - x)^2 + (y_o - y)^2\right]^2}{z_o^3}\right|_{\max} << 2\pi$$

and

$$\left|\frac{k}{8}\frac{\left[(x_2 - x)^2 + (y_2 - y)^2\right]^2}{z_2^3}\right|_{\max} << 2\pi, \tag{3.27}$$

the terms in the curly brackets and other higher-order terms can be neglected.

$$U(x_o, y_o, z_o) = \frac{-jA\mathrm{e}^{-jk(z_o - z_2)}\mathrm{e}^{-jk\left(\frac{x_o{}^2 + y_o{}^2}{2z_o}\right)}}{\lambda z_2 z_o} \iint\limits_{\Omega} \mathrm{e}^{j\frac{\pi}{\lambda}\left[\frac{1}{z_o} - \frac{1}{z_2}\right](x^2+y^2)} \mathrm{e}^{j2\pi\left[\frac{x_o}{\lambda z_o} - \frac{x_2}{\lambda z_2}\right]x} \mathrm{e}^{j2\pi\left[\frac{y_o}{\lambda z_o} - \frac{y_2}{\lambda z_2}\right]y} \mathrm{d}x\mathrm{d}y. \tag{3.28}$$

When $z_o \approxeq z_2$, the term in square brackets involving "$\frac{1}{z_o} - \frac{1}{z_2}$" in the above integral is 0. Therefore the radiation in the focal plane of the lens is a Fourier transform of the input with the limits of integration given by the aperture Ω. When Ω is very large, it approaches a δ function in the x–y plane centered at $z = z_2$.

Two conclusions can be drawn from this result. (1) Eq. (3.28) is the same as the result obtained for Fraunhofer diffraction in the far field expressed in Eq. (3.12), except for the constant A/z_2. The finite beam width implies that there will be a finite-sized focused spot for a plane wave input. From Eq. (3.23), the circular spot size is $0.61 f\lambda_o/nr'$. Note that, in the literature, $NA = f/nr'$ is commonly known as the numerical aperture of the lens, and n is the refractive index of the medium in the focal plane. $f/2r'$ is known as the F number of the lens. In other words, the resolution limit of a microscope is approximately λF. (2) If a thin transparent film with amplitude and phase transmission $t(x, y)$ is placed before the aperture and the lens at $z = 0$, then U at the focal plane for a normal incident uniform plane wave is:

$$U(x_o, y_o, z_2) = \frac{-jA\mathrm{e}^{-j2\pi\left(\frac{x_o{}^2 + y_o{}^2}{2\lambda z_o}\right)}}{\lambda z_0{}^2} \iint\limits_{\Omega} t(x, y)\mathrm{e}^{j2\pi\left(\frac{x_o}{\lambda z_o}\right)x}\mathrm{e}^{j2\pi\left(\frac{y_o}{\lambda z_o}\right)y}\mathrm{d}x\mathrm{d}y \tag{3.29}$$

This is an important result. It states that when the limits of integration are large, U at the focal plane $z_o = z_2$ is essentially the Fourier transform of t at $z = 0$. The spatial Fourier frequencies of the Fourier transform are $f_x = x_o/\lambda z_o$ and $f_y = y_o/\lambda z_o$.

The usefulness of this result can be illustrated by two simple applications:

(1) In the first application, a student wants to measure the far field radiation pattern of a laser. It is not necessary for him to actually do the measurement at a distance far away. All he needs to do is to use a camera focused to ∞. At the focal plane of the lens, he obtains the far field pattern.

(2) The second application is a spatial filter that can be described as follows. Let us consider two optical lenses with focal length f. Let the lenses be placed in series and perpendicular to the optical axis. They are separated from each other by a distance $2f$. If the size of the lens is sufficiently large then the integration limit in Eq. (3.28) can be approximated by ∞. Now consider the optical signal processing set up shown in Figure 3.5. Let U be a normally incident plane wave. The field at the focal plane of the first lens is the Fourier transform of the transmission of the transparent film, t, placed in front of the first lens. When this radiation is transmitted through an aperture placed at the focal plane of the first lens, the higher Fourier frequencies are blocked by the opaque portion of the aperture. Thus the U obtained after the second lens is $-tU$ filtered through a low-pass spatial frequency filter. Such a setup has many applications. For example, when a laser mode passes through optical instruments, it frequently is perturbed

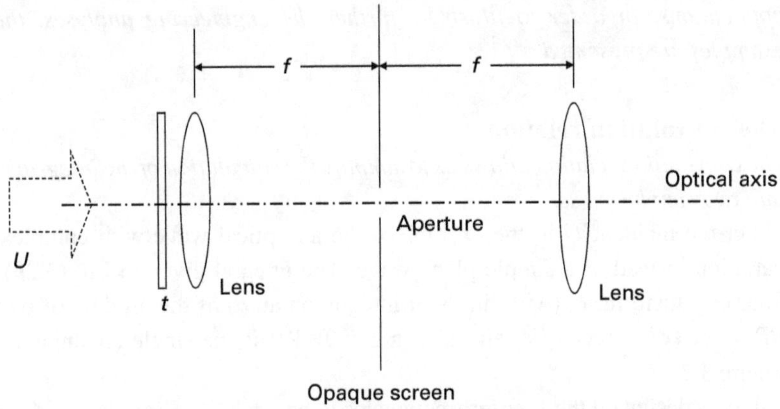

Figure 3.5 Spatial low-pass filtering of an optical wave. A transparent film with transmission function $t(x, y)$ is placed in front of an ideal lens with focal length f. A spatial filter which consists of an opaque screen with a pin-hole aperture is placed at the center, a distance f from the lens at the front. A second lens with focal length f is placed a distance f from the spatial filter.

because of imperfections or defects in the optical elements. A setup such as that shown in Figure 3.5 (without the transparent film t and with a pin-hole aperture) is commercially sold as a spatial filter to clean up the effects of perturbations or defects, which typically produce higher spatial frequencies than the laser mode.

(d) The lens viewed as a transformation element

A simple alternative way to consider a thin lens is to represent it by its transmission function t, as discussed in Section 1.3.5(b). Thus, for any U passing through a thin lens without any aperture, we can now multiply the incident U on the lens by a phase function,

$$t_l = t_o e^{j\frac{\pi}{\lambda f}(x^2+y^2)} \tag{3.30}$$

to obtain U immediately after the lens.

We emphasize that this is a thin lens approximation. Only an ideal lens can be represented by Eq. (3.30). A practical lens will have other higher-order phase shifts, which are considered to be distortions from an ideal lens. Although we have derived this result only for a thin spherical lens, it is used to represent any ideal compound lens where f is the focal length.

3.2.4 Convolution theory and other mathematical techniques

A major difference between the traditional optical analysis used for imaging and diffraction and engineering optical analysis of TEM waves presented here is the analysis of the transform relationship between the incident and the diffracted fields in various applications. In order to illustrate further the engineering analyses, the following examples are presented.

(a) The convolution relation

The convolution relation allows us to simplify the calculation of the integral in Eq. (3.29) *into two simpler parts.*

Let the incident U in the example to be an optical wave with complex functional variation instead of a simple plane wave. The integral given in Eq. (3.29) is then the Fourier transform, F, (with limits of integration at ∞) of the product of two functions, $RR = \text{rect}(x/l_x)\ \text{rect}(y/l_y)$ and $U(x, y, z = 0)$. RR is the single rectangular aperture in Figure 3.2.

Let us designate the Fourier frequencies to be:

$$f_x = \frac{x_o}{\lambda d} \quad \text{and} \quad f_y = \frac{y_o}{\lambda d}$$

$$\text{Let} \quad F_{RR}(f'_x, f'_y) = \int_{-\infty}^{+\infty}\int_{-\infty}^{+\infty} \text{rect}\left(\frac{x}{l_x}\right) \text{rect}\left(\frac{y}{l_y}\right) e^{+j2\pi f'_x x} e^{+j2\pi f_y' y}\, dx\, dy,$$

$$F_U(f'_x, f'_y) = \int_{-\infty}^{+\infty}\int_{-\infty}^{+\infty} U(x, y', z = 0) e^{+j2\pi f'_x x} e^{+j2\pi f'_y y'}\, dx\, dy' \tag{3.31}$$

Then, according to convolution theory,

$$U(\underline{r}_o) = \frac{-j}{\lambda d} e^{-jkd} e^{-jk\frac{x_o^2 + y_o^2}{2d}} \int_{-\infty}^{+\infty}\int_{-\infty}^{+\infty} F_{RR}(f_x', f_y')\, F_U(f_x - f_x', f_y - f_y')\, df_x'\, df_y' \tag{3.32}$$

F_{RR} and F_U are likely results that we already have. Thus we can obtain U at $\underline{r}_o$ from the known results by convolution.

(b) Double slit diffraction

Let us consider the diffraction pattern of a double slit, from $y = h - l_y$ to $y = h + l_y$ and from $y = -h - l_y$ to $y = -h + l_y$. The incident U is a plane wave ($U = A$) propagating in the $+z$ direction.

Using superposition theory and the convolution relation, we obtain immediately the diffraction pattern to be:

$$U(\underline{r}_o) = \frac{-2Aj}{\lambda d} e^{-jkd} e^{-jk\frac{x_o^2 + y_o^2}{2d}} \cos\left[2\pi\left(\frac{y_o}{\lambda d}\right)h\right]$$

$$\int_{-\infty}^{+\infty}\int_{-\infty}^{+\infty} \text{rect}\left(\frac{x}{l_x}\right)\text{rect}\left(\frac{y'}{l_y}\right) e^{+j2\pi\left(\frac{x_o}{\lambda d}\right)x} e^{+j2\pi\left(\frac{y_o}{\lambda d}\right)y'}\, dx dy'. \tag{3.33}$$

The cosine function expresses the interference effect of the diffracted radiation from the double slit.

(c) Diffraction by an opaque disk

Let us consider the diffracted field when the screen is an opaque obstacle such as a finite-sized disk Ω. We can express any opaque aperture Ω as Σ – (Σ – Ω), where Σ is the entire $z = 0$ plane. "Σ – Ω" is the complimentary aperture of Ω. Therefore we can rewrite Eq. (3.12) as

$$U(\underline{r}_o) = U_{\text{inc}}(\underline{r} = \underline{r}_o) - \left[\frac{-j}{\lambda d} \mathrm{e}^{-jkd} \mathrm{e}^{-jk\frac{x_o^2 + y_o^2}{2d}} \iint_{\Sigma - \Omega} U_{\text{inc}}(z = 0) \mathrm{e}^{+j2\pi\frac{x x_o}{\lambda d}} \mathrm{e}^{+j2\pi\frac{y y_o}{\lambda d}} \, \mathrm{d}x\mathrm{d}y \right] \quad (3.34)$$

(d) The Fresnel lens

Let us consider next a refractive Fresnel lens. This lens does not have a spherical surface. The configuration of a Fresnel lens is illustrated in Figure 3.6. It has a material structure that has a sectional continuous profile. For the first segment, the surface profile is such that from the center $r = 0$ to a radius r, the phase shift is described in Eq. (3.30). However, this surface stops at $r = r_1$ when the phase shift is 2π, i.e. when $r_1^2/\lambda f = 2$. A new segment of the surface starts at r_1 with 0 phase shift. This second segment of the surface will give a phase shift proportional to "$(r^2/\lambda f) - 2$." The second surface segment stops at r_2 when $r_2^2/\lambda f = 4$. The third segment starts at r_2 with 0 phase shift. These segments continue until the shortest length of segments, $r_j - r_{j-1}$, reaches the resolution limit of the fabrication technology. Figure 3.7 illustrates the phase shifts along the radial direction r.

When one calculates the diffraction pattern of the Fresnel lens, the Kirchhoff's integral will be performed over each segment of continuous phase shift zone separately. The sum of all the diffraction integrals gives $U(\underline{r}_o)$. The insertion of $e^{j2n\pi}$ (n = any integer) to any integrand does not change the value of the integral. We can easily show that for any normally incident plane wave, U given by the Kirchhoff's integrals for the Fresnel lens behaves identically to any thin spherical lens with the same focal length. The difference between the spherical lens and the Fresnel lens lies in the higher-order terms of the phase shifts, which we neglected in the first-order approximation. For oblique incident radiation, the diffraction pattern of the segments yields distortions.

(e) Spatial filtering

As a final example, let us consider an example in Goodman's book [3]. A plane wave with amplitude A is incident normally on a transparent film at $z = 0$, followed immediately by an ideal thin lens with focal length f, as shown in Figure 3.8. The film is placed in a square aperture ($d \times d$) centered at $x = y = 0$. The electric field transmission t of the transparent film is:

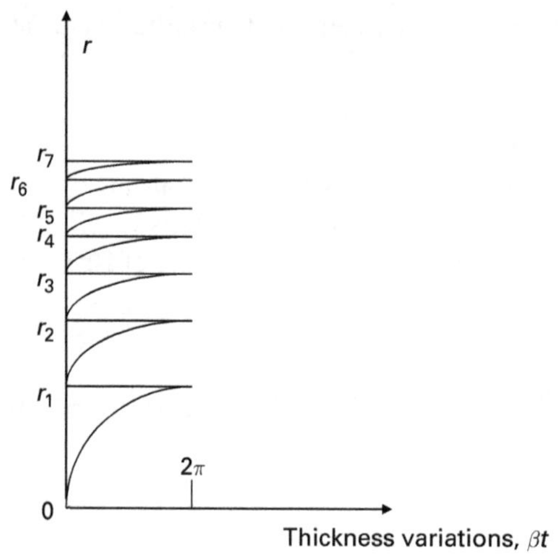

Figure 3.6 Thickness variation in a Fresnel lens.

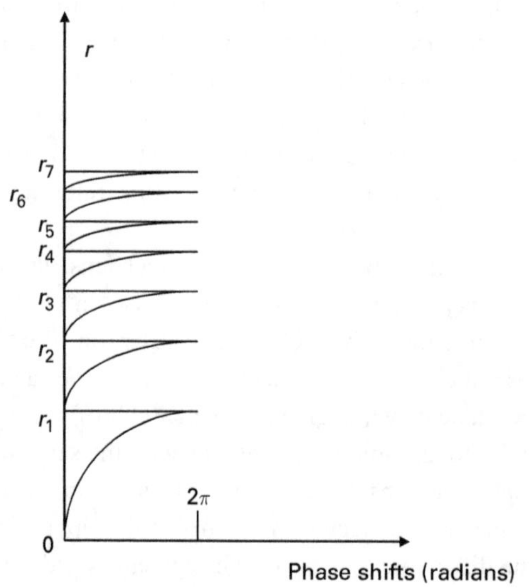

Figure 3.7 Phase shifts in a Fresnel lens.

$$t(x, y) = \frac{1}{2}[1 + \cos(2\pi Hx)]\,\mathrm{rect}\left(\frac{x}{d/2}\right)\mathrm{rect}\left(\frac{y}{d/2}\right). \tag{3.35}$$

Here $H >> (1/d)$. A spatial filter screen is placed at the focal plane of this lens. The screen is opaque in two regions: (1) $|x| < l/2$ and $|y| < l/2$ for the inside region and (2) $|x| > l$ and $|y| > l$ for the outside region. The spatial screen is shown in

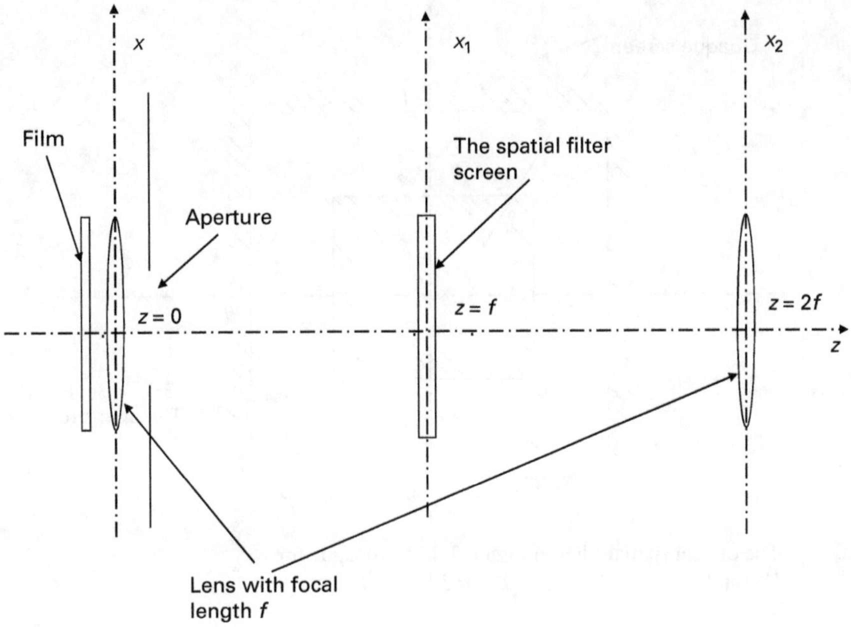

Figure 3.8 Illustration of an example of spectral filtering in the Fourier transform plane. A transparent film with transmission function $t(x, y)$ is placed in front of an ideal lens with focal length f at $z = 0$, followed by a square aperture ($d \times d$). A spatial filter is placed at $z = f$. A second lens with focal length f is placed at $z = 2f$ to reconstruct the filtered light.

Figure 3.9. A second lens with focal length f is placed at a distance f behind the screen.

At $z = f$, the incident field on the screen is:

$$U(x_1, y_1, f) = \frac{-jAe^{-jkf}e^{-jk\left(\frac{x_1^2+y_1^2}{2f}\right)}}{\lambda f}\int_{-d/2}^{d/2}\int_{-d/2}^{d/2}\frac{1}{2}\left[1+\frac{1}{2}e^{j2\pi Hx}+\frac{1}{2}e^{-j2\pi Hx}\right]e^{j2\pi\left(\frac{x_1}{\lambda f}\right)x}e^{j2\pi\left(\frac{y_1}{\lambda f}\right)y}\mathrm{d}x\mathrm{d}y$$

$$= \frac{-jAe^{-jkf}e^{-jk\left(\frac{x_1^2+y_1^2}{2f}\right)}}{2\lambda f}\left[\frac{\sin\left(\frac{\pi x_1 d}{\lambda f}\right)}{\pi\frac{x_1}{\lambda f}}+\frac{\sin\left(\pi\left(H+\frac{x_1}{\lambda f}\right)d\right)}{2\pi\left(H+\frac{x_1}{\lambda f}\right)}+\frac{\sin\left(\pi\left(H-\frac{x_1}{\lambda f}\right)d\right)}{2\pi\left(H-\frac{x_1}{\lambda f}\right)}\right]\frac{\sin\left(\frac{\pi y_1 d}{\lambda f}\right)}{\pi\frac{y_1}{\lambda f}} \tag{3.36}$$

Thus there are three radiation peaks in the x_1 direction, at $x_1 = 0$, $x_1 = \lambda fH$, and $x_1 = -\lambda fH$. The width (defined by the first zero of the field) of the peaks in the x direction is $\lambda f/d$ centered at the peaks. All radiation peaks in the y direction are centered about $y_1 = 0$ with width $\lambda f/d$. However, the transmission range of the screen in the x direction at $z = f$ is $l/2 < |x_1| < l$. Thus the peak centered about $x_1 = 0$ is always blocked by the screen. In order for the two side peaks to pass the screen, we

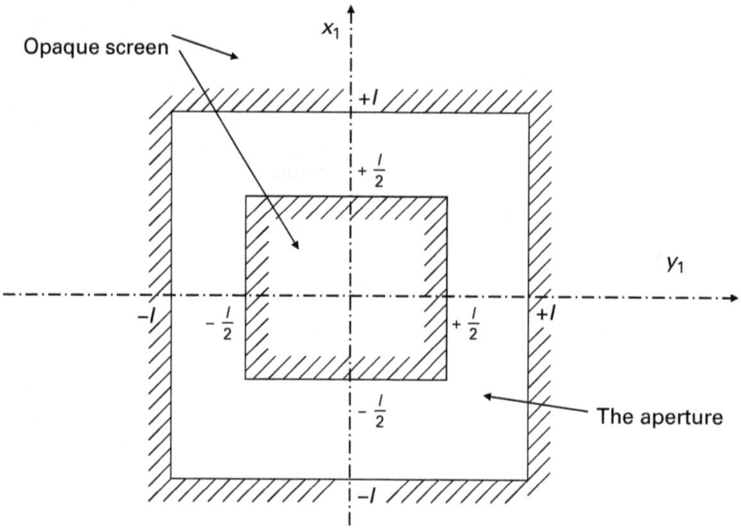

Fig. 3.9 The optical spatial filter in Figure 3.8. It is opaque for $|X_1|$ and $|Y_1| < l/2$ and $|X_1|$ and $|Y_1| > l$.

need $l/2 < f\lambda H < l$. In order for the main lobe of the two side peaks to pass through the screen, we need

$$\lambda Hf + \frac{\lambda f}{d} \langle l \langle 2\lambda Hf - \frac{2\lambda f}{d} \tag{3.37}$$

Since the peaks are centered in the y direction at $y_1 = 0$, the transmission of the screen is effectively from $y_1 = -l$ to $y_1 = l$. If we approximate the transmitted radiation field by deleting the term representing the peak centered about $x_1 = y_1 = 0$, we obtain the diffracted transmitted radiation after the screen to be:

$$U'(x,y,f) \cong \frac{-Ae^{-jkz}e^{-jk\left(\frac{x^2+y^2}{2(z-f)}\right)}}{2\lambda^2 f(z-f)} \int_{-l}^{l} \left[\frac{\sin\left(\frac{\pi y_1 d}{\lambda f}\right)}{\pi\frac{y_1}{\lambda f}} e^{-jk\frac{y_1^2}{2f}}\right] e^{-jk\frac{y_1^2}{2(z-f)}} e^{j2\pi\left(\frac{y}{\lambda(z-f)}\right)y_1} dy_1$$

$$\left\{ \begin{array}{l} \displaystyle\int_{l/2}^{l} \left[\frac{\sin\left(\pi\left(\frac{x_1}{\lambda f} - H\right)d\right)}{2\pi\left(\frac{x_1}{\lambda f} - H\right)} e^{-jk\frac{x_1^2}{2f}}\right] e^{-jk\frac{x_1^2}{2(z-f)}} e^{j2\pi\left(\frac{x}{\lambda(z-f)}\right)x_1} dx_1 \\ \displaystyle + \int_{-l}^{-l/2} \left[\frac{\sin\left(\pi\left(\frac{x_1}{\lambda f} + H\right)d\right)}{2\pi\left(\frac{x_1}{\lambda f} + H\right)} e^{-jk\frac{x_1^2}{2f}}\right] e^{-jk\frac{x_1^2}{2(z-f)}} e^{j2\pi\left(\frac{x}{\lambda(z-f)}\right)x_1} dx_1 \end{array} \right\} \tag{3.38}$$

When this diffracted field passes through the second lens at $z = 2f$, the exponential term in front of the integral, $e^{-jk\frac{x^2+y^2}{2(z-f)}}$, is canceled by the quadratic phase change of an ideal lens, $e^{jk\frac{x^2+y^2}{2f}}$, for $z \geq 2f$.

The integration is quite messy in the general case. However, the answer is simple for the following special case. Let $\lambda f/d$ be small (i.e. the width of radiation peaks is narrow), and let the dimension l be such that at least the main lobe of the two side peaks passes through the screen at $z = f$. Then $\sin(\pi y_1/\lambda f/d)/(\pi y_1/\lambda f)$ is large only for $y_1 < \lambda f/d$, and its peak value is proportional to d. Within such a small range of y_1, the three exponential terms in the above y_1 integral can be approximated by constant values at $y_1 \approxeq 0$. This means that $e^{-jky_1^2/2f}e^{-jky_1^2/2(z-f)}e^{jky_1/2(z-f)y} \cong 1$.

Similarly the three exponential terms in the two x_1 integrations can be approximated by $x_1 = \lambda Hf$ and by $x_1 = -\lambda Hf$, respectively.

Therefore, immediately after the second lens at $z = 2f$, we have the following field.

$$U''(x_2, y_2, 2f) \cong \frac{-Ae^{-jk2f}}{2\lambda^2 f^2} \int_{-l}^{l} \frac{\sin\left(\pi \dfrac{2\pi y_1 d}{\lambda f}\right)}{\pi \dfrac{y_1}{\lambda f}} dy_1$$
$$\left\{ \begin{array}{l} e^{-jk\lambda^2 H^2 f} e^{jk\lambda Hx_2} \displaystyle\int_{l/2}^{l} \frac{\sin\left[\pi\left(\dfrac{x_1}{\lambda f} - H\right)d\right]}{2\pi\left(\dfrac{x_1}{\lambda f} - H\right)} dx_1 \\ +e^{-jk\lambda H^2 f} e^{-jk\lambda Hx_2} \displaystyle\int_{-l}^{-l/2} \frac{\sin\left[\pi\left(\dfrac{x_1}{\lambda f} + H\right)\right]}{2\pi\left(\dfrac{x_1}{\lambda f} + H\right)} dx_1 \end{array} \right\}. \tag{3.39}$$

If the second lens has a sufficiently large size, the diffraction effect due to the finite size of the second lens can be neglected. The far field diffraction pattern will be given by two beams, one beam propagating as $e^{-jkz}e^{jk\lambda Hx}$ and the second beam propagating as $e^{-jkz}e^{-jk\lambda Hx}$. The incident beam propagating along the z axis has been filtered out.

The example is presented here to demonstrate the mathematical complexity of the use of Kirchoff's integral. It will be compared with the use of the Gaussian beam in Chapter 4.

Chapter summary

Diffraction analysis of optical waves is the forte of traditional optics. There is no need to present any extensive discussion of diffraction optics in this book. However, the limitation and the theoretical basis of diffraction analysis are not always clearly understood. It is shown here that the scalar wave equation can only be used to analyze TEM optical radiation. Under the TEM approximation, Kirchhoff's integral can be used to calculate the propagation of any incident radiation through various instruments, without solving Maxwell's equation separately each time.

In addition, it is shown here that, for single-frequency radiation such as laser radiation, the format of Kirchhoff's integral allows us to apply mathematical techniques to relate the diffracted field to the incident field. For example, under certain circumstances, the incident and diffracted fields are related by Fourier transform. Many engineering analytical techniques based on Fourier analysis, such as convolution, become applicable. Signal processing applications such as spatial filtering are based on the transform relations of the optical fields.

Natural light has many frequency components with random relative phases among them. Many effects, such as the Fourier transform relationship, convolution, spatial filtering, etc. depend on the phase relations. Thus, these effects will not be observed in general for natural light. On the other hand, techniques such as Kirchoff's integral are still applicable to each frequency component. Conclusions such as diffraction-limited focus spot size and far field beam width are valid. Even Fabry–Perot resonance could still be observed when the frequency range of the natural light is narrow. Since there are already many excellent books that discuss the optics of natural light, the diffraction of natural light is not presented extensively in this book.

References

[1] M. Born and E. Wolf, *Principles of Optics*, Pergamon Press, 1959.
[2] J. A. Stratton, *Electromagnetic Theory*, McGraw Hill, 1941.
[3] J. W. Goodman, *Introduction to Fourier Optics*, McGraw Hill, 1968, Chapters 5 and 6.

4 Optical resonators and Gaussian beams

Although diffraction analysis of many applications has already been presented extensively in the literature, laser and Gaussian beam analysis has not been emphasized much in traditional optical books. Laser analysis is the focus of the first part of this chapter; Gaussian beam analysis is the focus of the second part.

There are three reasons to present analysis of laser cavities and Gaussian beams. (1) Much of the optical applications today use laser radiation. Understanding of laser modes is very important. For example, analysis of laser cavities allows us to appreciate the difference between longitudinal and transverse modes. It shows us that laser modes are Gaussian. (2) In laser cavities, Kirchoff's diffraction analysis yields the characteristics of the modes. From that discussion, we can appreciate that diffraction and modal analysis are closely related. (3) The Gaussian modes can be used to represent any TEM wave. It is an important analytical technique in itself.

Although Gaussian modes were derived from the analysis of laser cavities in the first part of Chapter 4, they are also a direct solution of Maxwell's equations. Gaussian modes also constitute a complete set. It means that any TEM radiation can be represented as a summation of Gaussian modes. The propagation of a Gaussian beam through any component with a large enough aperture retains the Gaussian form. It simply suffers a loss of power due to diffraction. Therefore, its propagation through various components can be analyzed without carrying out the messy diffraction integrals used in Chapter 3.

It is well known that the basic solid-state and gas laser cavities consist of two concave end reflectors that have the transverse (or lateral) shape of a flat disk or a part of a large sphere. The reflectors are separated longitudinally by distances varying from centimeters to meters. The size of the end reflectors is small compared to the separation distance, but still very large compared to the optical wavelength. Thus, the cavity modes are resonant modes of TEM waves, bouncing back and forth between reflectors. They can be analyzed by means of Kirchhoff's integral. Laser cavities are sometimes called Fabry–Perot cavities because of their similarity to Fabry–Perot interferometers, discussed in Section 2.2.[1] This is the case for solid-state and gas lasers, but not for waveguide semiconductor lasers. Scalar wave equation analysis is not able to analyze

[1] However, Fabry–Perot interferometers have distances of separation much smaller than the size of the end reflectors; therefore, diffraction loss is negligible.

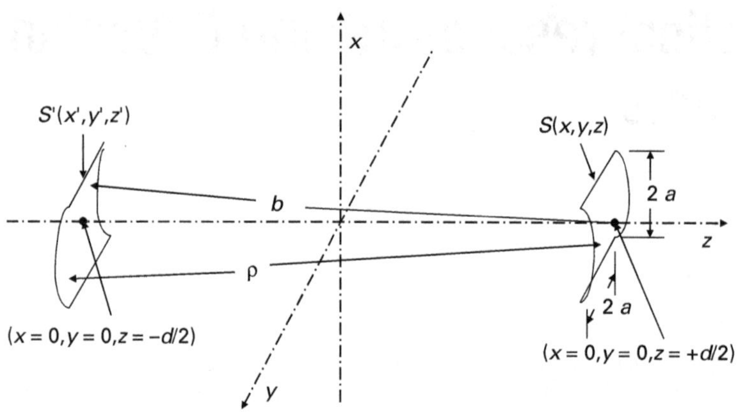

Figure 4.1 Illustration of a laser cavity. A confocal cavity has two spherical end reflectors, S and S'. The reflectors have a square aperture, $2a \times 2a$. The spherical center of the S surface is at the center of S', with radius b. The spherical center of the S' surface is at the center of S, also with radius b. The focus of both the S reflector and the S' reflector is at the center of the opposite mirror; ρ is the distance from a point on the S surface to a point on the S' surface.

waveguide laser cavities that have dimensions of the order of optical wavelength. The fields in waveguides are not transverse electric and magnetic.[2]

4.1 Integral equations for laser cavities

Consider a typical laser cavity as shown in Figure 4.1. Let the y-polarized electric field on the S' mirror be $E_y'(x', y')$ and the electric field on S be $E_y(x, y)$. The diffracted electric field $E_y''(x, y)$ on the S mirror can be calculated by Kirchhoff's integral from E_y' on S'. Similarly, the diffracted E_y''' on S' can be calculated from E_y on S.

$$E''_y(x,y) = \int_{S'} \frac{jk(1+\cos\vartheta')}{4\pi\rho} e^{-jk\rho} E'_y(x',y') ds'$$

and (4.1)

$$E'''_y(x',y') = \int_{S} \frac{jk(1+\cos\theta)}{4\pi\rho} e^{-jk\rho} E_y(x,y) d.$$

$P(x,y,z)$ is a point on S, $P'(x,y,z)$ is a point on S', and ρ is the distance between P and P', $\rho = \sqrt{(x-x')^2 + (y-y')^2 + (z-z')^2}$. If we have a symmetric pair of mirrors and if the cavity supports a stable mode, then E_y and E_y' must eventually reproduce each other, except by a complex constant γ, i.e.:

[2] Surface emitting semiconductor lasers also have TEM cavity modes. Their end reflectors are much smaller than solid-state and gas lasers. The distance between reflectors is comparable to wavelength. In this configuration, diffraction loss is negligible, so surface emitting laser cavities are also not analyzed by means of Kirchoff's integral.

$$\gamma E_y = E_y'' \quad \text{and} \quad \gamma E_y' = E_y'''$$

$$\gamma E_y(x,y) = \int_{S'} \frac{jk(1+\cos\theta')}{4\pi\rho} e^{-jk\rho} E_y'(x',y') ds', \qquad (4.2)$$

$$\gamma E'_y(x'y') = \int_{S} \frac{jk(1+\cos\theta)}{4\pi\rho} e^{-jk\rho} E_y(x,y) ds$$

E_y and E'_y have the same x and y functions.

Any stable resonant mode of the cavity must satisfy Eq. (4.2). Conversely, any solution of Eq. (4.2) is a resonant mode of the cavity. The field pattern of the resonant mode of the laser was found first by Fox and Li [1]. They calculated numerically the diffraction integral on a computer, starting from an assumed initial mode pattern on S. The resultant electric field pattern on the opposite mirror S' was then used as E_y in the diffraction integral to calculate the field on S after a round trip. This process was iterated back and forth many times. Eventually, stabilized mode patterns (i.e. mode patterns that differ from each other only by a complex constant after one diffraction) were found. In the next section, we will first discuss in detail the analytical solution of the integral equation for a specific cavity configuration, the confocal laser cavity. We will then extend the analysis of confocal cavities to other non-confocal cavity configurations. All the modes discussed in this chapter are "cold" or "passive" cavity modes, meaning that there is no gain in the material.

4.2 Modes of confocal cavities

Consider the resonator shown in Figure 4.1. In a confocal resonator, there are two identical spherical mirrors with radius b, symmetrically placed about the z axis at $z = \pm d/2$ ($d = b$ in confocal cavities). In order to take advantage of the simplicity of mathematical analysis in rectangular coordinates, both mirrors are assumed to have a square shape ($2a \times 2a$ in transverse dimension).[3] The size of the mirror is small compared to the separation distance, i.e. $d >> a$. While the centers of the spherical surfaces are located at $x = y = 0$ and $z = \pm d/2$, the focal point of both mirrors is at $x = y = 0$ and $z = 0$; hence it is called the confocal cavity. We will analyze the confocal cavities following Boyd and Gordon [2].

4.2.1 The simplified integral equation for confocal cavities

Since $a << d$, $\theta \cong 0$ and $\cos\theta \cong 1$ in Eq. (4.2). Thus Eq. (4.2) for an electric field polarized linearly in the y direction can be simplified as:

$$\gamma E_y(x,y)|_{on\,S} = \left(\frac{j}{\lambda d}\right) \int_{-a}^{a} \int_{-a}^{a} E_y(x',y')|_{on\,S'} e^{-jk\rho} dx'dy'. \qquad (4.3)$$

[3] The shapes of actual mirrors are round.

Here, ρ is the distance between P and P' on the S and S' surfaces. Clearly E_y on S and S' must be identical. Equation (4.3) is an integral equation for E_y. It is well known mathematically that, like differential equations with appropriate boundary conditions, such an integral equation has independent eigenfunctions and eigenvalues. If we can find these independent solutions, we have found the modes of the confocal cavity.

The S' and S surfaces are described by:

$$\begin{aligned} z' - \frac{d}{2} &= -\sqrt{d^2 - x'^2 - y'^2} \cong -d + \frac{x'^2 + y'^2}{2d}, \\ z + \frac{d}{2} &= \sqrt{d^2 - x^2 - y^2} \cong d - \frac{x^2 + y^2}{2d} \end{aligned} \tag{4.4}$$

When $e^{-jk\rho}$ is simplified by binomial approximation and when the higher-order terms are neglected, we obtain

$$\begin{aligned} \rho &= \sqrt{(z-z')^2 + (x-x')^2 + (y-y')^2} \cong (z-z') + \left[\frac{1}{2}\frac{(x-x')^2 + (y-y')^2}{(z-z')}\right] \\ &\cong \left(d - \frac{x'^2 + y'^2}{2d} - \frac{x^2 + y^2}{2d}\right) + \left[\frac{(x^2 + x'^2) + (y^2 + y'^2) - 2xx' - 2yy'}{2d}\right] \\ &\cong d - \frac{xx' + yy'}{d} \end{aligned} \tag{4.5}$$

When z and z' are on S and S', d is used to approximate the $(z - z')$ term in the denominator. Note that the quadratic terms $x'^2 + x^2/2d$ and $y'^2 + y^2/2d$ in the second square-bracketed term of the binomial series expansion are canceled by the quadratic terms $x'^2 + y'^2/2d$ and $x^2 + y^2/2d$ in the round brackets, created by the spherical surfaces of the confocal resonator. This coincidence gives us a simplified expression for ρ. When higher-order terms are neglected, E_y at (x, y, z) on S is related to E_y at (x', y', z') on S' by a simplified equation:

$$\gamma E_y(x,y,z)|_{on\ S} = \left(\frac{j}{\lambda d} e^{-jkd}\right) \int_{-a}^{+a}\int_{-a}^{+a} E_y(x',y',z')|_{on\ S'}\, e^{jk\left(\frac{xx'+yy'}{d}\right)}\, dx'dy' \tag{4.6}$$

Neglecting the higher-order terms in the binomial expansion is justified when $a^2/b\lambda << (b/a)^2$.

If we compare Eq. (4.6) with the diffraction integrals for Fraunhofer diffraction in the focal plane of a lens, we see that the relation between E_y on S and E_y on S' is again a Fourier transform with finite integration limits, $\pm a$. There are known mathematical solutions for such an integral equation. This is really the secret of the simplicity of a confocal cavity and the reason we started the cavity analysis with it.

4.2.2 Analytical solutions of the modes in confocal cavities

If we let E_y on S be described by $F(x)G(y)$, then the integral equation for F and G is:

$$\sigma_l\sigma_m F_l(x)G_m(y) = \int_{-a}^{+a}\int_{-a}^{+a} \frac{je^{-jkb}}{\lambda d} F_l(x')G_m(y')e^{jk\left(\frac{xx'+yy'}{b}\right)}dx'dy', \tag{4.7}$$

Here γ is represented by $\sigma_l\sigma_m$. When we make the following change of variables,

$$\Lambda = \frac{a^2k}{b}, \quad X = \frac{\sqrt{\Lambda}}{a}x, \quad \text{and} \quad Y = \frac{\sqrt{\Lambda}}{a}y$$

we obtain

$$\sigma_l\sigma_m F_l(X)\,G_m(Y) = \frac{je^{-jkb}}{2\pi}\int_{-\sqrt{\Lambda}}^{+\sqrt{\Lambda}} F_l(X')e^{jXX'}dX' \int_{-\sqrt{\Lambda}}^{+\sqrt{\Lambda}} G_m(Y')e^{jYY'}dY'. \tag{4.8}$$

This is a product of two well-known identical integral equations, one for X and one for Y. In order for both of them to be satisfied for all X and Y, each integral equation must be satisfied separately. Slepian and Pollak [3] have shown that the lth independent solution to

$$F_l(X) = \frac{1}{\sqrt{2\pi}\chi_l}\int_{-\sqrt{\Lambda}}^{+\sqrt{\Lambda}} F_l(X')e^{jXX'}dX' \tag{4.9}$$

is

$$F_l(X) = S_{0l}\left(\Lambda, \frac{X}{\sqrt{\Lambda}}\right)', \quad \text{and} \quad \chi_l = \sqrt{\frac{2\Lambda}{\pi}}j^l R_{0l}{}^{(1)}(\Lambda, 1), \qquad l = 0, 1, 2, \ldots \tag{4.10}$$

S_{0l} and R_{0l} are, respectively, the angular and radial wave functions in prolate spheroidal coordinates, as defined by Flammer [4]. Thus the eigenvalues and eigenfunctions of Eq. (4.9) are:

$$\sigma_l\sigma_m = j\chi_l\chi_m e^{-jkb} = \frac{2\Lambda}{\pi}R_{0l}{}^{(1)}(\Lambda, 1)R_{0m}{}^{(1)}(\Lambda, 1)j^{m+l+1}e^{-jkb}$$

and

$$E_y = U_{lm}(x, y) = S_{0l}\left(\Lambda, \frac{x}{a}\right)S_{0m}\left(\Lambda, \frac{y}{a}\right) \tag{4.11}$$

with l, $m = 0, 1, 2, 3\ldots$ According to Slepian and Pollak [3], the R and S functions are real. It confirms that the mirrors are surfaces of constant phase of E_y.

For each mode, as it propagates from one mirror to the other, its amplitude changes by $\chi_l\chi_m$ and its phase changes by je^{-jkb}. For a given transverse lth and mth mode, there are resonances at those frequencies when $kb = q\pi$.

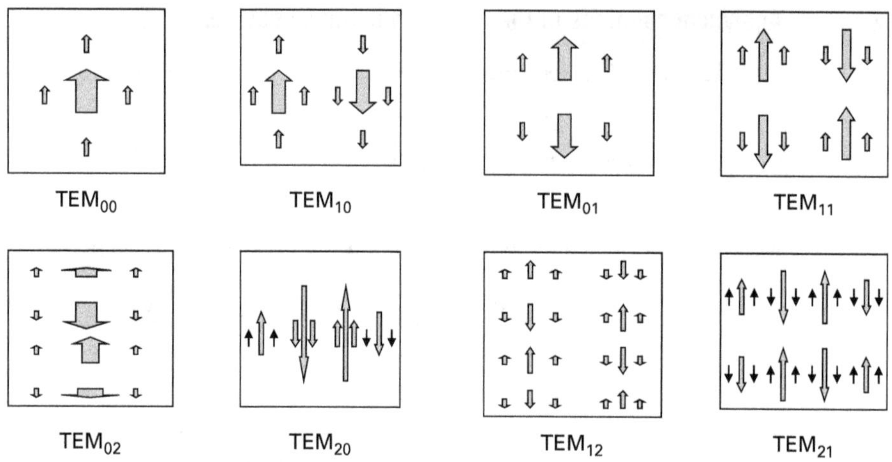

Figure 4.2 Sketch of transverse field distribution of lower-order modes in confocal resonators. The arrows are used to indicate the electric field patterns of various low-order TEM$_{lm}$ modes on the mirror. The direction of the field is shown by the direction of the arrows and the magnitude of the field is indicated by the size of the arrows.

It is interesting to recognize that we have obtained cavity modes as solutions of diffraction integrals without the use of any modal expansion concept. Yet, once we know the existence of these modes, we realize that the laser properties are best described by modes. In other words, diffraction analysis of laser cavities has led us directly into the modal concepts.

4.2.3 Properties of resonant modes in confocal cavities

Many conclusions can be drawn from the solution of the fields in the cavity discussed in Section 4.2.2 above. Seven properties of the resonant modes of the confocal cavities are presented below.

(a) The transverse field pattern

We normally designate the resonant modes as TEM$_{lm}$ modes, which have the transverse variation given by U_{lm}. Figure 4.2 illustrates the transverse field distribution of lowest order TEM$_{lm}$ modes in confocal resonators. There is also a set of TEM modes for an electric field polarized in the x direction.

Note that the lth order mode will have l nodes or zeros in the x direction, while the mth order mode will have m nodes in the y direction. This information is important. It allows us to experimentally identify the mode order by examining its intensity pattern. For a given transverse TEM$_{lm}$ mode, the cavity resonates at frequencies whenever $kb = q\pi$. The modes of the same transverse TEM$_{lm}$ pattern that have different resonance frequencies are known as the longitudinal modes.

(b) The resonance frequency

At resonance, the phase shift after each round trip of propagation in the z direction must be integers of 2π. Thus resonance in the z direction occurs only at discrete wavelengths λ_{lmq} that correspond to various values of q multiples of 2π.

$$\left|\pi - \frac{4\pi b}{\lambda_{lmq}} + (m+l)\pi\right| = 2q\pi \tag{4.12}$$

Here m, l, and q are all integers. From here on, we designate modes belonging to different l and m as transverse modes and modes belonging to different q longitudinal modes. Note that, lower-order transverse modes have small integers or 0 for m and l, while q may be a very large number, up to millions, for long cavities.

In summary, the resonance frequency f_{lmq} for a given order of mode, designated by l, m, and q, is:

$$f_{lmq} = \frac{c}{4b}(2q + l + m + 1) \tag{4.13}$$

where c is the velocity of light in the cavity.

From Eq. (4.12), we see that the TEM_{lm} modes are degenerate with respect to l and m. Degeneracy means independent modes with the same $l + m$ value, but different l and m values, will have the same resonance frequency. As we will show in the next section, such degeneracy does not exist in non-confocal cavities. In principle, degenerate modes may resonate at the same frequency. However, we usually do not want more than one mode to resonate at the same frequency because it creates uncertainty in the modal content of the total field. The mode degeneracy is a disadvantage of confocal resonators. Therefore confocal configuration is not used in practical lasers.

TEM_{lm} modes that have adjacent longitudinal mode orders, i.e. q and $q + 1$, will have resonance frequencies separated by $c/2b$. "$2b/c$" is the round trip propagation time for a wave front to travel around the cavity. Thus the frequency spacing of the longitudinal modes is controlled by the mirror separation between the reflectors and the velocity of light. For cavities filled with dielectric that has refractive index n, the resonance frequency separation of the adjacent longitudinal modes will be $2bn/c_o$ where c_o is the velocity of light in the free space.

(c) The orthogonality of the modes

U_{lm} is a set of orthogonal functions, i.e.

$$\int_{-a}^{a}\int_{-a}^{a} F_m\left(\frac{x\sqrt{\Lambda}}{a}\right) G_n\left(\frac{y\sqrt{\Lambda}}{a}\right) F_{m'}\left(\frac{x\sqrt{\Lambda}}{a}\right) G_{n'}\left(\frac{y\sqrt{\Lambda}}{a}\right) \mathrm{d}x\,\mathrm{d}y = \int_{-a}^{a}\int_{-a}^{a} U_{mn}U_{m'n'}\,\mathrm{d}x\,\mathrm{d}y = 0, \tag{4.14}$$

when $m \neq m'$ or $n \neq n'$. Therefore these modes are orthogonal modes.[4] Moreover, it can be shown mathematically that eigenfunctions of the integral equation of the form given in Eq. (4.3) always form a complete set.

[4] Orthogonality of modes can be proved in general only for cavity medium without loss.

The orthogonality relation is very helpful in expanding any arbitrary electric field $U(x, y)$ in terms of U_{lm}.[5] For example, for any field U in the cavity, we can write

$$U = \sum_{l.m} a_{lm}\, U_{lm} \tag{4.15}$$

Then, because of the orthogonality relation,

$$a_{lm} = \frac{\int_{-a}^{a}\int_{-a}^{a} U\, U_{lm} \mathrm{d}x\, \mathrm{d}y}{\int_{-a}^{a}\int_{-a}^{a} {U_{lm}}^2\, \mathrm{d}x\, \mathrm{d}y} \tag{4.16}$$

There are many important applications of the orthogonality properties of modes. When an input radiation is coupled into an optical component that has TEM_{nm} modes, the input radiation U_{in} can be written as a summation of TEM_{lm} modes. For example, if U is symmetric in x and y, the coupling between TEM_{lm} mode with an odd value of l or m and any symmetric input radiation will be zero. In another example, we may be interested in the power of a given U. The intensity of U is proportional to UU^*. Since U_{lm} is orthogonal, the total power is $\int_{-a}^{+a}\int_{-a}^{+a} UU^* \mathrm{d}x\mathrm{d}y = \sum_{lm} a_{lm}{a_{lm}}^* \int_{-a}^{+a}\int_{-a}^{+a} U_{lm}{U_{lm}}^* \mathrm{d}x\mathrm{d}y$. In other words, the total power is just the sum of the power in each mode.

(d) A simplified analytical expression of the field

For x and $y << a$, U_{lm} can be approximated by the product of a Hermite polynomial and a Gaussian envelope,

$$E_{y,lm} = U_{lm}(x,y) = \frac{\Gamma[(l/2)+1]\Gamma[(m/2+1)]}{\Gamma(l+1)\Gamma(m+1)} H_l\left(\frac{x\sqrt{\Lambda}}{a}\right) H_m\left(\frac{y\sqrt{\Lambda}}{a}\right) \mathrm{e}^{-\pi(x^2+y^2)/d\lambda} \tag{4.17}$$

Here, Γ is the usual gamma function, and Hermite polynomials are tabulated in many physics and mathematics books.

$$\begin{aligned} H_0(x) &= 1, \\ H_1(x) &= 2x, \\ H_2(x) &= 4x^2 - 2 \\ &\dots \\ H_n(x) &= (-1)^n e^{x^2} \frac{\partial^n}{\partial x^n} \mathrm{e}^{-x^2} \end{aligned} \tag{4.18}$$

[5] Any arbitrary TEM field polarized in the y direction can also be expressed as superposition of other complete set of modes, such as plane waves or spherical waves. The selection of what specific form of modal expansion to use will be based on the configuration of the device and mathematical convenience, such as the availability of expressions for the modes.

For $l = m = 0$, the lowest order Hermite polynomial is just 1. Thus the TEM_{00} mode is just a simple Gaussian function. An lth-order Hermite polynomial is an lth order algebraic polynomial function. Thus, it will have l zeros. Even-order modes will be even functions while odd-order modes will be odd functions. At large x and y values, polynomials are weakly varying functions, while the exponential function dominates the amplitude variation. The envelope of all TEM_{lm} modes is a Gaussian function that is independent of the mode order, l and m. Thus they are known as Gaussian modes.

(e) **The spot size**

The radius at which the exponential envelope term falls to 1/e of its maximum value at $x = 0$ and $y = 0$ is the spot size $\underline{\underline{\omega}}_s$ of the Gaussian modes on the mirror. At this distance from $x = 0$ and $y = 0$, the intensity falls to $1/e^2$ of its maximum value. Therefore, for all TEM_{lm} modes, the spot size on the mirror is:

$$\underline{\underline{\omega}}_s = \sqrt{b\lambda/\pi} \tag{4.19}$$

Note that the spot size on the mirror is independent of the mode order, l and m. It is controlled only by the radius of curvature of the confocal mirror.

(f) **The diffraction loss**

There is a fractional energy loss per reflection, γ_D. It is commonly called the diffraction loss per pass (i.e. the loss from propagation of the wave front from one reflector to the second reflector and back again) of the TEM_{lm} mode. It means that the diffracted field of the first mirror is only partially captured by the second mirror. Because of this loss, the magnitude of the eigenvalue χ_m is less than one. There are two ways to calculate γ_D.

(1)

$$\gamma_D = 1 - |\chi_l \chi_m|^2 \tag{4.20}$$

(2) We can calculate γ_D from the ratio of the energy captured by the mirror to the total energy in the E field at the mirror. i.e.:

$$\gamma_D = 1 - \frac{\iint\limits_{\Omega} |E(x,y,z)|^2 \mathrm{d}x\mathrm{d}y}{\int\limits_{-\infty}^{\infty}\int\limits_{-\infty}^{\infty} |E(x,y,z)|^2 \mathrm{d}x\mathrm{d}y} \tag{4.21}$$

Here E is given in Eqn. (4.17) and Ω is the aperture representing the mirror. Figure 4.3 shows γ_D for several lower-order modes of the confocal resonators, obtained by Boyd and Gordon, as well as the γ_D obtained by Fox and Li in their numerical calculation for two flat mirrors. This is a very important result. (1) Note that TEM_{lmq} and $TEM_{lmq'}$ modes have the same diffraction loss (i.e. the diffraction loss is independent of the longitudinal mode order). The diffraction loss increases, in general, for higher-order transverse modes. Note also that the diffraction loss varies rapidly as a function of $a^2/b\lambda$. In lasers, we like to have just a single TEM oscillating mode most of the time. If the

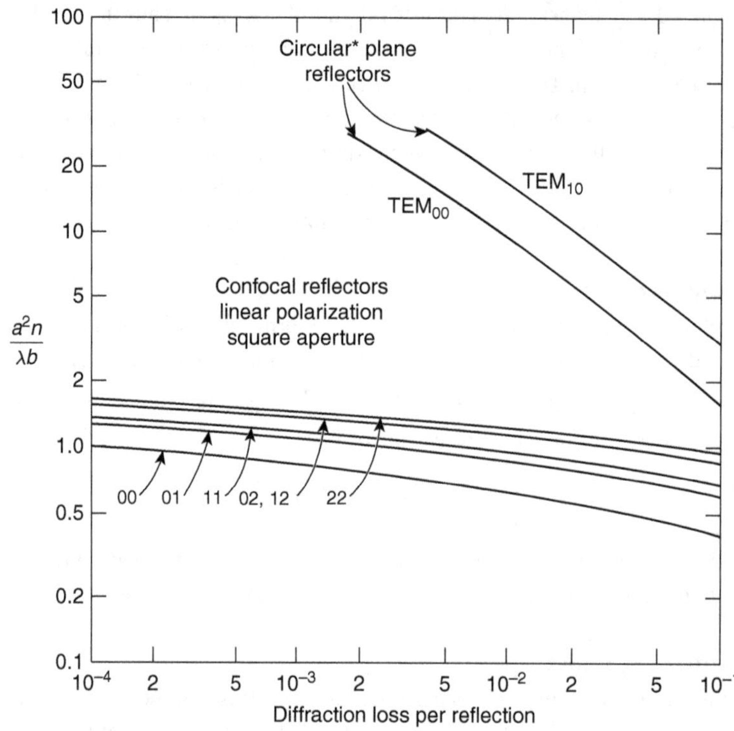

Figure 4.3 Diffraction loss per pass for the lowest-order mode of a plane parallel cavity and for lower-order modes of confocal cavities: a is the mirror size and b is the mirror spacing. The pairs of numbers under the arrows refer to the transverse mode order l and m of the confocal cavity; n is the refractive index of the material between reflectors.

diffraction loss is sufficiently high for higher-order modes, they will not oscillate. Controlling the diffraction loss by the aperture size is a very important technique in laser cavity design. (2) Note that, in conventional Fabry–Perot interferometers, $a^2/b\lambda$ is much bigger than those used in laser cavities shown in Figure 4.1. Therefore diffraction loss is insignificant for many modes in Fabry–Perot interferometers.

(g) The line width of resonances

In Section 2.2.2, we showed that the line width of Fabry–Perot resonance depends on the reflectivity of the mirror. In laser cavities, even for mirrors with perfect reflectivity, there will be diffraction loss. When we include the diffraction loss, the equivalent power reflectivity of a mirror is the material reflectivity of the mirror multiplied by γ_D. For a given total reflectivity, the line width of a given mode can be obtained from Eq. (2.41). The higher-order transverse modes will have much larger line width. Longitudinal modes of the same transverse order will have the same line width.

Knowing the properties of laser modes has many practical applications. For example: (1) It allows us to identify experimentally the modes that we are observing. (2) The minimum spot size of a laser beam is the beam waist of the Gaussian modes. (3) It allows us to understand the difference between transverse and longitudinal modes and their *resonance frequencies. (4) It shows us how to control the diffraction loss of different modes by varying $a^2/b\lambda$. Laser oscillation occurs in any mode whenever the gain exceeds the loss. (5) In order to have only a single-mode oscillation, one must control the diffraction loss and the mirror separation d such that only one transverse and one longitudinal mode has a diffraction loss profile that satisfies the oscillation condition within the gain profile. The higher-order modes do not oscillate because their loss exceeds the gain.*

Cavities for surface emitting semiconductor lasers also have TEM modes. Thus they can also be analyzed by scalar wave equations. However, they usually have transverse dimensions much larger than the separation of reflectors. Diffraction loss is not an important issue in these cavities. Instead, how to obtain high reflectivity in the reflectors becomes a major concern. Note that since the separation of reflectors is so small the Fresnel and Fraunhoffer approximations of Kirchhoff's integral are not applicable.

4.2.4 Radiation fields inside and outside the cavity

Inside the cavity, the internal field U can be obtained by applying Kirchhoff's diffraction formula to U on the mirror. If the mirror is partially transmitting, there will also be a radiation field outside the cavity.[6] Since U must be continuous across a partially transmitting surface, the propagation of U outside the cavity can also be calculated by Kirchoff's diffraction formula from U on the mirror. The result is

$$E_{ylm}(x, y, z) = A\frac{2}{1+\xi^2}\frac{\Gamma[(m/2)+1]\Gamma[(l/2)+1]}{\Gamma(m+1)\Gamma(l+1)}H_l\left(\frac{x}{\sqrt{\frac{b\lambda}{2\pi}}}\sqrt{\frac{2}{1+\xi^2}}\right)$$
$$\times H_m\left(\frac{y}{\sqrt{\frac{b\lambda}{2\pi}}}\sqrt{\frac{2}{1+\xi^2}}\right)\exp\left[-\frac{kr^2}{b(1+\xi^2)}\right]$$
$$\times \exp\left(-j\left\{k\left[\frac{b}{2}(1+\xi)+\frac{\xi}{1+\xi^2}\frac{r^2}{b}\right]-(1+l+m)\left(\frac{\pi}{2}-\phi\right)\right\}\right), \tag{4.22}$$

Here $r^2 = x^2+y^2$, $\xi = 2z/b$, $\tan\theta = (1-\xi)/(1+\xi)$, and A is the amplitude.

Eq. (4.22) implies that the amplitude spot size at any z is

[6] Since the transmission is usually low, the outside field will have much smaller amplitude than the internal field.

$$\underline{\underline{\omega}}(z) = \sqrt{\frac{b\lambda}{2\pi}(1+\xi^2)} \tag{4.23}$$

The intensity of the radiation is proportional to $E_y E_y^*$, thus the intensity falls to $1/e^2$ of its maximum value at the edge of the spot. Clearly the minimum spot size $\underline{\underline{\omega_o}}$ is at $z = 0$.

$$\underline{\underline{\omega_o}} = \sqrt{\frac{b\lambda}{2\pi}} \tag{4.24}$$

The Gaussian beam at $z = 0$ is known as the beam waist. Note again that, at large x and y, the amplitude variation will be dominated by the exponential function, instead of any polynomial function dependent on l and m. Thus the spot size is independent of the order of the mode.

Three important examples of how to use the above results are given here.

(a) The far field pattern of the TEM modes

From Eq. (4.23), we can calculate ω_s/z for very large z. This ω_s/z ratio is the radiation beam width θ_{rad} of the TEM modes in the far field,

$$\theta_{\text{rad}} = \tan^{-1}\left(\frac{\lambda}{\pi\underline{\underline{\omega_o}}}\right) \cong \frac{\lambda}{\pi\underline{\underline{\omega_o}}} = \sqrt{\frac{2\lambda}{\pi b}} \tag{4.25}$$

If we compare this far field beam width, $\lambda/\pi\underline{\underline{\omega_o}}$, with the beam width, $\lambda/2l_x$ or $\lambda/2l_y$, of a plane wave incident on a rectangular aperture, given in Eq. (3.22), we see immediately the similarity between them. However, in the case of Eq. (3.22), we defined the radiation intensity beam width by the first node of the radiation intensity; here we define the radiation beam width when the intensity falls to $1/e^2$ of its maximum.

(b) A general expression for the TEM_{lm} Gaussian modes

We can now rewrite $E_{y\,lm}$ given in Eq. (4.22) in a form that has clear physical meaning for different parts of the expression, as follows:

$$E_{ylm} = E_o \frac{\underline{\underline{\omega_o}}}{\underline{\underline{\omega}}(z)} H_l\left[\frac{\sqrt{2}x}{\underline{\underline{\omega}}}(z)\right] H_m\left[\frac{\sqrt{2}y}{\underline{\underline{\omega}}}(z)\right] e^{-\frac{r^2}{\underline{\underline{\omega}}^2(z)}} e^{-jk\left(\frac{r^2}{2R(z)}\right)} e^{-jkz+j(l+m+1)\eta} \tag{4.26}$$

Here, E_o is just the amplitude, a constant, and

$$\underline{\underline{\omega}} = \underline{\underline{\omega_o}}\left[1+\left(\frac{z}{z_o}\right)^2\right]^{1/2}, \qquad z_o = \frac{b}{2},$$
$$R(z) = \frac{1}{z}\left[z^2 + z_o^2\right],$$
$$\eta = \tan^{-1}(z/z_o) \tag{4.27}$$

The three exponential terms in the above expression have different physical meanings. (1) The first exponential term exhibits the Gaussian envelope amplitude variation at any z. This is the most commonly cited property of laser cavities. Because of this term, the laser modes are also known as the Gaussian modes. (2) The second exponential term exhibits the quadratic phase variation (i.e. the spherical wave front) with a specific radius of curvature $R(z)$ at each z value. Note that at $z = \pm d/2$, R is just the curvature of the confocal reflector, as we would expect. At $z = 0$, i.e. at the beam waist, the mode has a planar wave front, as well as the smallest spot size. (3) The third exponential term expresses the longitudinal phase shift in the z direction. The phase shift is important in determining the resonance frequency.

Note that the electric field distribution of any TEM_{lm} mode is independent of the size of the reflector. The Gaussian beam already includes the diffraction effect without explicitly invoking Kirchoff's formula. Only the diffraction loss is dependent on the reflector size.

The amplitude variation of the intensity is the main concern of conventional optics. Since U^*U is the intensity, we are not concerned with the phase variation of U in many applications. However, in laser optics, the quadratic phase variation is also important. For example: (1) high coupling efficiency between a specific laser mode and the mode of another optical component requires good phase matching, as well as amplitude matching of the two modes; (2) phase variations are important in analyzing the diffraction pattern; (3) as the laser light encounters a lens, the quadratic phase variation of the lens will control the focusing of the laser radiation.

(c) An example to illustrate confocal cavity modes

Consider a confocal cavity with end reflectors separated by 30 cm and $a = 0.5$ mm. The medium between the mirrors is air, i.e. $n = 1$. The wavelength is 1 μm. The reflectors are 99% reflection and 1% transmission in intensity. The confocal resonator modes will have a beam waist size on the mirror, $\sqrt{b\lambda/\pi} = 0.3$ mm, which is independent of and much smaller than the mirror size. The mode pattern in the x and y directions will not be dependent on the mirror size.

The mode pattern will depend only on the mode order, l and m, and $b\lambda$. According to Eq. (4.23), the radiation field of the mode assumes its far field pattern when $4z^2/b^2 >> 1$. The beam divergence angle at the far field is given by Eq. (4.25) as $\sqrt{2\lambda/\pi b}$, which is 1.5 m radians and independent of the mode order. Notice that the condition for the far field is different than the far field condition for Kirchhoff's diffraction given in Eq. (3.16). The Fraunhoffer condition requires a much longer distance to satisfy.

The diffraction loss per pass will depend on the mode order, l and m. For this cavity $a^2/b\lambda = 0.83$. From Figure 4.4 the diffraction loss for the TEM_{00} mode is approximately 10^{-3} per pass. The diffraction loss per pass for the TEM_{01} or TEM_{10} mode jumps to 2×10^{-2}, while the loss per pass for the TEM_{11} mode is 5×10^{-2}. The mirror size, a, is much larger than the spot size. The mode pattern in the x and y variations are the same. According to Eq. (4.21), the diffraction loss per pass will be independent of whether the mirrors are square or round in the cross section, as long as the area of the mirror is approximately the same. Since the transmission is 1%, the total loss per pass is 1.1×10^{-2} for the TEM_{00} mode, 3×10^{-2} for the TEM_{01} or TEM_{10} modes, and 6×10^{-2} for the TEM_{11} mode. Notice the sensitivity of the diffraction loss per pass to changes in $a^2/b\lambda$. In order to get much larger loss per pass for the TEM_{01}, TEM_{10}, or TEM_{11} mode, it is necessary to reduce the mirror size, a. At $a = 0.525$ mm, the total losses for these modes are: 1.02×10^{-2}, 1.5×10^{-2}, and 2×10^{-2}. The increase of total loss per pass for the higher-order modes will be much less significant for the larger mirrors. A favorite practical trick to increase the differential losses of the higher-order modes is to put an aperture in front of the mirror to reduce "a." In other cases, the effective "a" of the cavity may be limited by other considerations such as the size of the laser tube.

4.3 Modes of non-confocal cavities

In this section, modes of non-confocal cavities with arbitrary spherical end reflectors at a given distance of separation will be found by identifying them with the modes of a virtual equivalent confocal cavity as follows. (1) We will first show that the reflectors of any given confocal resonator can be replaced by other reflectors at various locations that have an appropriate radius of curvature. Such a replacement will not change the resonant mode pattern. We call this technique the formation of a new cavity for known modes of confocal resonators. (2) We will then solve the inverse problem: how to find the virtual equivalent confocal resonator for a given pair of non-identical spherical mirrors at a given distance of separation. (3) Once we have found the virtual equivalent confocal resonator we will obtain the properties of the modes of the original resonator, such as the field pattern, diffraction loss, resonance frequencies, etc. from the modes of the virtual resonator. (4) We will illustrate how to find the modes in non-confocal cavities via an example.

4.3.1 Formation of a new cavity for known modes of confocal resonator

Let us first examine closely the consequence of the confocal resonator modes found in Section 4.2. Eq. (4.26) implies that there is a constant phase surface for any resonator mode whenever x, y, and z satisfy the condition,

$$z + \frac{r^2}{2R(z)} = \text{constant} \tag{4.28}$$

It is clear that if a reflector with curvature $R(z)$ is placed at this z position to replace one of the confocal mirrors at $z = \pm d/2$, we will still have the same Gaussian transverse mode as the original confocal cavity. The frequency at which resonance will occur will be shifted because η is a function of z, and the

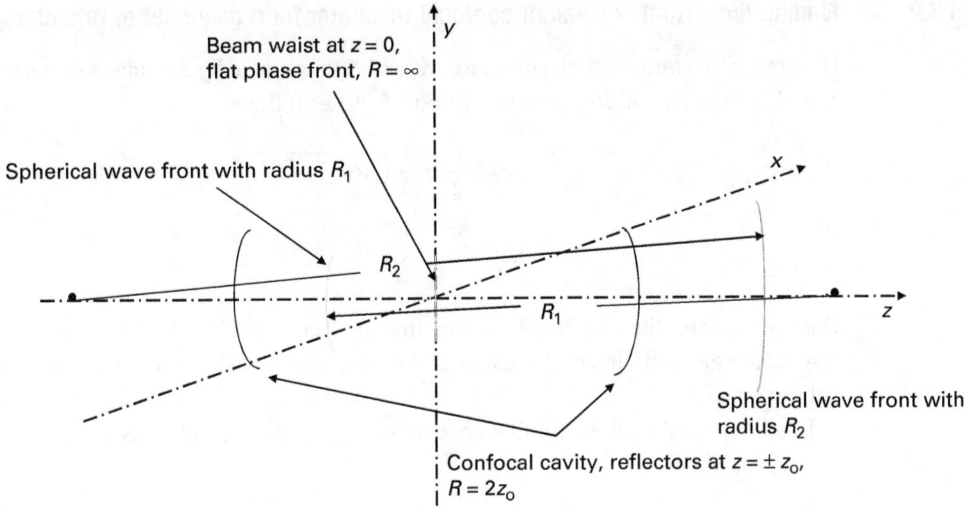

Figure 4.4 Illustration of constant phase fronts of the modes of confocal resonators. The confocal cavity is shown as two spherical reflectors at $z = \pm z_o$. The radius of these confocal spherical reflectors is $2z_o$. The modes of the confocal cavity have a spherical wave front inside and outside the cavity. Outside, a spherical wave front (dashed curve) is shown to have a radius of curvature R_2. Inside, a spherical wave front (dotted curve) is shown to have a radius of curvature R_1. The waist of the modes (solid line) is at $z = 0$; the modes have a flat wave front at this position.

round trip distance of propagation will be different than that of the original confocal resonator. However, the transverse mode variation will be the same. The spot size on this mirror at z is given by the $\underline{\underline{\omega}}$ in Eq. (4.27). The diffraction loss per pass will depend on the size of the reflectors.

In other words, a new optical cavity can be formed with mirrors at z_1 and z_2, provided

$$R_1 = z_1 + \frac{z_o^{\,2}}{z_1} \qquad \text{and} \qquad R_2 = z_2 + \frac{z_o^{\,2}}{z_2} \tag{4.29}$$

The transverse lm modes of the original confocal resonator are also modes of this new cavity. The resonant modes of the new cavity will have the same transverse field variation as the modes of the original cavity. The diffraction loss of the modes will be the same in the original cavity and in the new cavity when the mirror size varies proportional to $\underline{\underline{\omega}}(z)$. Figure 4.4 illustrates the surfaces of constant phase at two z positions. Note that one of z_1 and z_2 can have a negative value, producing a negative R, which means we have a curved mirror at $z < 0$ bending toward $z = 0$. As $|z_2|$ or $|z_1|$ becomes very large, $|R_1|$ and $|R_2|$ become approximately the same as $|z_1|$ or $|z_2|$, i.e. the surface of constant phase is approximately the same as a spherical wave originated from $z = 0$. As $|z_1|$ or $|z_2|$ becomes very small, $|R_1|$or $|R_2|$ becomes very much larger than $|z_1|$ or $|z_2|$. At $z = 0$, the surface of constant phase is a plane.

4.3.2 Finding the virtual equivalent confocal resonator for a given set of reflectors

If there are two mirrors with curvatures R_1 and R_2, separated by a distance D, we can find z_1 and z_2 to fit R_1 and R_2 according to Eq. (4.29) as follows:

$$z_1 = \frac{R_1}{2} \pm \frac{1}{2}\sqrt{R_1{}^2 - 4z_o{}^2}$$
$$z_2 = \frac{R_2}{2} \pm \frac{1}{2}\sqrt{R_2{}^2 - 4z_o{}^2} \qquad (4.30)$$

Here, $\pm z_o$ are the positions of the mirrors for the virtual equivalent confocal resonator that will have the same transverse modes. However, we still need to determine z_o.

In order to find z_o, we shall first observe some important conditions for z_o. Assuming $z_2 > z_1$, we obtain,

$$D = z_2 - z_1 = \frac{R_2}{2} - \frac{R_1}{2} \pm \frac{1}{2}\sqrt{R_2{}^2 - 4z_o{}^2} \mp \frac{1}{2}\sqrt{R_1{}^2 - 4z_0{}^2} \qquad (4.31)$$

Rearranging terms and squaring both sides to eliminate the square root, we obtain,

$$z_o{}^2 = \frac{D(-R_1 - D)(R_2 - D)(R_2 - R_1 - D)}{(R_2 - R_1 - 2D)^2} \qquad (4.32)$$

Clearly z_o must be a positive quantity in order to obtain real values for the equivalent confocal resonator position.

Eq. (4.32) allows us to calculate z_o with a real value only when the right-hand side is positive. The requirement for the right-hand side to be positive imposes also certain conditions on R_1, R_2, and D as follows. Let us assume that R_1 is negative at negative z_1. Then, we must have:

$$D(|R_1| - D)(|R_2| - D)(|R_1| + |R_2| - D) > 0 \qquad (4.33)$$

There are only two ways to satisfy this condition. (1) $0 < D < |R_1|$ or $|R_2|$, whichever is smaller. (2) $|R_1|+|R_2| > D > |R_1|$ or $|R_2|$, whichever is larger. Condition (1) can be expressed as $0 < (1 - D/|R_1|)(1 - D/|R_2|)$. Condition (2) can be expressed as $(1 - D/|R_1|)(1 - D/|R_2|) < 1$. Hence the criterion for the existence of a resonator mode, equivalent to a confocal resonator mode with z_o given in Eq. (4.32), is

$$0 < \left(1 - \frac{D}{|R_1|}\right)\left(1 - \frac{D}{|R_2|}\right) < 1 \qquad (4.34)$$

If we plot this equation in a rectangular coordinate system with the two axes as $D/|R_1|$ and $D/|R_2|$, then the boundary of the product to be zero consists of two straight lines, $D/|R_1| = 1$ and $D/|R_2| = 1$. On the other hand, the boundary of the product to be 1 is a hyperbola in this coordinate system. Figure 4.5 shows this plot. The shaded regions show the combinations of R_1, R_2, and D that satisfy the inequality in Eq. (4.34). Resonators with these combinations are called stable resonators. The regions

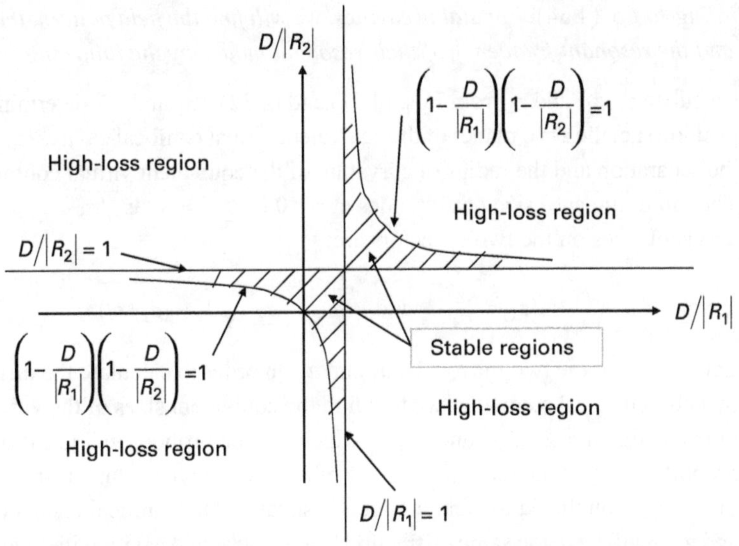

Figure 4.5 The stable and unstable regions of laser cavities. The straight lines are the plots of the lower limit of Eq. (4.34), and the hyperbola is the plot of the upper limit. The shaded region (i.e. the stable region) shows the $D/|R_1|$ and $D/|R_2|$values that satisfy Eq (4.34). In this region, modes have low or modest diffraction loss per pass. Cavities in the high loss region do not have $D/|R_1|$ and $D/|R_2|$ values that satisfy Eq. (4.34). It is called the unstable region.

outside of the shaded regions are called unstable resonator regions. The confocal resonator configuration has $D = |R_1| = |R_2|$. Thus the confocal resonator can easily be pushed into the unstable region by a slight misalignment of the cavity. In reality, the assumptions used in our diffraction loss calculation breakdown near the boundaries of stable and unstable regions. More precise calculations show the diffraction loss increases rapidly from the stable to the unstable configuration. There is no sudden change in diffraction loss from the stable to the unstable configuration. Unstable resonator modes not only exist, they are used often in very high-power lasers so that optical energy is not concentrated in a small physical region to avoid material damage by the high electric field.

In summary, when the given R_1, R_2, and D satisfy the stability criterion, Eq. (4.34), z_o, z_1, and z_2 are determined from Eqs. (4.31) and (4.32). z_o provides us with the specifications of the virtual equivalent confocal resonator. Note that the $\pm$ sign in Eq. (4.31) give us two answers for z_1 and for z_2. The correct choice is the one that gives the correct D.

4.3.3 A formal procedure to find the resonant modes in non-confocal cavities

A formal procedure can now be set up to find the resonant modes in non-confocal cavities for a given set of reflectors, according to the analysis presented in the previous subsection. We will first test the stability of the given R_1, R_2, and D

according to Eq. (4.34). *For stable cavities, we will find the field pattern, the diffraction loss and the resonant frequency of their resonant modes by the following seven steps:*

(1) Calculate z_o, z_1, and z_2 from Eqs. (4.31) and (4.32). "z_1 and z_2" determine the center position (i.e. the $z = 0$ plane) of the equivalent virtual confocal cavity. "z_0" determines the separation and the radius of curvature of the equivalent virtual confocal cavity.

(2) The minimum spot size of all modes at $z = 0$ is $\omega_o = \sqrt{\lambda z_o/\pi}$.

(3) The spot sizes on the two reflectors are:

$$\omega_{s1} = \omega_o\sqrt{1 + (z_1/z_o)^2}, \quad \text{and} \quad \omega_{s2} = \omega_o\sqrt{1 + (z_2/z_0)^2} \tag{4.35}$$

(4) Let the size of the two mirrors be a_1 and a_2. In order to calculate the diffraction loss of a non-confocal resonator, we first find the equivalent sizes of the virtual confocal mirrors with areas, ${a_{eq,1}}^2$ and ${a_{eq,2}}^2$, which will be proportional to ${a_1}^2$ and ${a_2}^2$. The proportionality is the ratio of the areas of the spots on the actual mirrors to the areas of the spots on the equivalent confocal resonator. The confocal resonator with $a_{eq,1}$ and $a_{eq,2}$ will have the same diffraction loss as the actual cavity with a_1 and a_2. Using Eq. (4.35) as the guide, we obtain

$$a_{eq,1} = \left(\sqrt{2}\omega_o\right)a_1/\omega_{s1} \ \textit{and}\ a_{eq,2} = \left(\sqrt{2}\omega_o\right)a_2/\omega_{s2} \tag{4.36}$$

(5) For symmetrical cavity, the diffraction loss per pass is calculated directly from the confocal resonator with size a_{eq}. For asymmetrical cavities, the diffraction loss per pass is the average of the diffraction losses. The averaged diffraction loss per pass for the cavity is ½ of the sum of the diffraction loss for the two different virtual confocal cavities, one with mirror size $a_{eq,1}$ and the second one with mirror size $a_{eq,2}$.

(6) In general, the resonance wavelength, λ_{lmq}, is determined by

$$(2\pi D/\lambda_{lmq}) = q\pi + (l + m + 1)(\tan^{-1}z_2/z_o - \tan^{-1}z_1/z_o) \tag{4.37}$$

The differences in resonance frequency for different longitudinal order q and transverse order l and m are:

$$\begin{aligned} f_{l,m,q+1} - f_{l,m,q} &= c/2D\,, \\ f_{l',m',q} - f_{l,m,q} &= \frac{c}{2\pi D}\left(\frac{\pi}{2} - \tan^{-1}\frac{z_2}{z_o} - \tan^{-1}\frac{z_1}{z_o}\right)(l' - l + m' - m) \end{aligned} \tag{4.38}$$

Note again that the difference in resonance frequency for two adjacent longitudinal orders is just $1/T$ where T is the round trip propagation time inside the cavity, $T = 2D/c$. If the cavity is filled with a dielectric that has an index of refraction n, $T = 2nD/c$. The transverse modes are still degenerate. All modes that have the same $l + m$ order will have the same resonance frequency.

(7) Practical resonators do not use end mirrors with square cross-sections. It is clear from the previous discussions that the mode patterns (i.e. the Hermite polynomials and the Gaussian envelope) will be affected only by the curvature and the position of the reflector, not by the shape of the cross-section, e.g. whether it is square or round.

Thus the modes derived for the square mirrors are equally applicable to the round mirrors. From Eq. (4.21), it is clear that the diffraction loss per pass depends primarily on the area Ω of the mirror. Round or square mirrors with the same size are likely to have the same diffraction loss per pass. Thus Figure 4.4 is also used to estimate the diffraction loss for round mirrors.

4.3.4 An example of resonant modes in a non-confocal cavity

Let us consider a semi-spherical cavity that has one flat reflector with $a_1 = 2$ mm and one spherical reflector with radius of curvature $R_2 = 0.7$ m and $a_2 = 0.6$ mm, separated by a distance of $D = 30$ cm. The wavelength is 1 μm. The medium between reflectors is air. Clearly, the stability criterion in Eq. (4.34) is satisfied so that we can find the modes and their diffraction losses by means of the virtual equivalent confocal cavity. Following the process outlined in Section 4.3.3, we obtain the following results.

For the equivalent virtual confocal cavity, $z_o = [D(R_2 - D)]^{1/2} = 0.346410$ m, $z_1 = 0$, $z_2 = 0.3$ m. Notice that there are two solutions for z_2 given in Eq. (4.32). The correct solution is the one that yields $z_2 - z_1 = D$.

The spot sizes are $\omega_o = 0.332063$ mm, $\omega_{s1} = 0.332063$ mm, and $\omega_{s2} = 1.32288 \times 0.332063 = 0.439278$ mm.

The appropriate sizes of the equivalent confocal reflectors for the calculation of diffraction loss are: $a_{eq,1} = 2.82843$ mm, $a_{eq,2} = 0.641427$ mm.

For reflector #1, $a^2/2z_o\lambda$ is 11.5. For reflector #2, $a^2/2z_o\lambda$ is 0.59. Therefore, the diffraction loss per pass of the TEM_{00} obtained from Figure 4.4 for the flat mirror is negligible, while the diffraction loss per pass for the second mirror is 5×10^{-3}. The averaged diffraction loss per pass for the cavity is 2.5×10^{-3}. The averaged diffraction loss per pass for the TEM_{01} mode will be approximately 5%.

In this section, we have not only shown how a non-confocal cavity can be analyzed, and designed from confocal cavity analysis. We have also shown important properties of non-confocal resonators, such as the stability diagram in Figure 4.5. The understanding of these properties allows us to determine the stability of any cavity configuration without detailed analyses.

From the discussion presented in Section 4.2.4, we observe that the Gaussian mode pattern also extends to the outside of non-confocal cavities. This is a very important result. It means that any Gaussian beam propagating through different environment remains a Gaussian beam.

4.4 The propagation and the transformation of Gaussian beams (the ABCD matrix)

There are many forms of the solutions of Maxwell's equations in the literature, such as plane waves, cylindrical waves, spherical waves, etc. These solutions have been used to analyze radiation fields propagating in different components whenever it is appropriate. The Gaussian beam is one of them.

Gaussian modes were shown in the last section to form a complete set. Any radiation field can be represented by a summation of Gaussian modes. The advantage of the use of a Gaussian beam to represent approximately any radiation field is that fields propagating through reasonably large apertures retain the same functional variations, except for a reduction in their amplitudes. Thus the wave equation is satisfied, and the diffraction effect is accounted for without the use of Kirchoff's diffraction integral. Furthermore, we will show in the following section that Gaussian modes are also natural mathematical solutions of Maxwell's equations without solving the scalar wave integral equation and without the existence of a cavity [5].

4.4.1 A Gaussian mode as a solution of Maxwell's equation

Consider, Maxwell's equations,

$$\nabla \times \underline{h} = \varepsilon^{\partial \underline{e}}/_{\partial t}, \quad \nabla \times \underline{e} = -\mu^{\partial \underline{h}}/_{\partial t}, \quad \nabla \bullet (\varepsilon \underline{e}) = 0 \tag{4.39}$$

In the most general case ε can be a function of (x,y,z). From $\nabla \times \nabla \times \underline{e}$, we obtain

$$\nabla^2 \underline{e} - \varepsilon\mu \frac{\partial^2 \underline{e}}{\partial t^2} = -\nabla\left(\frac{1}{\varepsilon}\underline{e} \cdot \nabla\varepsilon\right) \tag{4.40}$$

If $\nabla\varepsilon \perp \underline{e}$ (such as the ε variation in an optical fiber) or if $\nabla\varepsilon$ is small, we can then replace the right-hand side with 0. If we further assume the time variation to be $e^{j\omega t}$, then the equation for the electric field is:

$$\nabla^2 \underline{e} + k^2(\underline{r})\,\underline{e} = 0, \quad \text{where } k^2(\underline{r}) = \omega^2/\mu\varepsilon(\underline{r}) \tag{4.41}$$

When the medium is homogeneous, k is a constant. Note that Eq. (4.41) is similar to Eq. (3.4).

Let E be a linearly polarized field and,

$$E(x,y,z) = \psi(x,y,z)e^{-jkz} \tag{4.42}$$

We will now show below in five mathematical steps that, in a homogeneous medium, the circular symmetric ψ has a functional form identical to that of Gaussian modes.

(1) Substituting Eq. (4.42) into Eq. (4.41), we obtain in a cylindrical coordinate with $\partial\psi/\partial\theta = 0$:

$$\nabla_t^2 \psi - 2jk\frac{\partial \psi}{\partial z} = 0, \quad \text{where} \quad \nabla^2 = \nabla_t^2 + \frac{\partial^2}{\partial z^2} = \frac{\partial^2}{\partial r^2} + \frac{1}{r}\frac{\partial}{\partial r} + \frac{\partial^2}{\partial z^2} \tag{4.43}$$

(2) Let

$$\psi = \mathrm{e}^{-j\left[p(z)+\frac{k}{2q(z)}r^2\right]} \tag{4.44}$$

Substituting this functional form into the equation, we obtain

$$-\left(\frac{k}{q}\right)^2 r^2 - 2j\left(\frac{k}{q}\right) - k^2 r^2 \frac{\partial}{\partial z}\left(\frac{1}{q}\right) - 2k\frac{\partial p}{\partial z} = 0 \tag{4.45}$$

This equation must hold for all values of r. Thus, the terms involving different powers of r must vanish simultaneously, i.e.

$$\frac{1}{q^2} + \frac{\partial}{\partial z}\left(\frac{1}{q}\right) = 0, \quad \text{and} \quad \frac{\partial}{\partial z}p = \frac{-j}{q} \tag{4.46}$$

(3) Let $1/q = (\mathrm{d}S/\mathrm{d}z)/S$, then the equation for S is $\mathrm{d}^2 S/\mathrm{d}z^2 = 0$. The solution for S is obviously,

$$S = az + b, \; and \; q = S/(\mathrm{d}S/\mathrm{d}z) = z + b/a = z + q_o \tag{4.47}$$

Substituting this solution to the equation for $p(z)$, we obtain

$$\frac{\partial p}{\partial z} = -\frac{j}{z+q_o}, \quad p(z) = -j\ln\left(1 + \frac{z}{q_o}\right) \tag{4.48}$$

(4) The objective of finding the solutions for p and q is to show that ψ has the functional form of a TEM$_{00}$ Gaussian beam. Substituting q_o by a new constant $q_o = j\pi\omega_o^2/\lambda$, we obtain

$$\begin{aligned} \mathrm{e}^{-jp(z)} &= \mathrm{e}^{-\ln\left(1-\frac{j\lambda z}{\pi\omega_o^2}\right)} = \frac{1}{\sqrt{1+(\lambda^2 z^2/\pi^2\omega_o^4)}}\mathrm{e}^{j\tan^{-1}\left(\frac{\lambda z}{\pi\omega_o^2}\right)}, \\ \mathrm{e}^{\frac{-jkr^2}{2(z+q_o)}} &= \mathrm{e}^{\frac{-r^2}{\omega_o{}^2\left[1+\left(\frac{\lambda z}{\pi\omega_o^2}\right)^2\right]}} \mathrm{e}^{\frac{-jkr^2}{2z\left(1+\left(\frac{\pi\omega_o^2}{\lambda z}\right)^2\right)}} \end{aligned} \tag{4.49}$$

(5) Substituting the above results into the expression for ψ, we obtain an expression for E identical to the TEM$_{00}$ mode in Eq. (4.26) of Section 4.1,

$$E = \frac{1}{\sqrt{1+(\lambda^2\, z^2/\pi^2\omega_o^4)}}\mathrm{e}^{\frac{-r^2}{\omega_o^2\left[1+\left[\frac{\lambda z}{\pi\omega_o^2}\right]^2\right]}}\mathrm{e}^{\frac{-jkr^2}{2z\left[1+\left(\frac{\pi\omega_o^2}{\lambda z}\right)^2\right]}}\mathrm{e}^{-jkz}\mathrm{e}^{j\tan^{-1}\left(\frac{\lambda z}{\pi\omega_o^2}\right)} \tag{4.50}$$

In summary, a Gaussian beam is a natural solution of Maxwell's vector wave equations with $\nabla\varepsilon \perp \underline{e}$ or $\nabla\varepsilon \cong 0$. We have only derived the Gaussian mode for a homogeneous media. Yariv showed in his book that when $k^2(r) = k^2 - k\,k_2\,r^2$ in an inhomogeneous graded index medium, the solution of Eq. (4.43) for a circular symmetric mode is still a Gaussian beam [5].

4.4.2 The physical meaning of the terms in the Gaussian beam expression

We note that, for a given Gaussian mode, we can describe its functional variation at various values of z by

$$\begin{aligned} E &= A(x,y)\mathrm{e}^{-jkz}\mathrm{e}^{-jp(z)}\mathrm{e}^{-jk\frac{r^2}{2q(z)}}, \\ A(x,y) &= E_o H_l\left[\frac{\sqrt{2}x}{\underline{\underline{\omega}}(z)}\right] H_m\left[\frac{\sqrt{2}y}{\underline{\underline{\omega}}(z)}\right] \end{aligned} \tag{4.51}$$

Here, the coordinate z starts at the beam waist where the spot size is $\underline{\underline{\omega_o}}$. The E given here is taken from Eq. (4.26)

The first term, A, describes the x and y variation (i.e. the field pattern) of E. At two different z positions, z_1 and z_2, the A function will be the same. $A(x,y)$ is different for different l and m orders of the mode.

The second term, e^{-jkz}, and the third term, e^{-jpz}, are simple functions of z. They specify the phase of the beam as the beam propagates from one z position to another. They are independent of x and y. p is dependent on the mode order, l and m.

$$\mathrm{e}^{-jp(z)} = \frac{\underline{\underline{\omega_o}}}{\underline{\underline{\omega}}(z)}\mathrm{e}^{j(l+m+1)\tan^{-1}\left(\frac{\lambda z}{\pi \underline{\underline{\omega_o^2}}}\right)} \tag{4.52}$$

Thus, "$p + kz$" determines the phase of E at different z.

The $1/q$ term carries the most important physical meaning of the Gaussian beam. This term has a real part, which specifies the curvature of the phase front, and an imaginary part, which specifies the Gaussian variation of the amplitude at any z. To be more specific,

$$\frac{1}{q} = \frac{1}{R} - \frac{j2}{k\underline{\underline{\omega^2}}} \tag{4.53}$$

q is independent of the mode order, l and m.

q will be different at different z positions,

$$\frac{1}{q} = \frac{1}{z + q_o} \tag{4.54}$$

From Eq. (4.54), the q values at two z values are related to each other by

$$q(z_2) - q(z_1) = z_2 - z_1 \tag{4.55}$$

In the following sub-sections, 4.4.3 to 4.4.9, how a Gaussian beam propagates through various components will be presented. This will demonstrate clearly the advantage of using Gaussian beams to analyze propagation of waves with diffraction loss.

4.4.3 The analysis of Gaussian beam propagation by matrix transformation

It is important to note that as a Gaussian beam propagates, E is always given by Eq. (4.51). The relationship between $q(z_1)$, call it q_1, and $q(z_2)$, call it q_2, is linear. Instead of writing the Gaussian beam as a function of coordinates x, y, and z, we may write the relation between q_1 and q_2 in the formal form of a linear transformation,

$$q_2 = \frac{Aq_1 + B}{Cq_1 + D} \tag{4.56}$$

where $A = 1$, $B = z_2 - z_1$, $C = 0$, $D = 1$, $q_1 = q(z_1)$, and $q_2 = q(z_2)$. In other words, q_2 is transformed from q_1 by a linear transformation with the above ABCD coefficients.

A linear transformation relationship also exists between q values for Gaussian beams transmitting or reflecting from various optical components. When a Gaussian beam is incident on an ideal thin lens, we have learned from Eq. (3.30) that the transmitted field immediately after the lens, E_t, is related to the incident field E_{inc} by the transmission function of the lens, which is a quadratic phase shift, i.e.

$$E_t = E_{\text{inc}} e^{j\frac{\pi}{\lambda f}(x^2+y^2)} = A e^{-jkz} e^{-jp(z)} e^{-j\frac{\pi}{\lambda}\left(\frac{1}{q}-\frac{1}{f}\right)r^2} \tag{4.57}$$

Therefore, the transmitted beam will have the same form as given in Eq. (4.51). Let q_1 be the q parameter before the lens and q_2 the q parameter after the lens. q_2 is related to q_1 of the incident beam by

$$\frac{1}{q_2} = \frac{1}{q_1} - \frac{1}{f} \tag{4.58}$$

When we separate the imaginary and the real part of Eq. (4.58), we obtain

$$\omega_2 = \omega_1, \qquad \frac{1}{R_2} = \frac{1}{R_1} - \frac{1}{f} \tag{4.59}$$

It implies that the spot size is not changed by transmission through a thin lens. However, the radius of curvature of the phase front is changed according to Eq. (4.59). We conclude that q_2 and q_1 are again related by Eq. (4.56) with $A = 1$, $B = 0$, $C = -1/f$, and $D = 1$. p does not change when the beam propagates through a thin lens.

If the lens has a finite aperture, the transmitted Gaussian beam will have the same functional variation as for an infinite aperture. However, the amplitude

Table 4.1. The ABCD transmission matrix for some common optical elements and media.

Transformation description	Figure	Matrix
Homogeneous medium: length d	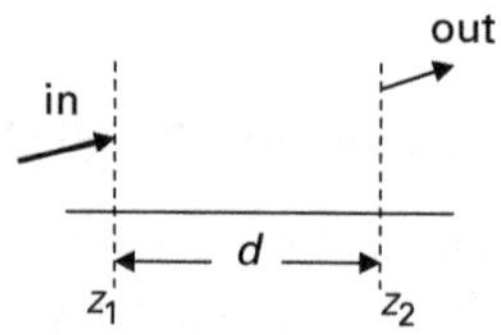	$\begin{bmatrix} 1 & d \\ 0 & 1 \end{bmatrix}$
Thin lens: Focal length f ($f > 0$, converging; $f < 0$, diverging)	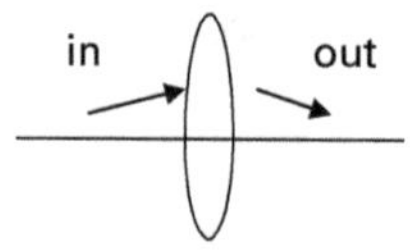	$\begin{bmatrix} 1 & 0 \\ \frac{-1}{f} & 1 \end{bmatrix}$
Dielectric interface: Refractive indices n_1, n_2	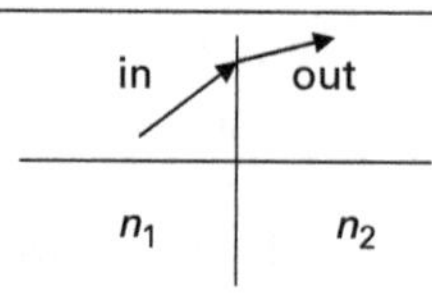	$\begin{bmatrix} 1 & 0 \\ 0 & \frac{n_1}{n_2} \end{bmatrix}$
Spherical dielectric interface: Radius R	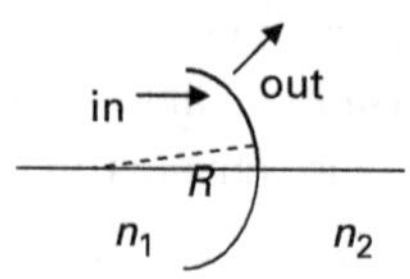	$\begin{bmatrix} 1 & 0 \\ \frac{n_2 - n_1}{n_2 \cdot R} & \frac{n_1}{n_2} \end{bmatrix}$
Spherical mirror: Radius of curvature R	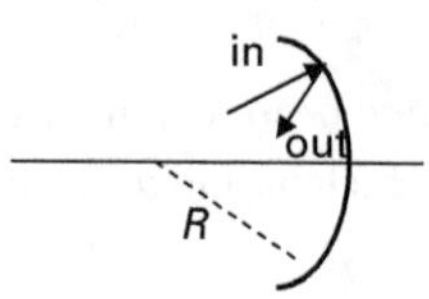	$\begin{bmatrix} 1 & 0 \\ \frac{-2}{R} & 1 \end{bmatrix}$

will be reduced. The reduction in amplitude will be identical to the amplitude reduction calculated from the diffraction loss per pass caused by the same aperture.

The ABCD transformation method is applicable to propagation of Gaussian modes through many optical elements. The ABCD transformation coefficients of various optical components, such as those shown in Table 4.1 are also given in other textbooks [5]. It does not include the diffraction loss.

If a Gaussian beam propagates through more than one optical component, the q parameters at various positions can be determined by ABCD transformations in succession. For two successive transformations:

$$q_3 = \frac{A_2 q_2 + B_2}{C_2 q_2 + D_2}, \qquad q_2 = \frac{A_1 q_1 + B_1}{C_1 q_1 + D_1}, \qquad \text{thus}$$
$$q_3 = \frac{(A_1 A_2 + B_2 C_1) q_1 + (A_2 B_1 + B_2 D_1)}{(A_1 C_2 + C_1 D_2) q_1 + (B_1 C_2 + D_1 D_2)} = \frac{A q_1 + B}{C q_1 + D} \tag{4.60}$$

The ABCD coefficients for q_3 in terms of q_1 in the above equation are simply the coefficients obtained by multiplying matrix $A_1\ B_1\ C_1\ D_1$ by matrix $A_2\ B_2\ C_2\ D_2$, as follows:

$$\left\| \begin{matrix} A & B \\ C & D \end{matrix} \right\| = \left\| \begin{matrix} A_2 & B_2 \\ C_2 & D_2 \end{matrix} \right\| \times \left\| \begin{matrix} A_1 & B_1 \\ C_1 & D_1 \end{matrix} \right\| \tag{4.61}$$

After the Gaussian beam has propagated through many optical components, this matrix multiplication process can be repeated many times to obtain the ABCD coefficients for the total transformation matrix.

q of the final output Gaussian beam is related to the input q by Eq. (4.56) where the ABCD coefficients are given by Eq. (4.61). Thus the ABCD coefficients are called the ABCD transformation matrices. It can be shown that any ABCD matrix is a unitary matrix, i.e. $AD - BC = 1$. It is important to keep in mind that the order of multiplication must follow the order in which the Gaussian beam is propagating through various elements. It cannot be taken for granted that permutation of the order of matrix multiplication will give the same result.

p changes only when the z position changes. Therefore when the TEM_{lm} mode passes through any component that has zero thickness, such as a thin lens, p does not change. After the mode has propagated through many elements and distances, the new p is obtained by using the distance of propagation as z. $A(x,y)$ does not change.

4.4.4 Gaussian beam passing through a lens

Consider a Gaussian beam at $\lambda = 1\ \mu\text{m}$ with $\omega_o = 0.4$ mm at $z = 0$. It propagates through a thin lens with $f = 2$ mm at $z = 0.1$ m. Let us find the field pattern at $z = 0.1$ m after the lens.

There are two ways to find the answer: (1) We can find the answer using Eq. (4.26). The given Gaussian beam has $z_o = \pi {\omega_o}^2/\lambda = 0.502665\,\text{m}$. From Eq. (4.26), we also know the field pattern for any TEM_{lm} mode incident on the lens at $z = 0.1$ m. It has a Gaussian amplitude variation with $\omega = 0.407839$ mm, a radius of curvature for the phase front $R = 2.62662$ m, and a phase shift given by $\eta = 0.1964$

radians. According to Eq. (4.57) the radiation field emerging from the thin lens will have the same phase and amplitude variation. However, the radius of curvature for the phase front will now be $Rf/(R-f)$, which is 2.00152×10^{-3} m. We would intuitively expect such an answer because the lens should create a focused spot near its focal plane. (2) The answer could also be obtained very quickly from the ABCD matrix transformation as follows:

$$\begin{Vmatrix} A & B \\ C & D \end{Vmatrix} = \begin{Vmatrix} 1 & 0 \\ \dfrac{-1}{0.002} & 1 \end{Vmatrix} \times \begin{Vmatrix} 1 & 0.1 \\ 0 & 1 \end{Vmatrix} = \begin{Vmatrix} 1 & 0.1 \\ -500 & -49 \end{Vmatrix}$$

At $z = 0$, q is $jk\omega^2{}_0/2$, which is $j0.502655$. Therefore at the exit plane of the lens,

$$\frac{1}{q} = \frac{Cq_1 + D}{Aq_1 + B} = \frac{-500(j0.502655) - 49}{j0.502655 + 0.1} = \frac{-131.231 - j0.502645}{0.262662}$$
$$= -499.619 - j1.91366\,.$$

Here the real part of $1/q$ is $1/R$, while the imaginary part $-\lambda/\pi\omega^2$. Note that the complete expression of the field is given in Eqs. (4.51) and (4.52) with this q value.

4.4.5 Gaussian beam passing through a spatial filter

Let us reconsider the example in Section 3.2.4(e) when the incident beam is a Gaussian beam. The ABCD transformation matrix method lets us find the main propagation characteristics of the incident beam without any integration. We will need to perform integration only when we want to know the diffraction loss.

Figures 3.8 and 3.9 have already illustrated the geometrical configuration of this spatial filtering set-up. Let the incident beam be a TEM_{00} Gaussian beam incident on the film at $z = 0$. The incident beam is:

$$E = E_o e^{-jkz} e^{-jp(z)} e^{-jk\frac{r^2}{2q_o}}$$

The beam waist is at $z = 0$ with spot size ω_o, $\omega_o << d$. Notice now the effective beam size is controlled by ω_o and not by d. Therefore,

$$\frac{1}{q_o} = -\frac{j2}{k\omega_o{}^2}$$

For $d > \omega_o$, the aperture size d does not change the functional form of the Gaussian beam. It introduces a reduction of the amplitude because of the diffraction loss caused by the aperture. At $z = 0$, immediately after the film with the transmission function t in Eq. (3.35), we obtain

$$E = \frac{1}{2}E_o e^{-jkz} e^{-jp(z)} e^{-jk\frac{r^2}{2q_0}} + \frac{1}{4}E_o e^{j2\pi Hx} e^{-jkz} e^{-jp(z)} e^{-jk\frac{r^2}{2q_o}} + \frac{1}{4}E_o e^{-j2\pi Hx} e^{-jkz} e^{-jp(z)} e^{-jk\frac{r^2}{2q_o}} \tag{4.62}$$

Each of the three terms in the above expression is still a Gaussian beam. The first term is the same as the incident Gaussian beam with half the amplitude. For $\lambda H << 1$, $e^{j2\pi Hx}e^{-jkz}$ is a propagating beam in the x–z plane at an angle $-\theta$ with respect to the z axis where $\sin\theta = \lambda H$. Similarly, the third term is a propagating beam in the x–z plane at an angle θ with respect to the z axis. For small θ, the three beams are still approximately Gaussian beams in their three respective directions of propagation, i.e. the z axis, the $+\theta$ axis and the $-\theta$ axis. Therefore, we will treat them as three separate Gaussian beams along those directions.

After the lens at $z = 0$, we have

$$E = \frac{1}{2}E_o e^{-jkz} e^{-jp(z)} e^{-jk\frac{r^2}{2q_1}} + \frac{1}{4}E_o e^{-jkz} e^{-jp(z)} e^{-j2\pi Hx} e^{-jk\frac{r^2}{2q_1}} + \frac{1}{4}E_o e^{-jkz} e^{-jp(z)} e^{+j2\pi Hx} e^{-jk\frac{r^2}{2q_1}} \tag{4.63}$$

where

$$\frac{1}{q_1} = \frac{1}{q_o} - \frac{1}{f}, \qquad q_1 = \frac{-(k\omega_o{}^2)^2 f + j2k\omega_o{}^2 f^2}{(k\omega_o{}^2)^2 + (2f)^2} \tag{4.64}$$

In front of the screen at $z = f$, the three beams are:

$$E = \frac{1}{2}E_o e^{-jkf} e^{-jp(z=f)} e^{-jk\frac{r^2}{2q_2}} + \frac{1}{4}E_o e^{-jkf} e^{-j2\pi Hx} e^{-jp(z=f)} e^{-jk\frac{r^2}{2q_2}} + \frac{1}{4}E_o e^{-jkf} e^{+j2\pi Hx} e^{-jp(z=f)} e^{-jk\frac{r^2}{2q_2}} \tag{4.65}$$

where

$$q_2 = q_1 + f = \frac{2jf^2}{k\omega_o^2 + 2jf}, \qquad \frac{1}{q_2} = +\frac{1}{f} - j\frac{k\omega_o^2}{2f^2} = \frac{1}{R_2} - \frac{j2}{k\omega_2^2} \tag{4.66}$$

Here R_2 is the curvature of the Gaussian beam and ω_2 is the spot size at $z = f$. Therefore,

$$R_2 = f \qquad \text{and} \qquad \omega_2 = \frac{\lambda f}{\pi \omega_o} \tag{4.67}$$

This means that the curvature of the beam is f and that the spot size is proportional to f/ω_o. This result fits our intuition since we expect an ideal lens to focus a plane wave into a spherical wave with a focused spot size proportional to the focal length and inversely proportional to the incident beam size. For small θ, we have approximated the distance along the respective directions of propagation by z in this calculation.

The centers of the three beams are at $z = 0$ and $z \cong \pm\theta f \cong \pm\lambda Hf$. The beam centered at $z = 0$ is always blocked by the screen. In order for the two beams in the $\pm\theta$ directions to pass, we need

$$\lambda Hf + \frac{f\lambda}{\pi\omega_o} \langle \ l \ \langle 2\lambda Hf - \frac{2f\lambda}{\pi\omega_o} \tag{4.68}$$

This is the same result we obtained in the example in Section 3.2.4(e).

When the two transmitted beams travels to $z = 2f$, in front of the second lens, the q parameter of the Gaussian beams is q_3, where

$$q_3 = q_2 + f, \quad q_3 = \frac{fk\omega_o^2}{k\omega_o^2 - 2jf}, \quad \text{and} \quad \frac{1}{q_3} = \frac{1}{f} - \frac{2j}{k\omega_o^2}. \tag{4.69}$$

After the lens, the parameter q_4 is

$$\frac{1}{q_4} = \frac{1}{q_3} - \frac{1}{f} = -\frac{2j}{k\omega_o^2} \tag{4.70}$$

Therefore we get back two original Gaussian beams, now propagating in the $\pm\theta$ directions with the same spot size. There will be some diffraction losses associated with the aperture and the screen.

Comparing the solution presented in this section with the solution presented in Section 3.2.4(e), the Gaussian beam analysis is much simpler.

4.4.6 Gaussian beam passing through a prism

A thin prism is illustrated in Figure 4.6. Let the prism be made of material with refractive index n at wavelength λ. Let the prism axis be the x axis and the base of the prism be

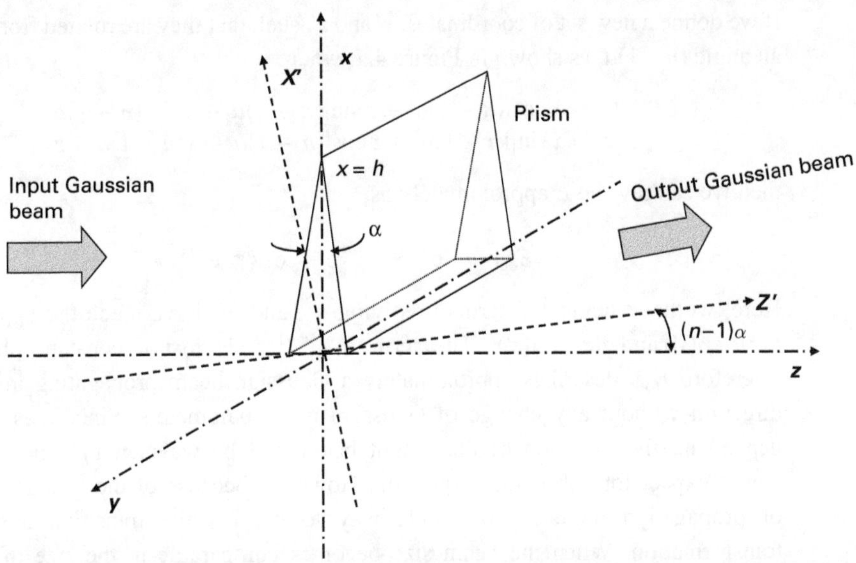

Figure 4.6 Illustration of a Gaussian beam passing through a prism. The phase shift of an optical beam passing through a thin prism can be represented as a phase shift equivalent to tilting the wave from the direction of propagation in xyz coordinates to a direction in $x'y'z'$ coordinates. The tilt angle is $(n-1)\alpha$, where n is the index of prism material at the wavelength and α is the vertex angle of the prism.

parallel to the y axis. The prism has a wedge angle α. The vertex of the prism is placed at $x = h$ and $z = 0$. Let a Gaussian beam

$$E_{\text{inc}} = A(x,y)\mathrm{e}^{-jkz}\mathrm{e}^{-jp(z)}\mathrm{e}^{-jk\frac{(x^2+y^2)}{2q(z)}} \tag{4.71}$$

be incident on the prism. The symbols in the expression for the incident E have already been defined and explained in Eq. (4.51).

Similar to a thin lens, discussed in Section 4.4.4, there is a phase change for any beam propagating through a thin prism. For the geometry shown in Figure 4.6, the phase change from any incident beam to the outgoing beam can be derived from phase changes of small optical rays passing through the prism at different x positions. The transfer function t for any beam passing through a thin prism was discussed in Section 1.3.4 (c) for plane waves. It is:

$$t = \mathrm{e}^{-jk(n-1)\alpha(h-x)} \tag{4.72}$$

Here we have assumed that the beams are located well below $x = h$ so that the diffraction from the prism vertex at $x = h$ can be neglected. α is small so that $\sin\alpha \cong \alpha$. Therefore, the output beam will be

$$E_{\text{out}} = A(x,y)\mathrm{e}^{-jk(n-1)\alpha h}\mathrm{e}^{-jkz}\mathrm{e}^{-jp(z)}\mathrm{e}^{jk(n-1)\alpha x}\mathrm{e}^{-jk\frac{(x^2+y^2)}{2q(z)}} \tag{4.73}$$

If we define a new set of coordinates, x' and z', such that they are rotated from x and z by an angle $(n-1)\alpha$, as shown in Figure 4.7, where

$$\begin{aligned} x' &= x\cos[(n-1)\alpha] - z\sin[(n-1)\alpha] \cong x - z(n-1)\alpha \\ z' &= x\sin[(n-1)\alpha] + z\cos[(n-1)\alpha] \cong x(n-1)\alpha + z \end{aligned} \tag{4.74}$$

then we can rewrite E approximately as:

$$E_{\text{out}} = A\mathrm{e}^{-jk(n-1)\alpha k}\mathrm{e}^{-jkz'}\mathrm{e}^{-jp(z')}\mathrm{e}^{-jk\frac{(x'^2+y^2)}{2q(z')}} \tag{4.75}$$

Here, we have neglected terms involving α^2, and we have made the approximation $p(z) \cong p(z')$ and $q(z) \cong q(z')$. The term $\mathrm{e}^{-jk(n-1)\alpha k}$ is just a constant phase factor. Therefore E_{out} describes approximately a Gaussian beam propagating in the new z' direction without any change of Gaussian beam parameters. Since n is wavelength dependent, the direction of the output beam will be wavelength dependent, as we would expect for chromatic dispersion. However, because of the change in direction of propagation there is no simple way to express the transition as an ABCD transformation. When the beam size becomes comparable to the size of the prism, there will be diffraction losses. The diffraction loss can be calculated according to Eq. (4.21). When there is diffraction loss the coefficient A in Eq. (4.75) will be reduced accordingly.

4.4.7 Diffraction of a Gaussian beam by a grating

Diffraction of a plane wave by a grating was discussed in Section 1.5. Analysis of the diffraction of a Gaussian beam is similar to that. However, the analysis now includes the effect of the finite size of the optical beam. Let there be a transmission grating with its transmission function t identical to that used in Section 1.5.

$$t = t_o(1 + \Delta t\cos 2\pi f_g y) = t_o + t_o\frac{\Delta t}{2}\mathrm{e}^{j2\pi f_g y} + t_o\frac{\Delta t}{2}\mathrm{e}^{-j2\pi f_g y} \tag{4.76}$$

Let there be an input Gaussian beam,

$$E_{\text{inc}} = A(x,y)\mathrm{e}^{-jkz}\mathrm{e}^{-jp(z)}\mathrm{e}^{-jk\frac{(x^2+y^2)}{2q(z)}} \tag{4.77}$$

The output beam is

$$E_{\text{out}} = t_o A'(x,y)\left(1 + \frac{\Delta t}{2}\mathrm{e}^{-j2\pi f_g y} + \frac{\Delta t}{2}\mathrm{e}^{+j2\pi_g f y}\right)\mathrm{e}^{-jkz}\mathrm{e}^{-jp(z)}\mathrm{e}^{-jk\frac{(x^2+y^2)}{2q(z)}} \tag{4.78}$$

There are three terms in Eq. (4.78). The first term is a Gaussian beam in the direction of the z axis. The second term is a diffracted Gaussin beam in the direction of θ^{+1}, called the +1 order diffracted wave where $\theta^{+1} = \sin^{-1}(2\pi f_g/\beta)$. The third term is a diffracted Gaussian beam in the direction of θ^{-1}, called the –1 order diffracted wave where

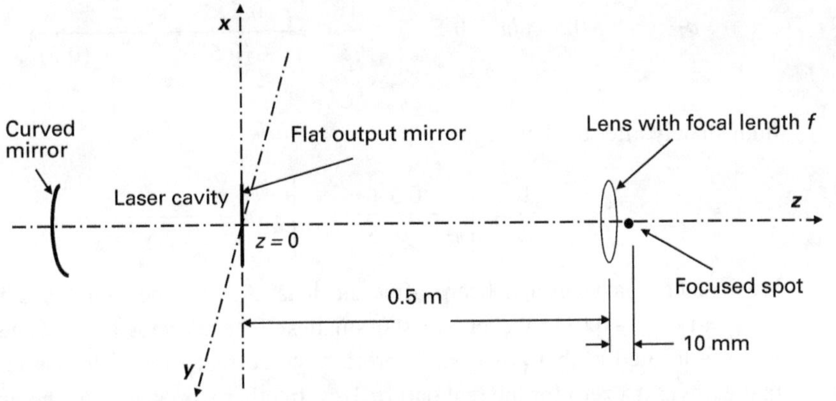

Figure 4.7 Illustration of a Gaussian beam focused by a lens. The laser is oscillating in the TEM_{00} mode of the cavity. The laser radiation from the flat reflector needs to be focused to a spot 10 mm beyond the lens. The Gaussian beam transformtion technique is used to find the optimal focal length of the lens.

$\theta^{-1} = \sin^{-1}(-2\pi f_g/\beta)$. A' will be proportional to A. The proportion will be determined by the diffraction loss, which can be calculated from Eq. (4.21). Diffraction by gratings of different groove periodicity and shape can be analyzed in a similar way to the discussion using Eq. (1.103).

4.4.8 Focusing a Gaussian beam

Intuitively, we know that in order to focus a beam to a distance d away from a lens or mirror, we use a lens or a mirror with a focal length of d. The smaller the value of d, the smaller the focused spot. However, we will always wonder whether the focusing will be affected by the characteristics of the Gaussian beam or by the location of the lens. It is also instructive to see how the focusing of a Gaussian beam can be analyzed by the ABCD transformation method. This analysis will allow us to calculate the value of f that will yield the smallest focused spot at a given distance away and the size of the focused spot.

Figure 4.7 shows a laser oscillating in the TEM_{00} mode and a lens focusing the laser mode. ω_o of the TEM_{00} oscillating mode is 1 mm on the flat mirror located at $z = 0$. Let the wavelength be 1 μm. A lens of focal length f is used to focus the laser beam to a distance 10 mm after the lens.

For a semi-spherical laser cavity the beam waist of the resonant mode is on the flat mirror. The Gaussian beam parameter, q_1, of this oscillating mode at $z = 0$ is

$$\frac{1}{q_1} = -j\frac{1}{\pi} \qquad 1/\text{meter} \tag{4.79}$$

The lens is located at $z = 0.5$ m away. The Gaussian beam parameter at $z = 0.5$ m, q_2, is

$$q_2 = q_1 + 0.5 = j\pi + 0.5, \qquad \frac{1}{q_2} = \frac{0.5}{\pi^2 + (0.5)^2} - j\frac{\pi}{\pi^2 + (0.5)^2} \tag{4.80}$$

Immediately after the lens, q_3 is

$$\frac{1}{q_3} = \left[\frac{0.5}{\pi^2 + (0.5)^2} - \frac{1}{f}\right] - j\frac{\pi}{\pi^2 + (0.5)^2} \tag{4.81}$$

We still have a Gaussian beam after the lens. At the intended focusing position, $1/q_4 = 1/(q_3 + 0.01)$. We obtain the smallest focused spot if the Gaussian beam waist is located at that position. Therefore the correct *f* for us to use is the *f* value that will yield a zero for the real part of $1/q_4$. In other words, q_4 must be imaginary, or the real part of q_3 should be –0.01. the numerical solution of that condition yields f= 0.00999516 m. In order to obtain the spot size at the focus, we need to find the imaginary part of $1/q_4$. Note that $Im[q_4] = Im[q_3]$. Substitution of the *f* value into $1/q_3$ yields a spot size of 9.88 μm at the focus. Clearly, a change in the position of the lens or in the Gaussian parameter q_1 will change very slightly the desired *f* value to use. On the other hand, if we reduce the distance of the focused spot to the lens, we will obtain a smaller focused spot size.

4.4.9 An example of Gaussian mode matching

Let there be a Gaussian beam with parameter q_a at location A. Let there be an optical instrument that requires a Gaussian beam with parameter q_b at location B, as illustrated in Figure 4.8. A lens with focal length *f* is placed at specific distance *d* from A to match the Gaussian beam with q_a at A with a Gaussian beam with q_b at B. We can find *f* and *d* by the ABCD transformation method as follows.

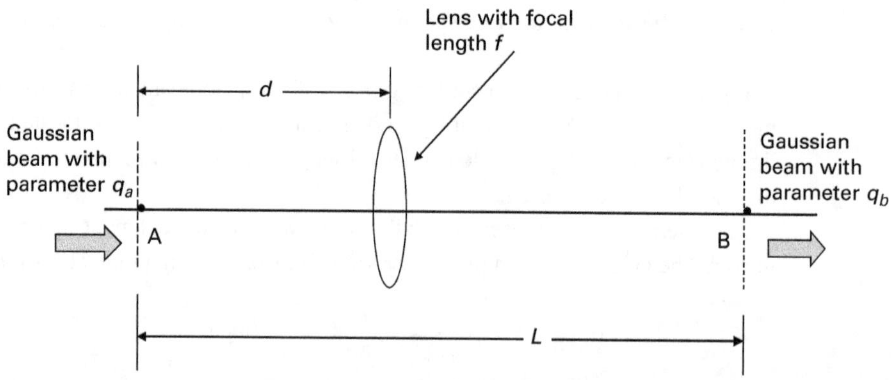

Figure 4.8 Matching a Gaussian beam at A to a Gaussian beam at B. A lens can be used to match a Gaussian beam at A to a different Gaussian beam at B. The Gaussian beam transformation technique can be used to determine the position and the proper focal length of the lens.

We know q_b is related to q_a as

$$q_b = \frac{(q_a + d)f}{f - (q_a + d)} + (L - d) \tag{4.82}$$

q_a and q_b have two differences, the difference in the real part (i.e. the curvature of the Gaussian beam wave front) and the difference in the imaginary part (i.e. the Gaussian spot size). We have two algebraic equations for f and d that can be easily obtained from Eq. (4.82) to match the two differences in q_a and q_b. Michael Spurr and Malcolm Dunn [6] have shown that high school geometry can be used to solve these algebraic problems arising from Gaussian beam optics.

4.4.10 Modes in complex cavities

When there are many optical elements in a cavity, the q parameter of the Gaussian beam at different positions in such a cavity can be found by considering the transformation of q after a round trip in the cavity.

Let the q parameter at any point in the cavity be q_s. The final q parameter after a round trip is $(Aq_s + B)/(Cq_s + D)$. For a stable mode in the cavity, it must also be the original q_s. Thus the equation for q_s is:

$$q_s = \frac{Aq_s + B}{Cq_s + D} \tag{4.83}$$

This is a quadratic algebraic equation for $1/q_s$. The solution is

$$\frac{1}{q_s} = \frac{D-A}{2B} \pm \frac{j\sqrt{1-[(D+A)/2]^2}}{B} = \frac{D-A}{2B} \pm \frac{j\sin\theta}{B}, \quad \text{where} \quad \cos\vartheta = \frac{D+A}{2} \tag{4.84}$$

We learned earlier that

$$\frac{1}{q_s} = \frac{1}{R} - j\frac{\lambda}{\pi\omega^2} \tag{4.85}$$

For a stable resonator, R is the radius of curvature of the spherical phase front, and ω is the spot size. Therefore the magnitude of $\cos\theta$ must be less than 1, or

$$\left|\frac{D+A}{2}\right| < 1 \tag{4.86}$$

For simple cavities, Eq. (4.86) is identical to Eq. (4.34). $|(D+A)/2| = 1$ is also represented by the boundary between stable and unstable regions shown in Figure 4.5.

Once q at various positions in the cavity is known, we can find the position at which q is purely imaginary. This position is that of the origin of the z axis, i.e. $z = 0$, for the virtual equivalent confocal resonator. At this position, the beam waist is ω_o. The lmth

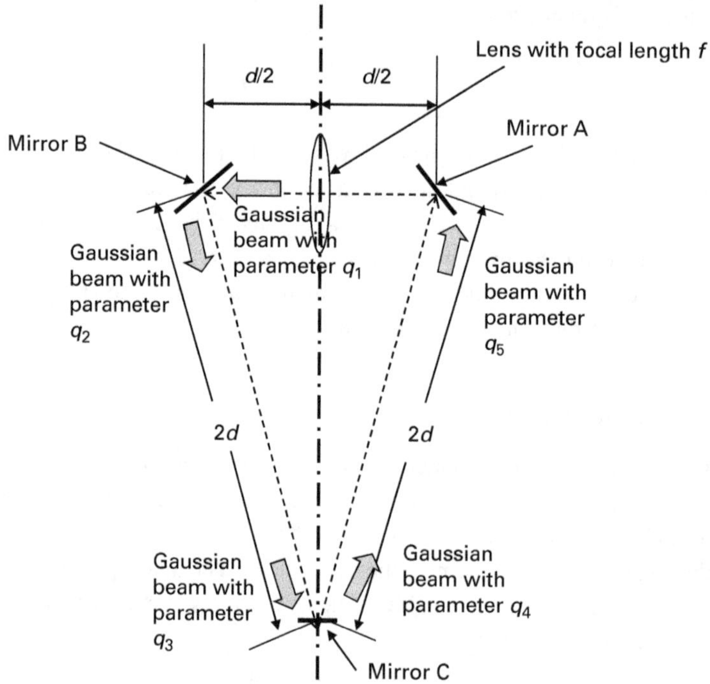

Figure 4.9 Illustration of a Gaussian mode in a ring cavity. In a ring cavity, the resonant mode is the recirculating mode that reproduces the field pattern with integer multiples of 2π phase shift after a round trip of multiple reflections. The optical path of the recirculating mode is shown by the block arrows. The Gaussian beam parameters values of q before and after each reflector are also shown.

mode of the equivalent virtual confocal resonator is given by Eqs. (4.51) and (4.52) in terms of these coordinates and the complex q values. The phase shift for the round trip propagation depends on the mode order, l and m, and the total distance of propagation from $z = 0$. The resonance frequency is determined by the wavelength at which the round trip phase shift is 2π. The diffraction loss per pass of each optical element encountered in the round trip path can be calculated by the same procedure as we have used for reflectors in non-confocal resonators at the end of Section 4.2.3.

4.4.11 An example of the resonance mode in a ring cavity

A ring cavity is illustrated in Figure. 4.9. There are three flat mirrors at A, B, and C, separated by a distance d between A and B. C is separated from A and B by $2d$. A lens with focal length 1 m is placed midway between mirrors A and B. The recirculating resonance mode is the mode that starts with Gaussian parameter q_2 at mirror B, is reflected by mirrors C and B, is transmited through the lens, and propagates back to mirror B. Let d = 1 m and λ = 1 μm. We can find the recirculating resonant modes and the diffraction loss per pass from the ABCD transformation matrix method as follows.

The transformation matrix M from q_2 to q_1 at mirror B through q_3, q_4, and q_5, in the counterclockwise direction in Figure 4.9 is:

$$M = \begin{Vmatrix} 1 & \frac{d}{2} \\ 0 & 1 \end{Vmatrix} \begin{Vmatrix} 1 & 0 \\ -\frac{1}{f} & 1 \end{Vmatrix} \begin{Vmatrix} 1 & \frac{9d}{2} \\ 0 & 1 \end{Vmatrix} \tag{4.87}$$

For $d = 1$ and $f = 1$,

$$M = \begin{Vmatrix} +\frac{1}{2} & \frac{11}{4} \\ -1 & -\frac{7}{2} \end{Vmatrix} \tag{4.88}$$

If we require that $q_s = q_2$ in a round trip, we have

$$q_2 = \frac{\frac{1}{2}q_2 + \frac{11}{4}}{-q_2 - \frac{7}{2}} \tag{4.89}$$

Therefore

$$\frac{1}{q_2} = -\frac{8}{11} \pm j\frac{\sqrt{20}}{11} \tag{4.90}$$

The values of $1/q$ at each mirror tell us the curvature and the spot size of the Gaussian beam at that mirror. We can obtain the diffraction loss per pass of each mirror from the mirror size and the spot size. In particular, q_3 is imaginary. Thus we know that the beam waist of the recirculating resonant mode is at mirror C. The size of the beam waist, ω_o, at mirror C is determined by the value of q_3. From ω_o we obtain z_o of the equivalent confocal resonator mode.

Chapter summary

Diffraction analysis presented in Chapter 3 *is the mathematical base used to analyze large laser cavities. The result of the analysis of laser cavities showed us various properties of laser modes. It is interesting to note that although the modes are the result of diffraction analysis, laser properties can much better be understood in terms of the modal description. There is not clear-cut boundary between diffraction analysis and modal analysis. The important considerations in analyzing any application are the geometrical configuration of the device and the most appropriate way to analyze the fields in that configuration.*

The laser cavity analysis yielded a set of Gaussian modes. Once we have the Gaussian modes, we can use them to represent any optical radiation, whenever it is appropriate. The advantage of representing the radiation beam by a Gaussian mode is that the diffraction loss through components with limited aperture is taken care of without the use of diffraction integrals. Several examples of how to use Gaussian mode analysis in applications have been demonstrated. Please note that the use Gaussian beam analysis is still limited to TEM waves.

References

[1] A. G. Fox and T. Li, Resonant modes in a maser interferometer, *Bell System Technical Journal*, **40**, 453, 1961.
[2] G. D. Boyd and J. P. Gordon, Confocal multimode resonator for millimeter through optical wavelength masers, *Bell System Technical Journal*, **40**, 496, 1961.
[3] D. Slepian and H. O. Pollak, Probate spheroidal wave functions, *Bell System Technical Journal*, **40**, 43, 1961.
[4] C. Flammer, *Spheroidal Wave Functions*, Stanford University Press, Stanford, CA, 1957.
[5] A. Yariv, *Quantum Electronics*, John Wiley & Sons Inc., New York, 1989, Chapter 6.
[6] M. Spurr and M. Dunn, Euclidian light: high-school geometry to solve problems in Gaussian beam optics, *Optics and Photonic News*, **13**, 40, 2002.

5 Optical waveguides and fibers

Kirchoff's integrals cannot be used to analyze optical waves in waveguides and fibers because they are not TEM waves and there are significant variations of the electromagnetic field in the transverse direction within distances comparable to or smaller than the wavelength. For electromagnetic analysis of guided-wave structures, Maxwell's vector equations plus appropriate boundary conditions need to be used to find the modes in these devices. Opto-electronic devices also function via the interaction of these modes. For these reasons, modal analyses are presented in Chapters 5 to 8. Chapter 5 focuses on the modes of optical fiber and channel waveguides. Chapter 6 presents the methods that analyze the mutual interactions of modes. Passive and active devices are discussed in Chapters 7 and 8. Many of the theoretical methods in optical guided-wave analysis are very similar to those used in microwaves.

From another perspective, modal analysis, plane wave, and diffraction integral analysis are all analyses based on Maxwell's equations. There are also other solutions of Maxwell's equations in the literature, such as cylindrical and spherical waves. They are just different ways to analyze optical fields demanded by different device configurations. The more complex the configuration, the more complex the mathematical analysis. What form of analyses should be used is determined by what is the most appropriate one to use.

5.1 Introduction to optical waveguides and fibers

Optical waveguides and fibers are made from dielectric materials. They have a high index core surrounded by lower index cladding or substrate. The transverse dimensions of the core are comparable to or smaller than the optical wavelength. Guided electromagnetic waves propagate in and around the core. A typical optical fiber and a typical channel waveguide are illustrated in Figure 5.1.

Guided-wave modes are solutions of homogeneous Maxwell's electromagnetic equations in waveguide structures that have a constant cross-section and infinite length in the direction of propagation. The modes of optical fibers and waveguides are the focus of discussion in this chapter. Homogeneous solution means that these are the propagating electric and magnetic fields that satisfy the differential equations and the appropriate boundary conditions in the absence of any radiation source. What modes are excited is determined by the input radiation.

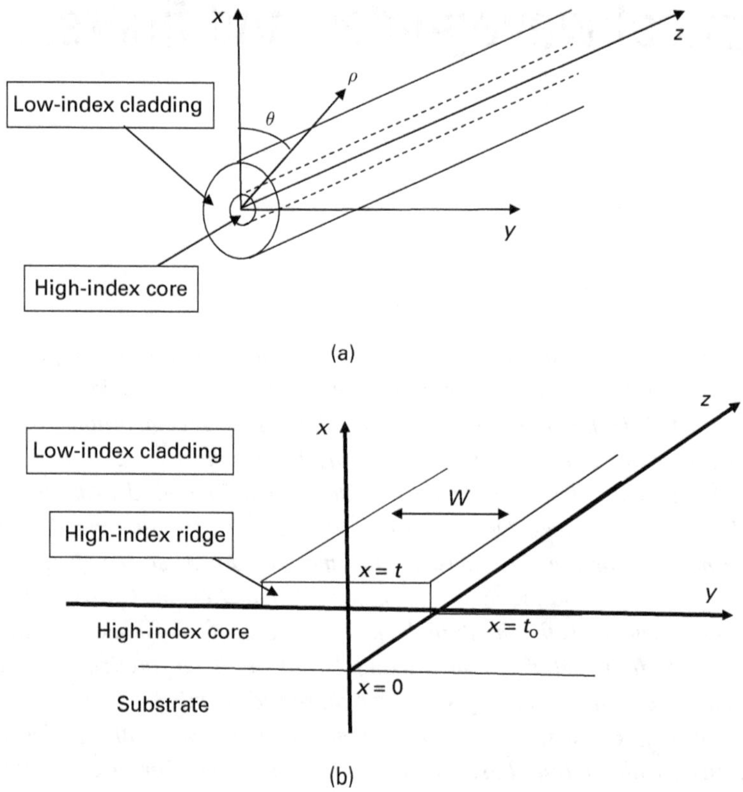

Figure 5.1 Illustration of a step-index optical fiber and a ridged channel waveguide. (a) The round optical fiber has a high-index core in the center, surrounded by a low-index cladding. It is shown in cylindrical coordinates. (b) The channel waveguide core consists of a high-index planar waveguide and a high-index ridge. The ridge plus the planar waveguide is t thick and W wide. There is a substrate under the core and a cladding surrounding the core on top.

Modal analyses are used in microwaves as well as optics. However, there are important differences between optical and microwave waveguides. In microwaves, we usually have closed waveguides inside metallic boundaries. Metals are considered to be perfect conductors at most microwave frequencies. Figure 5.2 illustrates a typical microwave rectangular waveguide, which is surrounded by metallic walls. The boundary condition at the metal surface is that the tangential electric field is zero. Microwaves propagate within the metallic enclosure. In such closed structures, we have only a discrete set of waveguide modes. At optical wavelengths, we avoid the use of metallic boundaries because of their strong absorption of radiation at optical frequencies. All optical waveguides are open dielectric waveguides. Two examples have been illustrated in Figures 5.1(a) and 5.1(b).

There are at least three differences between microwave and optical waveguide analysis.

(1) The mathematics of finding the modes is more complex for open waveguides. In fact, there exists no analytical solution for three-dimensional open-channel

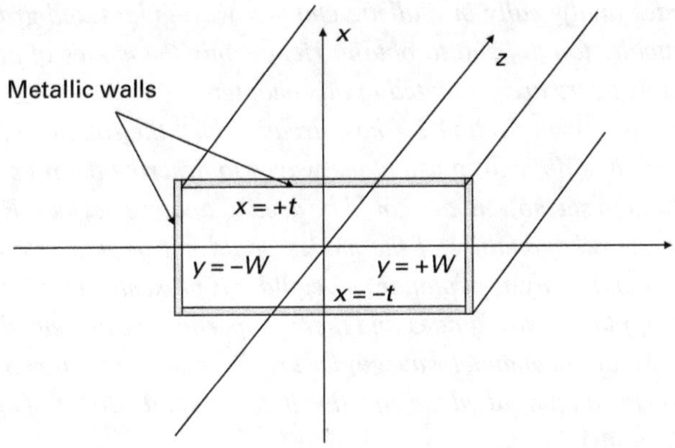

Figure 5.2 Illustration of a rectangular microwave waveguide. Within the metallic walls, it has lateral dimensions of $2t$ in the x direction and $2W$ in the y direction. The waveguide is oriented along the z direction.

waveguides or graded-index fibers. We only have analytical solutions for the modes of the round step-index fiber shown in Figure 5.2(a) and the planar waveguides shown in Figure 2.4. Numerical analysis or approximation methods must be used to analyze the modes of optical waveguides.

(2) Modes in microwave waveguides do not have an evanescent tail. In open dielectric waveguides, the discrete optical guided-wave modes have an evanescent field outside the core region (the core is often called vaguely the optical waveguide). The evanescent tail ensures that any perturbation of the mode from any structural change several decay lengths away is small. Since propagation loss of the guided-wave modes is caused often by scattering or absorption, it means the attenuation rate of the guided mode will be very low as long as there is little absorption or scattering in or near the high-index core. Thus optical fibers are used for long-distance communication. Yet, significant energy is still carried in the evanescent field near the core. This evanescent field may be used to achieve mutual interactions with other radiation fields. For example, the evanescent field is used to operate devices such as the dielectric grating, the distributed feedback laser, and the directional coupler.

(3) In addition to the guided modes that have discrete eigenvalues, there is an infinite set of continuous modes in open waveguides in optics. Only the sum of both the discrete and continuous modes constitutes a complete set of functions. It means that, rigorously, any arbitrary incident field should be expanded mathematically as a summation of this complete set of modes. At any dielectric discontinuity, the boundary conditions of the continuity of electric and magnetic fields are satisfied by the summation of both the guided-wave modes and the continuous modes. In other words, continuous modes are excited at any discontinuity in optics. Energy is radiated away from the discontinuity by the continuous modes. In microwaves, only discrete modes are excited at any discontinuity.

Because of the differences between the optical and microwave waveguide structures, the calculation of their modes also differs.

Finding the modes analytically in realistic channel waveguides and graded optical fibers is mathematically too difficult to obtain. Hence only the modes of planar waveguides and step-index fibers are presented in this chapter.

The plane wave analysis in Section 2.4 has already given the solutions for a planar waveguide. However, it is difficult to use plane waves to describe the properties of the waveguide modes and to see how modes could be used to analyze devices. For example, properties such as the othogonalilty of the modes could not be understood easily by plane waves. Thus modal analysis of planar waveguides is presented in this chapter. The TE and TM modes of planar waveguides and their properties are presented in detail in Sections 5.2 to 5.5. Modes of channel waveguides are discussed in Section 5.6 by means of an approximation technique called the effective index method. Modes of optical fibers are discussed in Section 5.7.

Although the configurations of planar waveguides and step-index fibers are very simple compared to those of the actual devices, the properties of the modes of these simple structures illustrate clearly the properties of optical guided waves in general. The modes of these simple structures also serve as the basis for the approximate analyses to be presented later to analyze realistic devices such as directional couplers, resonators, modulators, and filters in channel waveguides.

Modal analysis of planar and channel waveguides has been presented in many existing books. However, these discussions do not emphasize the implications of the approximation methods, the mutual interaction properties of the modes, and the effect of the excitation of modes.

5.2 Electromagnetic analysis of modes in planar optical waveguides

5.2.1 The asymmetric planar waveguide

A typical uniform dielectric planar waveguide has been shown in Figure 2.4. For planar waveguides, the core, the cladding, and the substrate are all uniform and infinitely wide in the y and the z directions. The core typically has a thickness, t, of the order of a wavelength or less, supported by a substrate and covered by a cladding (or air) many wavelengths (or infinitely) thick. The refractive index of the waveguide core, n_2, is higher than the indices of the surrounding layers, n_c of the cladding ($n_c = 1$ for air) and n_s of the substrate. All layers have the same magnetic permeability μ, and the time variation is $e^{j\omega t}$.

Since the structure is identical in any direction in the y–z plane, we could choose the $+z$ axis as the direction of propagation in our analysis without any loss of generality. For planar modes, we further assume $\partial/\partial y \equiv 0$. This assumption on the y variation applies in Sections 5.2, 5.3, and 5.4.

5.2.2 Equations for TE and TM modes

When we substitute $\partial/\partial y = 0$ into $\nabla \times \underline{E}$ and $\nabla \times \underline{H}$ in Maxwell's equations, we obtain two separate groups of equations.

$$\frac{\partial E_y}{\partial z} = \mu \partial H_x/\partial t,\ \frac{\partial E_y}{\partial x} = -\mu \partial H_z/\partial t,\ \frac{\partial H_z}{\partial x} - \frac{\partial H_x}{\partial z} = -\varepsilon \partial E_y/\partial t$$

and (5.1)

$$\frac{\partial H_y}{\partial z} = -\varepsilon \partial E_x/\partial t,\ \frac{\partial H_y}{\partial x} = \varepsilon \partial E_z/\partial t,\ \frac{\partial E_z}{\partial x} - \frac{\partial E_x}{\partial z} = \mu \partial H_y/\partial t$$

Clearly, E_y, H_x, and H_z are related only to each other, and H_y, E_x, and E_z are related only to each other. Since the direction of propagation is z, the solutions of the first group of equations are called the TE, transverse electric, modes. The solutions of the second group of equations are the TM, transverse magnetic, modes. In other words, all planar waveguide modes can be divided into TE and TM types.

Since ε is only a function of x, the z variation of the fields must be the same in all layers. This is the consequence of the continuity of E_y or H_y for all z. For TE modes, the transverse electric E_y in Eq. (5.1) can now be written as a product of a function in y and a function in z, i.e. $E_y(x,z) = E_y(x)E_y(z)$. When all these considerations are taken into account, we obtain:

$$\left[\frac{\partial^2}{\partial x^2} + \left(\omega^2 \mu\varepsilon(x) - \beta^2\right)\right] E_y(x) = 0 \tag{5.2a}$$

$$\left[\frac{\partial^2}{\partial z^2} + \beta^2\right] E_y(z) = 0 \tag{5.2b}$$

Similar equations exist for TM modes.

Mathematically, Eq. (5.2) and its equivalent for TM modes are second-order differential equations. All the TE modes form a complete set of TE eigenfunctions, meaning that any arbitrary electric field polarized in the y direction with $\partial/\partial y = 0$ can be represented as a summation of TE modes. Similarly, all the TM modes form a complete set of TM eigenfunctions, meaning at any arbitrary electric field polarized in the x direction with $\partial/\partial y = 0$ can be represented as a summation of TM modes. Any radiation field with arbitrary polarization needs to be decomposed first into TE and TM components, and then analyzed.

5.3 TE modes of planar waveguides

The planar TE modes (i.e. modes with $\partial/\partial y = 0$) in the planar waveguides are eigen-solutions of the equation

$$\left[\frac{\partial^2}{\partial x^2} + \frac{\partial^2}{\partial z^2} + \omega^2\mu\varepsilon(x)\right] E_y(x)E_y(z) = 0$$

$$\begin{aligned} \varepsilon(x) &= {n_c}^2\varepsilon_o & x &\geq t \\ &= {n_2}^2\varepsilon_o & t &> x > 0 \\ &= {n_s}^2\varepsilon_o & 0 &\geq x \end{aligned} \tag{5.3}$$

$$H_x = -\frac{j}{\omega\mu}\frac{\partial E_y}{\partial z}, \qquad H_z = \frac{j}{\omega\mu}\frac{\partial E_y}{\partial x}$$

Here, ε_o is the free space electric permittivity. The boundary conditions are the continuity of the tangential electric and magnetic fields, E_y and H_z, at $x = 0$ and at $x = t$. The continuity of H_z is equivalent to the continuity of $\partial E_y/\partial x$. Note that when E_y is known, H_x and H_z can be calculated directly from E_y. Thus only E_y is shown explicitly in the following sections.

5.3.1 TE planar guided-wave modes

Eqs. (5.2) and (5.3) suggest that the solution of $E_y(x)$ is either a sinusoidal or an exponential function, and the solution of $E_y(z)$ is $e^{\pm j\beta z}$. Guided by the discussion in Section 2.4, we look for solutions of $E_y(x)$ with sinusoidal variations for $t > x > 0$ and with decaying exponential variations for $x > t$ and $x < 0$. In short, we obtain the following functional form for a forward propagating $E_y(x,z)$. The subscript m stands for the mth order solution of Eq. (5.3).

$$
\begin{aligned}
E_m(x,z) &= E_m(x)E_m(z) = A_m\left\{\sin(h_m t + \varphi_m)e^{-p_m(x-t)}\right\}e^{-j\beta_m z} && x \geq t \\
E_m(x,z) &= E_m(x)E_m(z) = A_m\left\{\sin(h_m x + \varphi_m)\right\}e^{-j\beta_m z} && t > x > 0 \\
E_m(x,z) &= E_m(x)E_m(z) = A_m\left\{\sin\varphi_m e^{q_m x}\right\}e^{-j\beta_m z} && 0 \geq x
\end{aligned}
$$

where, in order to satisfy Eq. (5.2)

$$
\begin{aligned}
(\beta_m/k)^2 - (p_m/k)^2 &= n_c^2 \\
(\beta_m/k)^2 + (h_m/k)^2 &= n_2^2 \\
(\beta_m/k)^2 - (q_m/k)^2 &= n_s^2
\end{aligned}
\tag{5.4}
$$

Eq. (5.3) is clearly satisfied by E_m in all the individual regions. Note that the continuity of E_y is automatically satisfied at $x = 0$ and $x = t$. In order to satisfy the H_z magnetic boundary conditions at $x = 0$ and $x = t$, h_m, q_m, and p_m must be the mth set of the root of the transcendental equations, which are also called the characteristic equations,

$$
\tan[(h_m/k)kt + \phi_m] = -h_m/p_m \quad \text{and} \quad \tan\phi_m = h_m/q_m \tag{5.5}
$$

For a given normalized thickness kt, there are only a finite number of roots of the characteristic equations yielding a discrete set of real values for h_m, p_m, and q_m. For this reason, the guided-wave modes are also called the discrete modes. They are labeled by the integer subscript m ($m = 0, 1, 2, \ldots$). The lowest-order mode with $m = 0$ has the largest β value, $\beta_0 > \beta_1 > \beta_2 > \beta_3\ldots$ and $h_0 < h_1 < h_2\ldots$ Moreover, one can show that the number of times that $\sin(h_m x + \varphi_m)$ is zero is m. Thus, we could identify experimentally the order of the mode by the number of zeros in its intensity pattern. The β_m/k value is called the effective index, $n_{eff,m}$, of the mode. The velocity of light in the free space divided by effective index n_{eff} is the phase velocity of the mth-order guided-wave mode. The exponential decay rate of any guided-wave mode in the cladding and the substrate is determined by the index of the surrounding layer (either at $x > t$ or at $x < 0$) and the β_m/k value of the mode. Lower-order modes will have larger effective index and faster exponential decay.

The lossless TE planar guided-wave modes are orthogonal to each other and to any other TE or TM modes of the same waveguide [1,2]. It is customary to normalize the constant A_m so that a unit amount of power (1 W) per unit length in the y direction is carried out by a normalized mode. Thus,

$$\frac{1}{2}\,\mathrm{Re}\left[\int_{-\infty}^{+\infty} E_{yn}H_{xm}{}^{*}\mathrm{d}x\right] = (\beta_m/2\omega\mu)\int E_n E_m{}^{*}\mathrm{d}x = \delta_{nm} \tag{5.6}$$

From this condition, we obtain

$$A_m{}^2 = \frac{4\omega\mu}{\beta_m}\left[\frac{1}{p_m}+\frac{1}{q_m}+t\right]^{-1} \tag{5.7}$$

5.3.2 TE planar guided-wave modes in a symmetrical waveguide

In order to visualize more easily why there should be only a finite number of modes, let us consider the example of a symmetrical waveguide. In this case, $n_c = n_s = n$ and $p_m = q_m$. The quadratic equations for h_m and β_m and the transcendental equation in Eq. (5.5) now become

$$\left(\frac{h_m}{k}\right)^2 + \left(\frac{p_m}{k}\right)^2 = n_2{}^2 - n^2 \tag{5.8}$$

and

$$\tan\left[\left(\frac{h_m}{k}\right)kt\right] = \frac{-2\dfrac{h_m}{p_m}}{1-\dfrac{h_m{}^2}{p_m{}^2}} \tag{5.9}$$

Since

$$\tan\left[2\left(\frac{h_m}{k}\right)\frac{kt}{2}\right] = \frac{2\tan\left[\left(\dfrac{h_m}{k}\right)\dfrac{kt}{2}\right]}{1-\tan^2\left[\left(\dfrac{h_m}{k}\right)\dfrac{kt}{2}\right]} \tag{5.10}$$

Eq. (5.9) can be reduced to two equations,

$$\tan\left[\left(\frac{h_m}{k}\right)\frac{kt}{2}\right] = \frac{p_m/k}{h_m/k}, \qquad \text{hence} \qquad \frac{h_m}{k}\tan\left[\left(\frac{h_m}{k}\right)\frac{kt}{2}\right] = \frac{p_m}{k} \tag{5.11}$$

or

$$\tan\left[\left(\frac{h_m}{k}\right)\frac{kt}{2}\right] = -\frac{h_m/k}{p_m/k}, \qquad \text{hence} \qquad -\frac{h_m}{k}\cot\left[\left(\frac{h_m}{k}\right)\frac{kt}{2}\right] = \frac{p_m}{k} \tag{5.12}$$

If we seek graphical solutions in the coordinate system of p_m/k and h_m/k, they are given by the intersections of one of the two curves described by the equivalent tangent equations, either Eq. (5.11) or Eq. (5.12), and the quadratic equation, $(h_m/k)^2 + (p_m/k)^2 = n_2{}^2 - n^2$.

In summary, there are two sets of equations. The solutions for the first tangent equation *(5.11) and the quadratic equation (5.8) are known as the even modes because they lead to field distributions close to a cosine variation in the film. They are symmetric with respect to x = t/2. The solutions from the second tangent equation (5.12) and the quadratic equation (5.8) are called the odd modes because the fields in the film have distributions in sine variations. They are anti-symmetric with respect to x = t/2.*

It is instructional to examine the even modes in detail. If we plot the quadratic equation of h_m/k and p_m/k in Eq. (5.8), it is a circle with radius $(n_2{}^2 - n^2)^{1/2}$. The curve describing the tangent equation in Eq. (5.11) as functions of h_m/k and p_m/k will be obtained whenever the left-hand side (LHS) equals the right-hand side (RHS) of the tangent equation. The RHS is just p_m/k; the LHS has a tangent which is a multi-valued function. It starts from 0 whenever $(h_m/k)kt/2$ is 0, π, or $m\pi$. It approaches + or – infinity when $(h_m/k)kt/2$ approaches $+\pi/2$ or $-\pi/2$, or $(m\pi + \pi/2)$ or $(m\pi - \pi/2)$, where m is an integer. The curves representing these two equations are illustrated in Figure 5.3. Clearly there is always a solution, as long as $n_2 > n$, i.e. there is an intersection of the two curves, no matter how large (or how small) the circle (i.e. the n_2 value). This is the fundamental mode, labeled by $m = 0$. However, whether there will be a solution for $m \geq 1$ depends on whether the radius is larger than $2\pi/kt$. There will be $m = j$ solutions when the radius is larger than $2j\pi/kt$. When the radius of the circle is just equal to $2j\pi/kt$, the value for p/k is 0. This is the cut-off point for the jth ($j > 1$) mode. Notice that $h_0 < h_1 < h_2 \ldots$ and $\beta_0 > \beta_1 > \beta_2 > \ldots$ m is called the order of the mode.

The odd modes are solutions of Eqs. (5.8) and (5.12). The solutions of odd modes can be obtained similarly to the even modes in Figure 5.3. However, since Eq. (5.12) contains a cotangent function, p_m/k is zero when $h_m/k = \dfrac{(2m+1)\pi}{kt}$ and $+\infty$ when $h_m/k = \dfrac{2m\pi}{kt}$. Therefore there is a minimum value of $n_2{}^2 - n^2$ below which the circle given in Eq. (5.8) does not intercept the curve representing Eq. (5.12).

There are two conclusions that can be made: (1) There is a minimum $n_2{}^2 - n^2$, below which there is no solution of the odd mode for a given t. (2) For a given t and $n_2{}^2 - n^2$, the value of h_m/k for the mth odd mode is larger than the mth even mode. Therefore the value of β_m/k (or $n_{eff,m}$) for the mth even mode is larger than for the mth odd mode.

If we list all the modes in descending order of the values of n_{eff}, then the lowest-order mode that has the largest n_{eff} is the $m = 0$ even mode, followed by the $m = 0$ odd mode, the $m = 1$ even mode, the $m = 1$ odd mode, etc. Note that the number of x positions at which the electric field E is zero for the first mode ($m = 0$, even) is zero, one for the second mode ($m = 0$, odd), two for the third mode ($m = 1$, even), three for the fourth mode ($m = 1$, odd), etc. If $n_2{}^2 - n^2$ and t are sufficiently small, there is only a single mode, the $m = 0$ even mode.

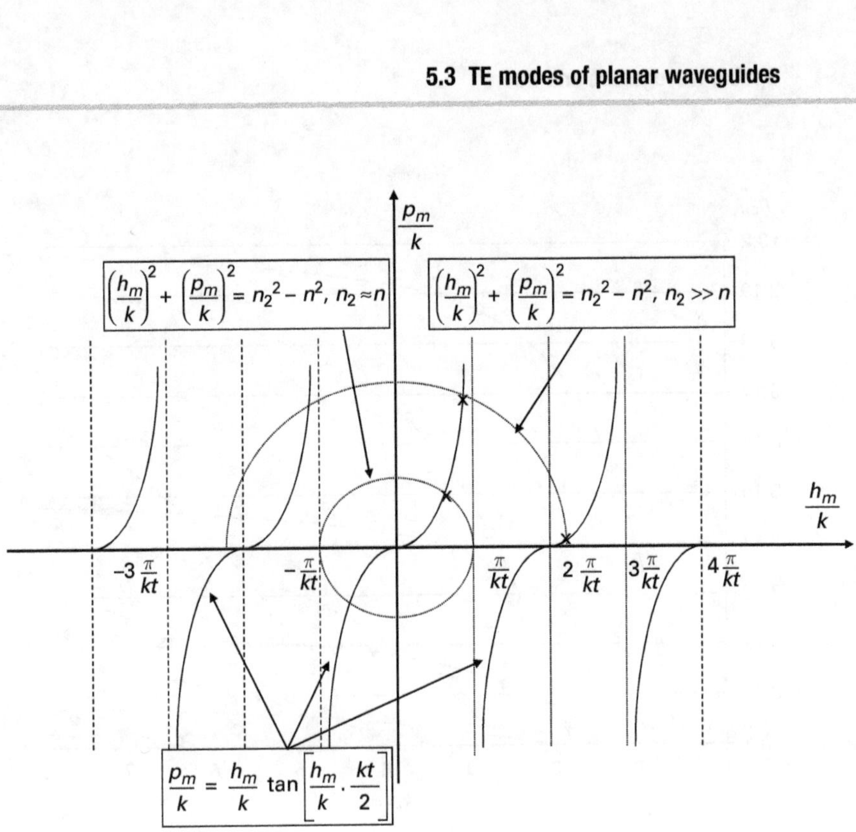

Figure 5.3 Illustration of the graphical solution for h_m and p_m, for even TE guided-wave modes in a symmetrical planar waveguide.

The symmetric waveguide is not a realistic waveguide. Its analysis is presented here because the mathematics is simple to understand, and the consequence of the various aspects of the modal characteristic can be easily shown.

5.3.3 The cut-off condition of TE planar guided-wave modes

In order to have an mth-order mode in an asymmetric planar waveguide, there are two conditions that need to be satisfied. The first condition is: $n_2 > n_s$ and n_c. Let us assume $n_2 > n_s \geq n_c$. For a given set of n, the second condition is that there is a minimum thickness, called the cut-off core thickness t_m, that will permit the mth solution of Eq. (5.5) to exist.

At the cut-off thickness of the mth mode, $q_m = 0$, $\beta_m/k = n_s$, $h_m/k = (n_2^2 - n_s^2)^{1/2}$, and $\phi_m = \pm(m+1/2)\pi$. Thus the cut-off thickness can be calculated from Eq. (5.5) to be:

$$kt_m = \left\{\left(m + \frac{1}{2}\right)\pi - \tan^{-1}[(n_2^2 - n_s^2)/(n_s^2 - n_c^2)]^{1/2}\right\}(n_2^2 - n_s^2)^{-1/2} \quad (5.13)$$

The thicker the core, the larger the number of guided-wave modes the waveguide can support. For all guided-wave modes above the cut-off, $n_2 \geq |\beta_m/k| > n_s$.

Note that, in symmetric waveguides, the cut-off condition is different from the condition shown in Eq. (5.13) for asymmetric waveguides discussed above. For symmetric waveguides, there is always an even $m = 0$ mode. There is no cut-off condition for the even $m = 0$ mode. In asymmetric waveguides, there is a cut-off condition below which no mode

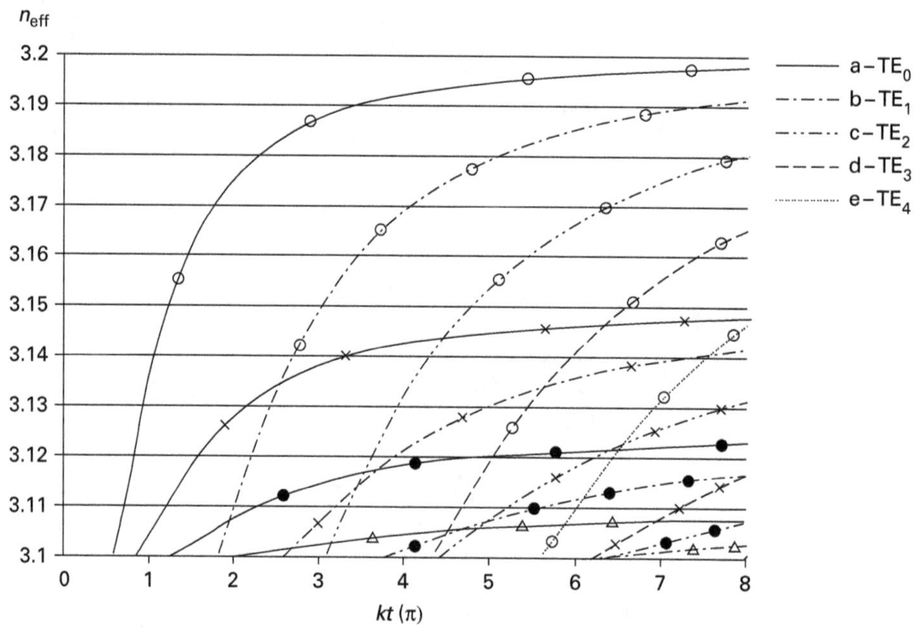

Figure 5.4 n_{eff} values of TE_m modes in epitaxially grown waveguides on InP substrates.

exists. In many applications, a single-mode waveguide is required. In that case, t and the indices of the layers are controlled so that only one mode exists in the waveguide.

5.3.4 An example of TE planar guided-wave modes

Figure 5.4 shows the effective index, $n_{eff} = \beta_m/k$, of TE_m planar guided-wave modes in epitaxially grown waveguides on InP substrates as a function of the waveguide thickness t, where $n_s = 3.10$, $n_2 = n_s + \Delta n$, and $n_c = 1$. The abscissa is kt in units of π. Δn, i.e. $n_2 - n_s$, depends on the alloy composition of the epitaxially grown layer. Curves with circles, o, are for $\Delta n = 0.10$; curves with crosses, x, are for $\Delta n = 0.05$, curves with solid dots, •, for $\Delta n = 0.025$, and curves with triangles, Δ, for $\Delta n = 0.01$. The (a) curves are for TE_0 modes, (b) curves are for TE_1, (c) curves for TE_2, (d) curves for TE_3, and (e) curves for TE_4. These curves are taken from Figure 1.5 of another book of mine [3]. At large kt, n_{eff} increases monotonically toward n_2. At the cut-off, all modes have $n_{eff} = n_s$. n_{eff} for higher-order modes is always smaller than n_{eff} for lower-order modes. For a given thickness t, there are more modes for waveguides that have a larger Δn. For $kt < 1.8\pi$, the waveguide has only the TE_0 mode for $\Delta n = 0.1$, 0.05, and 0.025. Notice that we have real eigenvalues for β, h, p, and q. Since β is real, these modes propagate in the z direction without attenuation. The fields of these modes are evanescent in the air and in the substrate.

Physically, as we have discussed in Section 2.4, the electric field of the mth TE guided-wave mode inside the core is just a plane wave (with the electric field polarized in the y direction), totally internally reflected back and forth from the two boundaries at $x = 0$ and $x = t$. Its propagation direction in the x–z plane makes an angle θ_m with respect to the x axis.

$$\beta_m = n_2 k \sin\theta_m,\ h_m = n_2 k \cos\theta_m \tag{5.14}$$

Since θ_m is a very small angle, the magnetic field of TE modes is polarized predominantly in the x direction with a small component in the z direction.

5.3.5 TE planar substrate modes

As we have discussed in Section 2.4, when $n_s > |\beta/k| > n_c$, the electric field has an exponential decay for $x > t$, and sinusoidal variation within the core and in the substrate. The plane waves in the core are totally internally reflected at the boundary $x = t$. The plane waves in the substrate are propagating. These are the substrate modes. From Eq. (5.3), we obtain the following expression for TE substrate modes:

$$\begin{aligned} E^{(s)}(x,z;\beta) &= A^{(s)}\sin(ht+\varphi)\mathrm{e}^{-p(x-t)}\mathrm{e}^{-j\beta z} && x \geq t \\ E^{(s)}(x,z;\beta) &= A^{(s)}\sin(hx+\varphi)\mathrm{e}^{-j\beta z} && t \rangle x \rangle 0 \\ E^{(s)}(x,z;\beta) &= [C^{(s)}\mathrm{e}^{-j\rho x} + C^{(s)*}\mathrm{e}^{+j\rho x}]\mathrm{e}^{-j\beta z} && 0 \geq x \end{aligned} \tag{5.15}$$

with

$$\begin{aligned} (h/k)^2 + (\beta/k)^2 &= n_2^2 \\ (\beta/k)^2 - (p/k)^2 &= n_c^2 \\ (\rho/k)^2 + (\beta/k)^2 &= n_s^2 \end{aligned} \tag{5.16}$$

$$\tan[(h/t)kt + \varphi] = -h/p \tag{5.17}$$

and

$$C^{(s)} = A^{(s)}[\sin\varphi + j(h\cos\varphi/\rho)]/2. \tag{5.18}$$

$C^{(s)}$ and $A^{(s)}$ are normalized so that

$$(\beta/2\omega\mu)\int_{-\infty}^{\infty} E^{(s)}(x,z;\beta)E^{(s)*}(x,z;\beta')\mathrm{d}x = \delta(\rho-\rho') \tag{5.19}$$

which requires

$$C^{(s)}C^{(s)*} = \frac{\omega\mu}{\beta\pi} \tag{5.20}$$

Unlike guided-wave modes, which have $n_2 > |\beta_m/k| > n_s > n_c$ and discrete values of β_m, the β, p, h, ρ, and φ of the substrate modes have a continuous range of values that satisfy the above equations within the range $n_s > |\beta/k| > n_c$. Thus, these modes are continuous modes; they are orthogonal to each other and to the guided-wave modes.

5.3.6 TE planar air modes

In Section 2.4, we have shown that there are two equivalent independent solutions of Maxwell's equations corresponding to either waves incident from the cladding with incident angle $\theta_c < \pi/2$, or waves incident from the substrate with incident angle $0 < \theta_s < \sin^{-1}(n_c/n_s)$. At any specific θ_c and θ_s such that $n_c \sin\theta_c = n_s \sin\theta_s$, the two independent solutions are equivalent in θ_c and θ_s. By linearly combining the two independent equivalent solutions one can obtain two orthogonal independent modes for each set of propagation constants. These orthogonal modes propagate in both substrate and cladding. They are known as air modes in the literature because the cladding medium is often the air.

For symmetrical structures, i.e. $n_s = n_c$, these two orthogonal modes represent odd and even linear combinations of two equivalent solutions, one solution consists of plane waves incident from the cladding and the second solution consists of plane waves incident from the substrate. For asymmetrical structures, such as the one shown in Figure 2.4, the x variations are more complex than just odd and even combinations. Nevertheless, there are still two orthogonal modes for each set of propagation constants, these two modes differ from each other by a $\pi/2$ phase shift of the sinusoidal variations in the x direction in the film, which has the index n_2.

The mathematical expressions for E_y of the air modes that satisfy Eq. (5.3) are:

$$\begin{aligned} E'(x,z;\beta) &= \{D' e^{-j\sigma(x-t)} + D'^* e^{+j\sigma(x-t)}\} e^{-j\beta z} && x \geq t \\ E'(x,z;\beta) &= A' \sin(hx+\varphi) e^{-j\beta z} && t > x > 0 \\ E'(x,z;\beta) &= [C' e^{-j\rho x} + C'^* e^{+j\rho x}] e^{-j\beta z} && 0 \geq x \end{aligned} \tag{5.21}$$

for the first set

and

$$\begin{aligned} E''(x,z;\beta) &= \{D'' e^{-j\sigma(x-t)} + D''^* e^{+j\sigma(x-t)}\} e^{-i\beta z} && x \geq t \\ E''(x,z;\beta) &= A'' \sin\left(hx+\varphi+\frac{\pi}{2}\right) e^{-j\beta z} && t > x > 0 \\ E''(x,z;\beta) &= [C'' e^{-j\rho x} + C''^* e^{+j\rho x}] e^{-j\beta z} && 0 \geq x \end{aligned} \tag{5.22}$$

for the second set

with

$$\begin{aligned} (\beta/k)^2 + (\sigma/k)^2 &= n_c{}^2 \\ (\beta/k)^2 + (h/k)^2 &= n_2{}^2 \\ (\beta/k)^2 + (\rho/k)^2 &= n_s{}^2 \end{aligned} \tag{5.23}$$

Imposing the boundary conditions at $x = 0$ and $x = t$, we obtain:

$$\begin{aligned} C' &= A'[\sin\varphi + j(h\cos\varphi/\rho)]/2 \\ D' &= A'\left[\sin(ht+\varphi) + j\frac{h}{\sigma}\cos(ht+\varphi)\right]/2 \end{aligned} \tag{5.24}$$

For the second set of modes, A'', C'', and D'' are obtained when φ is replaced by $\varphi + \pi/2$ in the above equation. For both sets of modes, a continuous range of solutions of ρ, σ, β, and h exist, where $n_c \geq |\beta/k| \geq 0$. All modes form an orthogonal normalized set,

$$(\beta/2\omega\mu)\int_{-\infty}^{\infty} E'(x,z;\beta^a)E'^*(x,z;\beta^b)\mathrm{d}x = \delta(\rho^a-\rho^b) \quad (5.25)$$

$$(\beta/2\omega\mu)\int_{-\infty}^{\infty} E''(x,z;\beta^a)E''^*(x,z;\beta^b)\mathrm{d}x = \delta(\rho^a-\rho^b) \quad (5.26)$$

$$(\beta/2\omega\mu)\int_{-\infty}^{\infty} E'(x,z;\beta^a)E''^*(x,z;\beta^b)\mathrm{d}x = 0 \quad (5.27)$$

Air modes are also orthogonal to substrate and guide wave modes.

5.4 TM modes of planar waveguides

The planar TM modes are eigensolutions of the wave equation (with $\partial/\partial y = 0$ and $e^{j\omega t}$ time variation):

$$\begin{aligned}
&\left[\frac{\partial^2}{\partial x^2}+\frac{\partial^2}{\partial z^2}+\omega^2\varepsilon(x)\mu\right]H_y(x)H_y(z) = 0\\
&E_x = \frac{j}{\omega\varepsilon(x)}\frac{\partial H_y}{\partial z}, \quad E_z = \frac{-j}{\omega\varepsilon(x)}\frac{\partial H_y}{\partial x}
\end{aligned} \quad (5.28)$$

where $\varepsilon(x)$ is the same as given in Eq. (5.3). Or, in a manner similar to Eq. (5.2a), we can write,

$$\left[\frac{\partial^2}{\partial x^2}+\left(\omega^2\mu\varepsilon(x)-\beta^2\right)\right]H_y(x) = 0 \quad (5.29)$$

TM modes are similar to TE modes. The main difference between TE and TM modes is the polarization. In lossless waveguides, all TM modes are orthogonal to each other and to TE modes [2,3].

5.4.1 TM planar guided-wave modes

Like the TE modes, the y component of the magnetic field for the nth TM planar guided-wave mode propagating in the $+z$ direction is:

$$\begin{aligned}
H_{yn}(x,z) = H_{yn}(x)H_{yn}(z) &= B_n\left\{\sin(h_n t+\varphi_n)e^{-p_n(x-t)}\right\}e^{-j\beta_n z} && x\geq t\\
H_{yn}(x,z) = H_{yn}(x)H_{yn}(z) &= B_n\sin(h_n x+\varphi_n)e^{-j\beta_n z} && t>x>0\\
H_{yn}(x,z) = H_{yn}(x)H_{yn}(z) &= B_n\left\{\sin\varphi_n e^{q_n x}\right\}e^{-j\beta_n z} && 0\geq x
\end{aligned} \quad (5.30)$$

with

$$\begin{aligned}(\beta_n/k)^2-(p_n/k)^2&=n_c^{\,2}\\(\beta_n/k)^2+(h_n/k)^2&=n_2^{\,2}\\(\beta_n/k)^2-(q_n/k)^2&=n_s^{\,2}\end{aligned}\tag{5.31}$$

Continuity of the tangential electric field requires that h_n, q_n, and β_n also satisfy the transcendental equation,

$$\tan[(h_n/k)kt+\varphi_n]=-\frac{n_c^{\,2}h_n}{n_2^{\,2}p_n}\quad\text{and}\quad\tan\varphi_n=\left(\frac{n_s}{n_2}\right)^2\frac{h_n}{q_n}.\tag{5.32}$$

TM$_n$ modes are given by the nth solutions of Eq. (5.29). The magnetic field is in the y direction. The dominant electric field is in the x direction.

TM modes are orthogonal and normalized:

$$\frac{1}{2}\mathrm{Re}\left[\int_{-\infty}^{+\infty}H_{yn}E_{xm}{}^{*}\,\mathrm{d}x\right]=\frac{\beta_n}{2\omega}\int_{-\infty}^{+\infty}H_{yn}H_{ym}{}^{*}\frac{1}{\varepsilon(x)}\,\mathrm{d}x=\delta_{nm}\tag{5.33}$$

From this condition, we obtain:

$$B_n^{\,2}=\frac{4\omega\varepsilon_o}{\beta_n}\cdot\left[\frac{n_2^{\,2}}{n_s^{\,2}p_n}\cdot\frac{p_n^{\,2}+h_n^{\,2}}{h_n^{\,2}+\left(\frac{n_2^{\,2}}{n_s^{\,2}}\right)^2p_n^{\,2}}+\frac{n_2^{\,2}}{n_c^{\,2}q_n}\cdot\frac{q_n^{\,2}+h_n^{\,2}}{h_n^{\,2}+\left(\frac{n_2^{\,2}}{n_s^{\,2}}\right)q_n^{\,2}}+t\right]^{-1}\tag{5.34}$$

All TM modes are orthogonal to all TE modes [1,2].

5.4.2 TM planar guided-wave modes in a symmetrical waveguide

It is instructive to see what happens to the TM modes in a symmetrical waveguide, i.e. $n_c=n_s=n$. In this case, $p_n=q_n$. The quadratic equation for h_n and β_n and the transcendental equations now becomes

$$\left(\frac{h_n}{k}\right)^2+\left(\frac{p_n}{k}\right)^2=n_2^{\,2}-n^2\quad\text{and}\quad\tan\left[\left(\frac{h_n}{k}\right)kt\right]=-\frac{2\dfrac{n^2h_n}{n_2^{\,2}p_n}}{1-\left(\dfrac{n^2h_n}{n_2^{\,2}p_n}\right)^2}\tag{5.35}$$

As we have seen in the case of TE guided-wave modes in symmetrical waveguide structures, the above tangent equation is equivalent to two equations,

$$\tan\left[\left(\frac{h_n}{k}\right)\frac{kt}{2}\right]=-\frac{n^2h_n/k}{n_2^{\,2}p_n/k}\quad\text{and}\quad\tan\left[\left(\frac{h_n}{k}\right)kt\right]=\frac{n_2^{\,2}p_n/k}{n^2h_n/k}\tag{5.36a}$$

or,

$$-\frac{n^2}{n_2^2}\left(\frac{h_n}{k}\right)\cot\left[\left(\frac{h_n}{k}\right)\frac{kt}{2}\right]=\frac{p_n}{k}\quad\text{and}\quad\frac{n^2}{n_2^2}\left(\frac{h_n}{k}\right)\tan\left[\left(\frac{h_n}{k}\right)\frac{kt}{2}\right]=\frac{p_n}{k}. \tag{5.36b}$$

These equations again point to the existence of two orthogonal sets of modes. They are either even or odd with respect to $t/2$. The $n=0$ even TM mode has no cut-off thickness t. These equations are very similar to the equations for the TE modes, except for the ratio, $(n/n_2)^2$, which is always smaller than 1. Therefore, for the same order (i.e. $m=n$), the p_n values of the TM modes are slightly smaller than the p_m values of the TE modes for the same thickness t and indices.

5.4.3 The cut-off condition of TM planar guided-wave modes

Again, for a given normalized thickness kt, there is only a finite number of discrete modes, labeled by the subscript n ($n=0, 1, 2,\ldots$), where $h_0 < h_1 < h_2 < h_3\ldots$ and $n_2 > \beta_0 > \beta_1 > \beta_2 > \beta_3\ldots > n_s$. At the cut-off of each mode, $\beta_n = n_s k$ and $q_n = 0$. The cut-off thickness for the nth TM mode for the asymmetric waveguide is:

$$kt_n = \left\{n\pi + \tan^{-1}\left[\frac{n_2^2}{n_c^2}\sqrt{\frac{n_s^2-n_c^2}{n_2^2-n_s^2}}\right]\right\}(n_2^2-n_s^2)^{-1/2} \tag{5.37}$$

Note that the cut-off thickness t_n for TM modes is always larger than the cut-off thickness t_m for TE modes of the same order.

In many applications we want a single-mode waveguide so that there cannot be any conversions into higher-order modes. Since TM modes have a larger cut-off thickness, it is possible to design the asymmetric waveguide with appropriate n_2, n_s, n_c, and t so that only the lowest-order TE mode can exist.[1] *On the other hand, in other applications, we may want to have two or more modes interacting with each other. In that case the indices of the layers and the thickness t are controlled to yield the desired number of guided-wave TE and TM modes. Note that TE and TM modes have perpendicular polarizations. Thus the total electric field in the TE polarization direction is not affected by the TM modes. Conversely, the total electric field in the TM polarization direction is not affected by the TE modes. Please also note that, in a multi-mode waveguide, the properties of the component are governed only by the modes excited in the waveguide.*

5.4.4 An example of TM planar guided-wave modes

Figure 5.5 shows the effective index n_{eff}, i.e. β_m/k, of TM_m planar guided-wave modes in epitaxially grown waveguides on InP substrates as a function of the waveguide thickness, t, where $n_s=3.1$, $n_2=n_s+\Delta n$, and $n_c=1$. This figure is taken from Figure 1.6 of my book, published by Cambridge University Press [3]. The abscissa is kt in units of π. Δn, i.e. n_1-n_2, depends on the alloy composition of the epitaxially grown layer. Curves with circles, o, are for $\Delta n=0.10$; curves with crosses, x, are for $\Delta n=0.05$; curves with solid dots, •, for

[1] Notice the difference between the symmetric and the anti-symmetric waveguides.

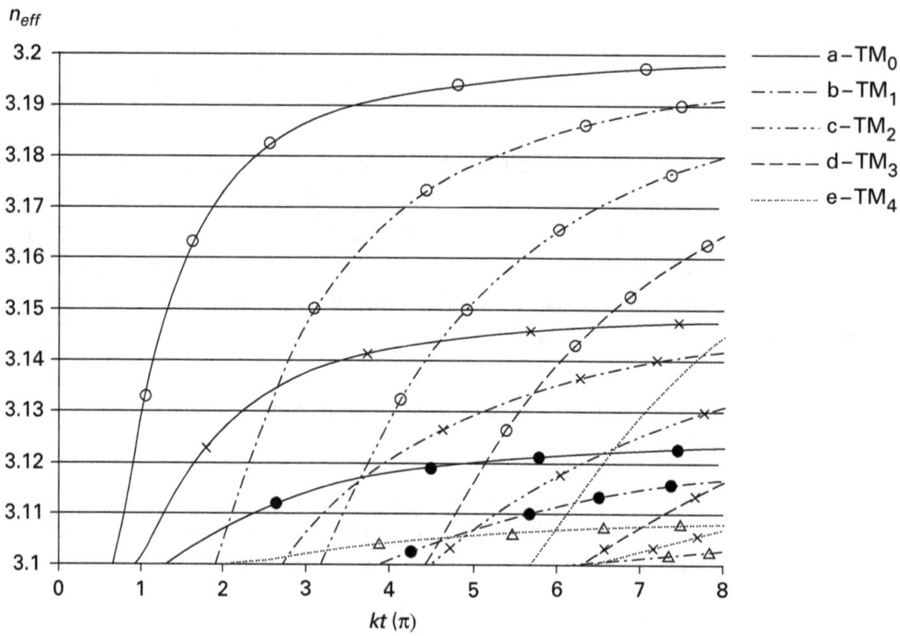

Figure 5.5 n_{eff} values of TM_m modes in epitaxially grown waveguides on InP substrates.

$\Delta n = 0.025$; and with triangles, Δ, for $\Delta n = 0.01$. The a curves are for TM_0 modes, b curves are for TM_1, c curves are for TM_2, d curves are for TM_3, and e curves are for TM_4. At the cut-off, all modes have $n_{eff} = n_s$. n_{eff} for the higher-order modes is always smaller than n_{eff} for lower-order modes. For a given thickness, t, there are more modes for waveguides that have a larger Δn. For $kt < 0.6\pi$, there is no TM guided wave. Thus a single-mode waveguide has only the TE_0 mode. For $kt < 1.9\pi$, the waveguide has only the TM_0 mode for $\Delta n = 0.1$, 0.05, and 0.025. Because of the dependence on $(n_s/n_2)^2$ and $(n_c/n_2)^2$, which are always smaller than 1, β/k of the TM modes is usually slightly smaller than the corresponding TE modes. Like the TE modes, the exponential decay rate of any guided-wave mode is determined only by the index of the layer (either at $x > t$ or at $x < 0$) and the β_n/k value of the mode. The velocity of light in free space, c, divided by n_{eff} is the phase velocity of the guided-wave mode. For the same polarization, lower-order modes will have a larger effective index and faster exponential decay. The most important difference between TM and TE modes is, of course, the polarization of the optical electric field.

5.4.5 TM planar substrate modes

For the substrate TM modes, the y component of the magnetic field is:

$$\begin{aligned} H^{(s)}(x,z;\beta) &= B^{(s)}\sin(ht+\phi)e^{-p(x-t)}e^{-j\beta z} && x \geq t \\ H^{(s)}(x,z;\beta) &= B^{(s)}\sin(hx+\phi)e^{-j\beta z} && t > x > 0 \end{aligned} \tag{5.38a}$$

$$H^{(s)}(x,z;\beta) = [D^{(s)}\mathrm{e}^{-jpx} + D^{(s)*}\mathrm{e}^{+jpx}]\mathrm{e}^{-j\beta z} \qquad 0 \geq x$$
$$D^{(s)} = \left(\frac{B^{(s)}}{2}\right)\left[\sin\phi + j\left(\frac{n_s{}^2 h\cos\phi}{n_2{}^2 p}\right)\right] \tag{5.38b}$$

$$\tan[(h/k)kt + \phi] = -\frac{n_c{}^2 h}{n_2{}^2 p} \tag{5.38c}$$

D and B are obtained from the orthogonalization and normalization conditions,

$$\frac{\beta}{2\omega}\int_{-\infty}^{+\infty} H^{(s)}(\beta)H^{(s)*}(\beta')/\varepsilon(x)\,\mathrm{d}x = \delta(\rho - \rho') \tag{5.39}$$

From Eq. (5.39), we obtain,

$$D^{(s)}D^{(s)*} = \frac{\omega\varepsilon_o n_s{}^2}{\beta\pi} \tag{5.40}$$

β, p, h, ρ, and ϕ have a continuous range of solutions within the range, $n_s > |\beta/k| > n_c$.

5.4.6 TM planar air modes

There are again two orthogonal TM air (or cladding) modes for each set of propagation constants. For the first set of modes,

$$\begin{aligned} H'(x,z;\beta) &= \{E'\mathrm{e}^{-j\sigma(x-t)} + E'^{*}[\mathrm{e}^{+j\sigma(x-t)}]\}\mathrm{e}^{-j\beta z} && x \geq t \\ H'(x,z;\beta) &= B'\sin(hx+\varphi)\mathrm{e}^{-j\beta z} && t > x > 0 \\ H'(x,z;\beta) &= [F'\mathrm{e}^{-jpx} + F'^{*}(\mathrm{e}^{+jpx})]\mathrm{e}^{-j\beta z} && 0 \geq x \end{aligned} \tag{5.41}$$

And, for the second set of modes,

$$\begin{aligned} H''(x,z;\beta) &= \{E''\mathrm{e}^{-j\sigma(x-t)} + E''^{*}[\mathrm{e}^{+j\sigma(x-t)}]\}\mathrm{e}^{-j\beta z} && x \geq t \\ H''(x,z;\beta) &= B''\sin\left(hx + \varphi + \frac{\pi}{2}\right)\mathrm{e}^{-j\beta z} && t > x > 0 \\ H''(x,z;\beta) &= [F''\mathrm{e}^{-jpx} + F''^{*}(\mathrm{e}^{+jpx})]\mathrm{e}^{-j\beta z} && 0 \geq x \end{aligned} \tag{5.42}$$

For both sets of orthogonal modes, a continuous range of solutions of ρ, σ, β, and h, exist where $n_c \geq |\beta/k| \geq 0$. For the first set of modes, the continuity of the electric and magnetic fields at $x = 0$ and $x = t$ requires:

$$\begin{aligned} E' &= \frac{1}{2}B'\left\{\sin(ht+\phi) + j\frac{h\,n_c{}^2\cos(ht+\phi)}{\sigma\,n_2{}^2}\right\} \\ F' &= \frac{1}{2}B'\left\{\sin\phi + j\frac{hn_s{}^2\cos\phi}{\rho\,n_2{}^2}\right\} \end{aligned} \tag{5.43}$$

For the second set of modes, ϕ is replaced by $\phi + \pi/2$ in Eq. (5.43).

Similar to Eqs. (5.24) to (5.27) in Section 5.3 for TE modes, TM air modes are orthogonal and normalized. They are also orthogonal to TM substrate and guided-wave modes, and to all TE modes.

5.4.7 Two practical considerations for TM modes

(1) The effect of metal electrodes. Often, metallic electrodes are fabricated on top of the n_c layer intended for applying a DC or RF electric field to the opto-electronic device. Since the electric field is polarized predominantly in the y direction for TE modes and in the x direction for TM modes, the difference in the polarization of the optical electric field may make a difference to the attenuation of the guided-wave mode in the z direction caused by the metal electrode. For example, when there is metallic absorption, the TM modes have higher attenuation.

(2) Scattering or absorption losses. Scattering or absorption loss in cladding or substrate usually does not significantly affect the mode pattern. However, at the scattering centers, radiation modes are excited. The radiation loss will cause attenuation as the mode propagates. Figures 5.4 and 5.5 demonstrated clearly that the higher-order modes have lower β/k values, and the evanescent decay in the cladding and substrate layers will be slower for higher-order modes. When there are scattering centers or absorption losses in the substrate or cladding, the evanescent decay is slower, and the attenuation rate larger. For this reason, higher-order guided-wave modes often have larger attenuation rate. Thus TM modes may have higher scattering loss.

The distinction between TE and TM modes is very important in applications. TE modes are excited by input radiation that has an electric field polarized in the y direction. TM modes are excited by input radiation that has an electric field polarized in the x direction. Most waveguide structures support both types of modes. The performance of the devices depends which modes have been excited.

In ideal straight waveguides, the electric fields of TE and TM modes do not interact with each other except at discontinuities or defects in the waveguide. At each defect or discontinuity, we need the sum of all TE and TM modes to satisfy the boundary condition. Thus TM as well as TE modes may be excited by the incident mode (or modes) at any defect or discontinuity. The scattered modes constitute the scattering loss.

5.5 Guided waves in planar waveguides

There are various applications, such as the Star coupler, acousto-optical scanner and RF spectral analyzer [4], that use planar waveguides. It is important to learn about the properties of generalized planar guided waves, how they focus or collimate and how to excite them in various configurations.

5.5.1 The orthogonality of modes

The orthogonality condition is important to analyze how modes function and interact in a waveguide.

When there are several modes propagating in the waveguide,

$$\underline{E} = E_{x}\underline{i_x} + E_{y}\underline{i_y}, \quad E_y = \sum_m \eta_m E_{ym}, \quad E_x = \sum_n \eta_n E_{xn} \tag{5.44}$$

η_m and η_n can be obtained from the orthogonality condition by

$$\eta_m = \frac{\beta_m}{2\omega\mu}\int_{-\infty}^{+\infty} E_y E_m^*(x)\mathrm{d}x \qquad \eta_n = \frac{\omega}{2\beta_n}\int_{-\infty}^{+\infty} \varepsilon(x) E_x E_n^*(x)\mathrm{d}x \tag{5.45}$$

The total power carried by the modes is:

$$\frac{1}{2}\mathrm{Re}\left[\int_{-\infty}^{+\infty} (E_y H_x^* + E_x^* H_y)\mathrm{d}x\right] = \sum_m \eta_m \eta_m^* + \sum_n \eta_n \eta_n^*. \tag{5.46}$$

In other words, the total power is just the sum of the powers carried in each mode.

5.5.2 Guided waves propagating in the *y–z* plane

In Sections 5.3 and 5.4, we have presented the analysis of the planar modes when they propagate in the direction of the z axis. In reality, planar modes for a waveguide structure such as that as shown in Figure 2.4 can propagate in any direction in the y–z plane with the same x functional variation, $E_{ym}(x)$ in Eq. (5.4) for TE modes and $H_{yn}(x)$ in Eq. (5.30) for TM modes. For a planar guided-wave mode propagating at an angle θ with respect to the z axis, $E_{ym}(y,z)$ or $H_{yn}(y,z)$ will have a z variation of $\mathrm{e}^{-jn_{eff}k(\cos\theta)z}$ and a y variation of $\mathrm{e}^{-jn_{eff}k(\sin\theta)y}$. For such a planar guided wave, there is no variation of the field in the direction perpendicular to the direction of propagation in the y–z plane.

5.5.3 Convergent and divergent guided waves

There can be superposition of TE_m modes propagating at different θ angles to form diverging or focusing waves in the y–z plane with identical x variations. Similarly, there can be superposition of TM_n modes propagating at different θ angles to form diverging or focusing waves in the y–z plane with the same x variation. Notice that, for TE modes, the electric fields are polarized in the y–z plane perpendicular to their direction of propagation and the dominant magnetic field is polarized in the x direction. Conversely, for TM modes, the magnetic fields are polarized in the y–z plane perpendicular to their directions of propagation, while the dominant electric field is polarized in the x direction. What TE or TM mode will be excited depends on the polarization and the xyz variation of the incident field.

Superposition of planar guided waves with the same $E_m(x)$ or $H_n(x)$ that propagate in different θ directions in the y–z plane can yield very complex field variations in the y–z plane. For example, a planar $E_m(x)$ beam propagating in the z direction with a finite beam width $2l_y$ can be written as

$$E_y = AE_m(x)\mathrm{rect}\left(\frac{y}{l_y}\right)\mathrm{e}^{-jn_{eff,m}kz} \tag{5.47}$$

$$\mathrm{rect}(\tau) = 1 \quad \text{for} \quad |\tau| \leq 1 \quad \text{and} \qquad \mathrm{rect}(\tau) = 0 \quad \text{for} \quad [\tau[> 1$$

The rect function can be expressed by Fourier transform as

$$F_y(f_y) = \int_{-\infty}^{+\infty} \mathrm{rect}\left(\frac{y}{l_x}\right)\mathrm{e}^{-j2\pi f_y y}\mathrm{d}y \tag{5.48}$$

$$\mathrm{rect}\left(\frac{y}{l_x}\right) = \int_{-\infty}^{+\infty} F_y(f_y)\mathrm{e}^{+j2\pi f_y y}\mathrm{d}f_y \tag{5.49}$$

Substituting Eq. (5.49) into Eq. (5.47), we obtain

$$E_y = AE_m(x)\int_{-\infty}^{+\infty} F_y(f_y)\left[\mathrm{e}^{+j2\pi f_y y}\mathrm{e}^{-jn_{eff,m}kz}\right]\mathrm{d}f_y \tag{5.50}$$

This means that E_y is made up of a summation of planar $E_m(x)$ modes propagating in θ directions with amplitude $F_y(f_y)$ and $e^{-jn_{eff,m}k \sin \theta y}$ variation in the y direction where

$$\theta = \sin^{-1}\left(\frac{2\pi f_y}{n_{eff,m}k}\right) \tag{5.51}$$

In other words, the beam will diverge as it propagates. The beam divergence will be determined by $F_y(f_y)$.

5.5.4 Refraction of a planar guided wave

There are refractions in planar waveguides. Let there be a straight junction of two waveguides at $z = z_o$. When a planar TE guided-wave mode E_m at the θ_{in} direction of propagation is incident on the second planar waveguide, it excites transmitted TE discrete and continuous modes in the second waveguide. The mode, E_k, that has the x variation closest to the $E_m(x)$ variation will be the dominant mode excited in the second waveguide. The direction of the propagation of the E_k guided-wave mode θ_{out} will be determined by the direction of the incident radiation beam through a relationship similar to Snell's law in free-space optics,

$$n_{eff,m} \sin \theta_{in} = n_{eff,k} \sin \theta_{out}. \tag{5.52}$$

In other words, when continuous modes are neglected, Snell's law is directly applicable using the effective indices. For example, a prism for a planar waveguide can be made by simply depositing an extra high-index layer on top of the waveguide cladding in the shape of a triangle. However, the change in direction of propagation is small because the difference between $n_{eff,m}$ of the original waveguide and $n_{eff,k}$ of the guided-wave mode under the prism is small.

5.5.5 Focusing and collimation of planar waveguide modes

Similar to the focusing of a plane wave by a lens discussed in Section 1.3.5, any guided wave that has an x variation of the mth order mode and a variation $e^{-jn_{eff,m}k\sqrt{y^2+z^2}}$ in the y–z plane is an outgoing wave cylindrically radiating away from $z=y=0$. Conversely, an $e^{+jn_{eff,m}k\sqrt{y^2+z^2}}$ variation will represent an incoming cylindrical wave focused at $z=y=0$. When z is large,

$$\sqrt{z^2+y^2} \cong z + \frac{y^2}{2z} \tag{5.53}$$

Let there be a planar mth-order guided-wave mode propagating in the forward z direction that has $e^{-jn_{eff,m}kz}$ variation at $z<-f$. When its phase front is modified by the factor, $e^{+jn_{eff,m}k(y^2/2f)}$ at $z=-f$, it will be converted into a cylindrical guided wave focused at $y=z=0$. In other words, an ideal lens with focal length f would transform any input guided wave by multiplying its amplitude variation by a phase factor $e^{+jn_{eff,m}k(y^2/2f)}$.

Conversely, when the phase factor $e^{+jn_{eff,m}k(y^2/2f)}$ is applied at $z=f$ to an outgoing cylindrically divergent guided wave originating from $z=0$, the resultant amplitude variation is $e^{-jn_{eff,m}kz}$, which is a planar guided wave in the $+z$ direction. In other words, an outgoing divergent cylindrical guided wave is also collimated by a lens. Needless to say, for any lens or guided-wave beam of finite size, there will be diffraction effect due to the limited aperture, such as those discussed in Sections 3.2.3 and 5.5.3.

There are several ways to obtain a guided-wave lens, including the Luneberg lens, the geodesic lens, and the Fresnel diffraction lens.

(a) The Luneberg lens

A generalized Luneberg lens in three dimensions is a variable-index, circular, symmetric refracting structure that reimages two objects to each other. Luneberg and other researchers have analytically determined the refractive index distribution that will give a diffraction limited performance. Using the dispersion relation of the waveguide (i.e. n_{eff} vs thickness), the analysis has been extended to the required variation of the thickness profile of the waveguide that will yield a waveguide lens [4]. A Luneberg lens has been fabricated by depositing lens material on a planar waveguide through a shaped mask. However, it is difficult to achieve the prescribed effective index distribution. Consequently it has not been used in practice.

(b) The geodesic lens

When a planar waveguide is fabricated on a substrate with a contoured surface, propagation of a guided-wave beam will follow the contour. Let there be a contoured depressed area. Guided-wave beams propagating through the depressed area in different paths will experience different phase shifts produced by the different path lengths. Figure 5.6

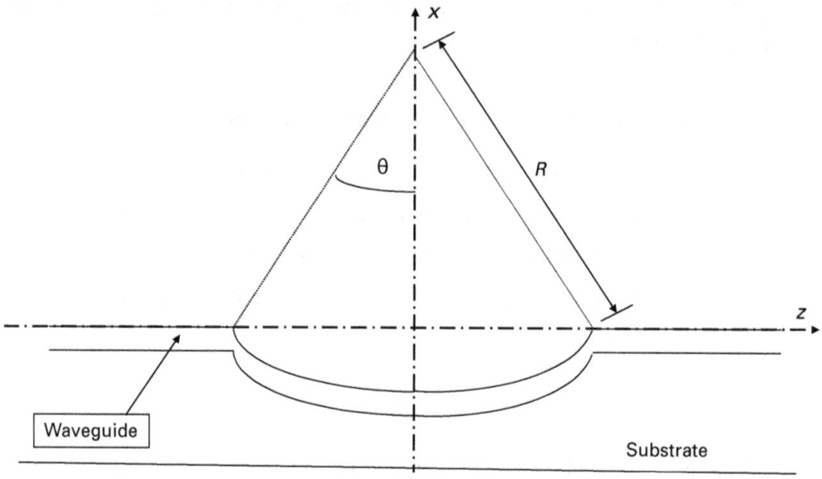

Figure 5.6 Cross-sectional view of a geodesic lens.

shows a waveguide on a substrate that has a spherical depression on its surface. R is the radius of curvature of the surface depression and 2θ is the vertex angle subtended by the arc of depression. It has been shown that a guided wave propagating in the z direction through the depression will have an additional quadratic phase variation $e^{+jn_{eff,m}k\frac{y^2}{2f}}$, where

$$f = \frac{R\sin\theta}{2(1-\cos\theta)} \tag{5.54}$$

Therefore, a plane guided wave will be focused at a distance f after the lens.[2] Conversely, a cylindrical guided wave originating at a distance f before the lens will be collimated. This is known as a geodesic lens. Since all spherical lenses have spherical aberrations, research has been conducted to use aspheric rotationally symmetric depression to correct them [4]. A numerically controlled, precision lathe has been used for diamond tuning the required surface contour on a y-cut $LiNbO_3$ substrate, followed by Ti-diffusion, to make a geodesic lens on a $LiNbO_3$ waveguide with $f = 2$ cm [4].

(c) The Fresnel diffraction lens

In Luneberg and geodesic lenses, the argument in the expression of the quadratic phase shift for a lens, $e^{+jn_{eff,m}k\left(\frac{y^2}{2f}\right)}$, exceeds multiples of 2π as $|y|$ increases. It is well known that a phase shift of $2n\pi$ is identical to a 0 phase shift. Curve (a) in Figure 5.7 shows the normal quadratic phase shift for a lens. Curve (b) shows only the value of the phase shift that exceeds $2n\pi$. Clearly, the multiplication of the amplitude and phase of a

[2] Note that f is independent of effective index or wavelength. It depends only on the geometry.

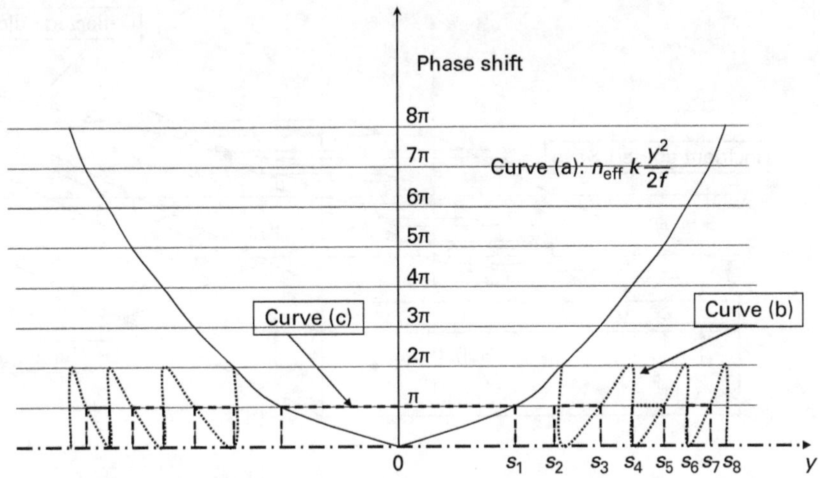

Figure 5.7 Digital approximation of the quadratic phase shift – the Fresnel lens. (a) Quadratic phase shift of an ideal lens. (b) Phase shift of an analog Fresnel lens. (c) Phase shift of a digital Fresnel lens.

guided wave as a function of y by a phase shift shown in either (a) or (b) has the same effect. In other words, a component that provides the phase shift shown in (b) will also serve as a lens with a focal length f. Curve (c) shows a digitalized approximation of (b) in which any phase shift from 0 to π is approximated by π, and any phase shift from π to 2π is approximated by 0. The zones in which the sectional change of phase shift is applied to an incoming guided wave are called the Fresnel zones. The digitalized change of phase for an incident planar TE_0 guided wave has been obtained by depositing rectangular pads of high-index materials with length L in the zone pattern on a planar waveguide [5]. The focusing effect of such a lens could also be viewed as the diffraction effect of the zone pads. Thus it is also known as a Fresnel diffraction lens.

A Fresnel lens is much shorter than a Luneberg or a geodesic lens. However, for large-angle oblique incident or divergent waves, the zone structure gives a phase shift distorted from that described in (c) of Figure 5.7.

5.5.6 Grating diffraction of planar guided waves

Gratings can be fabricated on waveguides by etching the grating pattern either onto the cladding layer or onto the core. It can also be obtained by depositing a material that has the grating pattern onto the waveguide. An ideal etched or deposited grating would have a periodic rectangular spatial profile for the grooves, which have permittivity ε', periodicity T, groove width δ, and thickness d (illustrated in Figure 5.8). The grooves are located from $x = H - d/2$ to $x = H + d/2$, with length, W. It can be described mathematically by $\Delta\varepsilon$, which is the spatial variation of the permittivity on top of a waveguide with a core from $x = 0$ to $x = t$ and cladding from $x = t$ to $x = H - d/2$.

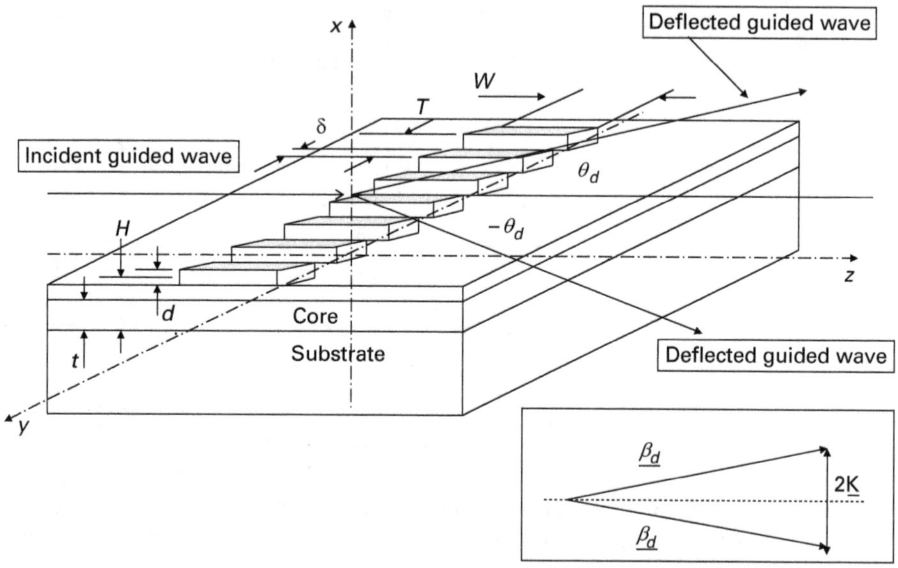

Figure 5.8 Illustration of a deflection grating on a planar waveguide. The input wave is incident at the normal direction. The deflected waves are at angle $\pm\theta_d$. The inset shows the matching of $\underline{\mathbf{K}}$, $\underline{\boldsymbol{\beta}_i}$ and $\underline{\boldsymbol{\beta}_d}$.

$$\Delta\varepsilon(x,y,z) = \Delta\varepsilon(x,y)\Delta\varepsilon(z) = \left[\sum_k (\varepsilon' - \varepsilon_0)\,\mathrm{rect}\left(\frac{kT-y}{\frac{\delta}{2}}\right)\mathrm{rect}\left(\frac{x-H}{\frac{d}{2}}\right)\right]\mathrm{rect}\left(\frac{z+\frac{W}{2}}{\frac{W}{2}}\right) \tag{5.55}$$

where

$$\mathrm{rect}(\tau) = 1 \quad \text{for} \quad |\tau| \le 1 \quad \text{and} \qquad \mathrm{rect}(\tau) = 0 \quad \text{for} \quad [\tau[>1$$

Here ε_0 is the free-space permittivity. For gratings with rectangular grooves, $\Delta\varepsilon(x,y,z) = \Delta\varepsilon(x)\Delta\varepsilon(y)\Delta\varepsilon(z)$.

$\Delta\varepsilon(y)$ causes a y variation in which the mth planar waveguide mode under the grating grooves has $n'_{eff,m}$ different than that of the waveguide without the grating groove. Thus, we have

$$\Delta n_{eff}(y) = \sum_k (n'_{eff,m} - n_{eff,m})\,\mathrm{rect}\left(\frac{kT-y}{\frac{\delta}{2}}\right) \tag{5.56}$$

For a forward mth guided-wave mode in the $+z$ direction incident on the grating, $Ae^{-jn_{eff,m}\beta z}$, the phase is modified as it transmits through the grating.[3] The phase

[3] From a strictly theoretical point of view, the incident mth mode will excite the mth mode in the grating section, plus other substrate and air modes of the same polarization. When the groove depth, d, and the refractive index of the grooves are not very large, very few substrate or air modes are excited.

change for the part of the guided wave without the groove is $jn_{eff,m}\beta W$, while the phase change for the part of the guided wave with the groove is $jn'_{eff,m}\beta W$. Therefore the transmitted wave will have a phase variation that is a function of y,

$$E_{out} = Ae^{+jn_{eff,m}kW}e^{-j\Delta n_{eff}(y)kW}e^{-jn_{eff,m}kz} \tag{5.57}$$

It is well known that any periodic function of y can always be represented by its Fourier series. Since the grating in Figure 5.8 is an even function of y, we have

$$\begin{aligned}\Delta n_{eff}(y) = {} & \frac{1}{T}\int_0^T \Delta n_{eff}(y)\mathrm{d}y + \frac{2}{T}\left(\int_0^T \Delta n_{eff}(y)\cos\frac{2\pi y}{T}\mathrm{d}y\right)\cos\frac{2\pi}{T}y + \\ & \frac{2}{T}\left(\int_0^T \Delta n_{eff}(y)\cos\frac{4\pi y}{T}\mathrm{d}y\right)\cos\frac{4\pi}{T}y + \\ & \frac{2}{T}\left(\int_0^T \Delta n_{eff}(y)\cos\frac{6\pi y}{T}\mathrm{d}y\right)\cos\frac{6\pi}{T}y + \text{higher orders}\end{aligned} \tag{5.58}$$

Since $\Delta n_{eff}kW$ is small, we can express $e^{-j\Delta n_{eff}(y)kW}$ by its Taylor's series,

$$e^{-jkW\Delta n_{eff}(y)} = 1 - jkW\left\{\begin{aligned} & \frac{1}{T}\int_0^T \Delta n_{eff}(y)\mathrm{d}y + \frac{2}{T}\left(\int_0^T \Delta n_{eff}(y)\cos\frac{2\pi y}{T}\mathrm{d}y\right)\cos\frac{2\pi}{T}y \\ & +\frac{2}{T}\left(\int_0^T \Delta n_{eff}(y)\cos\frac{4\pi y}{T}\mathrm{d}y\right)\cos\frac{4\pi}{T}y \\ & +\frac{2}{T}\left(\int_0^T \Delta n_{eff}(y)\cos\frac{6\pi y}{T}\mathrm{d}y\right)\cos\frac{6\pi}{T}y + \text{higher orders}\end{aligned}\right\} \tag{5.59}$$

The first term of the Fourier series is a constant. The cosine in the second term can be written as:

$$jkW\left[\Delta n_0\, e^{-j\frac{2\pi}{T}y} + \Delta n_o e^{+j\frac{2\pi}{T}y}\right]. \tag{5.60}$$

$$\Delta n_o = \frac{1}{T}\int_0^T \Delta n_{eff}(y)\cos\frac{2\pi y}{T}\mathrm{d}y \tag{5.61}$$

Therefore the transmitted wave has many terms. The first term is an mth guided-wave mode in the $+z$ direction. The second term has $e^{-jn_{eff,m}kz}e^{\pm j\frac{2\pi}{T}y}$ variation in the y–z plane. This second term represents two planar guided waves propagating in the $\pm\theta_d$ directions with respect to the z axis where $\theta_d = \sin^{-1}(2\pi/Tn_{eff,m}k)$.There are also higher-order terms with declining magnitudes at angles $\pm\theta_d = \sin^{-1}(2K\pi/Tn_{eff,m}k)$, where $K = 2,3$ 4... The analytical results presented here are applicable to both TE and TM modes.

5.5.7 Excitation of planar guided-wave modes

Planar guided waves can be excited by two different ways:

(1) End excitation. In this case an abruptly terminated planar waveguide with polished end surface is illuminated by an external incident beam. At the vertical end surface of the waveguide the incident tangential electric field $\underline{E_{inc}}$ excites various modes in the waveguide via the continuity condition,

$$\underline{E_{inc}} = E_{inc,x}(x)E_{inc,x}(y)\underline{i_x} + E_{inc,y}(x)E_{inc,y}(y)\underline{i_y} = \sum_n A_n E_n(x)\int_{-\infty}^{+\infty} F_{inc,x}(f_y)e^{+j2\pi f_y y}df_y\underline{i_x}$$

$$+\sum_m A_m E_m(x)\int_{-\infty}^{+\infty} F_{inc,y}(f_y)e^{+j2\pi f_y y}df_y\underline{i_y} + \text{TM and TE substrate and air modes} \tag{5.62}$$

where $F_{inc}(f_y)$ is the Fourier transform of $E_{inc}(y)$ in the y direction similar to Eq. (5.45). A_n and A_m are determined from the orthogonality relations similar to those shown in Eqs. (5.44) and (5.45).

(2) Prism excitation. In this case a prism with refractive index larger than the refractive index of the core is placed close to the core. The bottom prism surface is parallel to the top surface of the core with a low-index gap between them. Because of the gap, an incident wave in the prism at the appropriate angle will be internally reflected at the bottom surface of the prism with a propagation constant in the z direction equal to $n_{eff,m}k$. When mechanical pressure is applied so that the gap between the prism and the cladding is less than the decay length of evanescent tails of both the incident beam and the mth-order mode of the waveguide, optical energy is transferred from the prism to the mth mode of the waveguide. When the size of the input beam illuminating the bottom surface of the prism is adjusted appropriately, the energy transfer from the incident beam to the mth order mode is maximized [6,7]. Note that planar guided waves can be selectively excited by the prism, while several modes are often excited simultaneously by the end excitation method.

5.5.8 Multi-layer planar waveguides

In practice, epitaxial growth of III–V semiconductor materials on InP or GaAs is often used to create multiple-layer waveguides. These layers have different refractive indices. Let the substrate with index n_o at $x \leq x_o$ be labeled as the zeroth layer. There are N layers on top. Each layer is located from $x_{j-1} < x < x_j$. It is labeled as the jth layer, with index n_j. For planar waveguide modes, Eqs. (5.1) and (5.2) still apply. Thus we still have TE or TM modes with $e^{-j\beta_m z}$ variation in the z direction. For TE modes, $E(x)$ in the jth layer for the mth mode in the x direction is:

$$E_{m,j}(x) = \left[A_{m,j}e^{jh_{m,j}x} + B_{m,j}e^{-jh_{m,j}x}\right] \tag{5.63}$$

At the boundary $x = t_{j-1}$, the boundary condition requires:

$$A_{m,j}e^{jh_{m,j}t_{j-1}} + B_{m,j}e^{-jh_{m,j}t_{j-1}} = A_{m,j-1}e^{jh_{m,j-1}t_{j-1}} + B_{m,j-1}e^{-jh_{m,j-1}t_{j-1}}$$

$$jh_{m,j}A_{m,j}e^{jh_{m,j}t_{j-1}} - jh_{m,j}B_{m,j}e^{-jh_{m,j}t_{j-1}} = jh_{m,j-1}A_{m,j-1}e^{jh_{m,j-1}t_{j-1}} - jh_{m,j-1}B_{m,j-1}e^{-jh_{m,j-1}t_{j-1}} \tag{5.64}$$

The TE modes are obtained from the solution of these equations. Note that when the index n_j of the jth layer is low, $h_{m,j}$ may be imaginary. It means the field in the jth layer is exponentially decaying in the x direction. Similar comments apply to TM modes.

Modes of multi-layer planar waveguides are usually calculated by numerical methods.

5.6 Channel waveguides

Channel waveguides are used in devices such as directional couplers, Y-branch splitters, waveguide lasers, guided-wave modulators, waveguide photo-detectors, waveguide demultiplexers, ring resonators, and waveguide filters. Most channel waveguides are microns wide and a few centimeters long. Because of the complexity of the geometry of the dielectric boundaries, there is no analytical solution of the modes of a channel waveguide. There are only approximate solutions [8] and computer programs such as Rsoft BeamProp© or the Finite Element Method that can simulate the modes[9]. The guided-wave modes can also be obtained by an approximation method called the effective index method. Discussions of channel waveguides using the effective index analysis will be the focus of discussion in this section.

The properties of the channel guided-wave mode that are most important in these applications are n_{eff}, the attenuation rate, the polarization of the modes that have been excited by the incident radiation, and the decay rate of the evanescent tails.

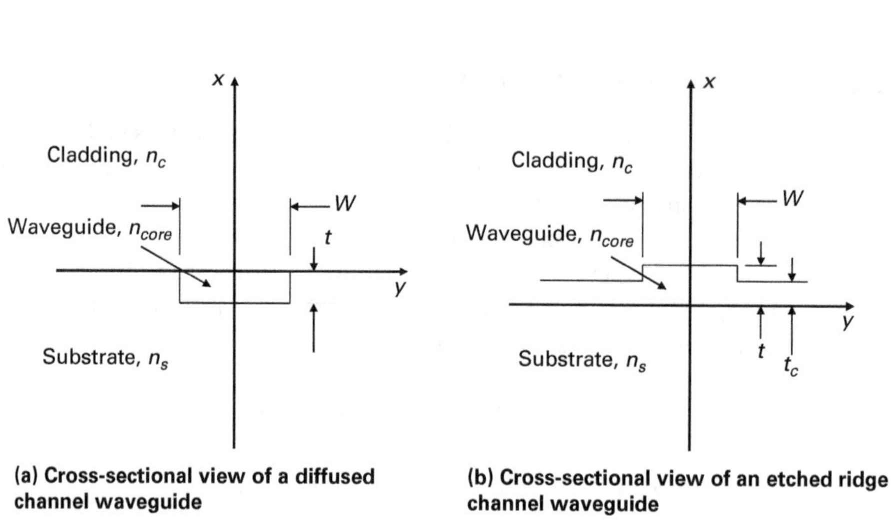

Figure 5.9 Illustration of the index profile of two examples of channel waveguides. (a) A diffused waveguide. The illustration shows only a uniform index variation in the core instead of a graded index variation. (b) An etched channel waveguide. The ridge is W wide. The thickness of the planar waveguide and the ridge is t.

Channel waveguide devices often involve one guided-wave mode interacting with another guided-wave mode through their evanescent tails. These interactions and the applications resulting from these interactions will be analyzed in detail in the next three chapters.

Channel waveguides are often fabricated either by diffusion or by microfabrication procedures, such as etching, from planar waveguides. A realistic diffused waveguide would have a graded variation of index from the core to the cladding. Comparing to etched waveguides, the advantage of diffused channel waveguides is that there is very little scattering loss at the boundaries. The cross-sectional index variation of an idealized diffused waveguide is illustrated in Figure 5.9(a). It shows only a core with a constant index. Figure 5.9(b) illustrates an etched channel waveguide in which the thickness of the core in the cladding region has been reduced. The waveguide shown in Figure 5.9 (b) is also called a ridge waveguide. Ridge waveguides are used because the roughness of the etched surface produces a large scattering loss. The thinner the ridge, the lower the total scattering loss. Sometimes, a ridge waveguide is also formed by depositing a ridge on top of the core. In both cases, the center core of the channel waveguide is located at $W/2 \geq |y|$.

5.6.1 The effective index analysis

Consider the rectangular channel waveguides in Figure 5.9(a) and (b), where there is a rectangular core region in the y direction, $|y| \leq W/2$, and a cladding region, $|y| \geq W/2$. Let us assume that the planar waveguide in the core region has only one mode in the x direction, the TE_0 mode. In Section 5.3.1, the propagation of the TE_0 planar guided-wave mode in the core region along its propagation direction z is given by $e^{\pm j\beta_o z}$ where β_0/k is its effective index, n_{e1}. In the ridge waveguide shown Figure 5.9(b), there is also a planar TE_0 guided-wave mode in the cladding region. Let the effective index of the TE_0 planar guided-wave mode of the structure in the cladding region be n_{e2}. Since the high-index layer is thicker at $|y| \leq w/2$, $n_{core} > n_{e1} > n_{e2} > n_s$. In the channel waveguide shown Figure 5.9(a), there is no guided-wave mode in the cladding region. There are only continuous substrate and air propagating modes for $|y| \geq W/2$.

In the channel waveguide shown in Figure 5.9(b), the planar guided-wave mode in the $|y| \leq W/2$ core region can propagate in any direction in the y–z plane. Let us consider a planar guided-wave mode TE_0 propagating in a direction that makes an angle δ with respect to the z axis. Its propagation constant in the z direction is $n_{e1}k \cos \delta$. If δ is so small that $n_{e1} \cos \delta > n_{e2}$, where n_{e2} is the effective index of the TE_0 mode outside the ridge, this planar guided wave will be totally internal reflected repeatedly at the $|y| = W/2$ boundaries. At any arbitrary δ, the sum of all the reflected waves is zero because of phase cancellations. The sum of all the reflected core planar guided waves would only yield a non-zero solution when the round-trip phase shift of total internal reflection is a multiple of 2π. It happens only at specific values of δ. Thus the allowed specific values of δ depend on n_{e1}, W, and the phase of the reflection coefficient at the $|y| = W/2$ boundaries, which in turn depend on n_{e1}, n_{e2}, and the polarization. These totally internally reflected planar waveguide modes in the core constitute the channel guided-wave modes. The lowest-order mode (i.e. the zeroth-order mode) in the y direction has a round trip phase shift of 2π, and the nth order mode has a round-trip phase shift of $2(n+1)\pi$. Consequently the field of the zeroth-order mode ($n = 0$) has no node in the y direction in the core. The nth-order mode has n nodes. If W is sufficiently small, then we would have only a single mode in the y direction in the core. Since the lowest order mode in the x direction is the TE_0 mode, the lowest-order mode of the channel waveguide is called the TE_{00} mode. Similarly, the nth totally internal reflected TE_m mode yields the TE_{mn} modes in the channel waveguide.

The effective index method is just a simplified method to match the boundary conditions, thereby determining approximately δ for the discrete modes of the channel waveguides.

Consider now the mathematical details of effective index analysis. Let the TE_0 electric field of the reflected planar waveguide mode have amplitude A. At the $|y| = W/2$ boundaries, the tangential electric and magnetic fields need to be matched. The dominant component of the electric field is approximately perpendicular to the boundaries. A small component is in the z direction. This component has an amplitude $A \sin \delta$. It is tangential to the $|y| = W/2$ boundaries. The magnetic field has two components, H_x and H_z. The dominant tangential field of the core planar guided wave is H_x. At the $|y| = W/2$ boundary, we need to match the magnetic field H_x and the z component of the electric field of the core and cladding modes of the planar waveguides.

Let the x variation of the mode in the core region be the TE_0 mode. The mode in the cladding region that matches closely the x variation of H_x and E_z in the core at the $|y| = W/2$ boundary, is the H_x and E_z of the cladding TE_0 planar guided-wave mode. If we neglect the continuous modes in the x direction, which will be excited at the boundaries, we only need to match the amplitude and phase of the tangential components of the TE_0 mode on both sides of the boundary as a function of y and z.

In order to satisfy the boundary condition for all z values, the z variation of this cladding guided-wave mode must be equal to $e^{-jn_{e1}k\cos\delta z}$. If we let the y variation of the TE_0 cladding guided-wave be $e^{-jy\frac{k}{\gamma}}$, γ must now satisfy the equation,

$$\gamma^2 = n_{e2}{}^2 - n_{e1}{}^2\cos^2\delta \tag{5.65}$$

γ is imaginary when $n_{e1}\cos\delta > n_{e2}$. An imaginary γ represents an exponentially decaying cladding guided-wave in the y direction, not a propagating wave.

The equation for H_x and the boundary conditions of the continuity of the amplitude and phase of H_x and the z component of the electric field here is similar to the equation and boundary conditions of an equivalent TE plane wave polarized in the y direction (with H_x in the x direction and $\partial/\partial x = 0$) bouncing back and forth between the y boundaries and propagating in the z direction. The equivalent TE plane waves and the configuration of the boundaries are illustrated in Figure 5.10. In this case, the equivalent plane wave in the core has index n_{e1}, and the equivalent plane wave in the cladding has index n_{e2}.

In short, the mathematics used here for analyzing the total internal reflection of a core planar guided wave in the y direction is approximately equivalent to analyzing a total reflection of the equivalent TM plane wave propagating in the y–z plane at angle δ with respect to the z axis where the magnetic field H_x is polarized approximately in the x direction and electric field E_z is in the z direction. The equivalent material refractive

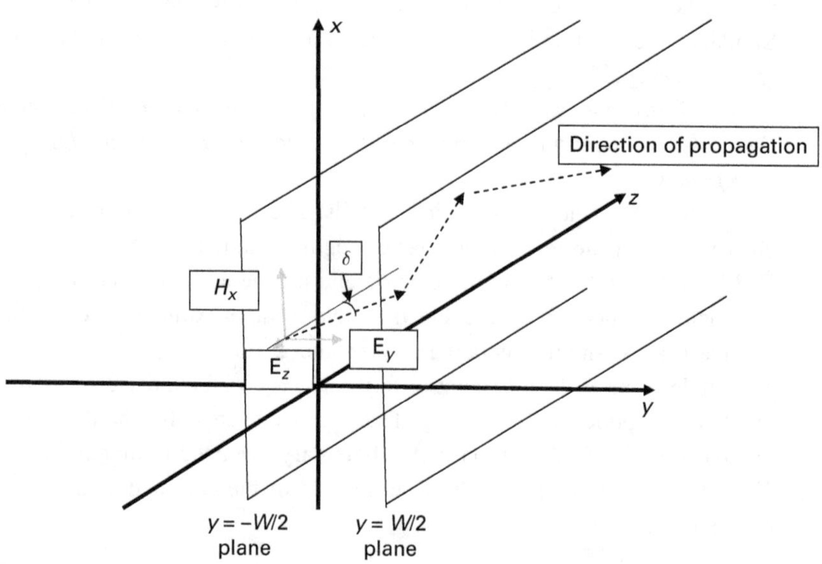

Figure 5.10 Illustration of the TM wave used in the effective index approximation.

indices are n_{e1} and n_{e2}. In other words, we can use the TM planar guided-wave mode equation for a symmetric waveguide in the y direction, i.e.

$$
\begin{gathered}
H_x(z) = A e^{-j\beta_o \cos\delta z} \\
\left[\frac{\partial^2}{\partial y^2} + \omega^2 \varepsilon(y)\mu - \beta_0{}^2 \cos^2\delta\right] H_x(y) = 0 \\
\varepsilon(y) = \varepsilon_o n_{ej}{}^2 \qquad j = 1 \; or \; 2 \\
E_y = \frac{j}{\omega\varepsilon(y)} \frac{\partial H_x(z)}{\partial z} \quad \text{and} \quad E_z = \frac{-j}{\omega\varepsilon(y)} \frac{\partial H_x}{\partial y}
\end{gathered}
\tag{5.66}
$$

Note that the equivalent plane wave has no x variation.[4] The boundary conditions are the continuity of H_x and E_z (or $\partial H_x / \partial y$) at $y = \pm W/2$. The nth solution of this equation will yield the effective index and the y variation of even channel guided-wave mode $TE_{0\,n}$ that we are looking for.[5] Note that the channel waveguide is now symmetric with respect to $y = 0$. As we have discussed in Section 5.4.2, there are even and odd solutions of Eq. (5.36). The total of all the even and odd modes constitutes all the channel waveguide modes. This is the effective index method.

Although the TE_0 mode is used in the preceding discussion, the result is applicable to any TE_m mode in the core. The most important quantity to be obtained is the effective index, i.e. the $\beta_{m,n}/k$ or $n_{m,e1}\cos\delta_n$, of the TE_{mn} mode of the channel waveguide in the z direction. Knowing this effective index, we know both the δ in the core and the exponential decay constant γ in the cladding. Since δ is very small, the channel guided-wave mode obtained from the TE core planar guided mode is still approximately a y-polarized TE mode propagating in the z direction. The x variation of E_y for $|y| < W/2$ is approximately the same as the core planar guided-wave mode TE_m.

Similarly, a channel guided-wave mode with approximately TM polarization can be obtained from TM planar guided-wave modes in the core and in the cladding region. In that case the equivalent TE guided-wave equation will be used to find the effective index of the channel waveguide mode and the y variation.

Notice that we no longer have pure TE or TM modes. We have basically TE- or TM-like modes. These modes are called hybrid modes. Note also that the effective index approximation did not give us a complete solution for the x variation of the electric field near the boundaries. In order to satisfy the boundary conditions accurately, many other modes, especially the substrate and air modes, need to be involved. The electric and magnetic fields of these substrate modes will exponentially decay in the y direction, even faster than the planar guided-wave mode in the cladding.

The n_{eff} of the channel waveguide mode calculated by the effective index method is reasonably accurate when modes are well above the cut-off and when the field variations of the mth order modes inside and outside of the core are close. No matter how accurate

[4] The x variations of the core and cladding modes have been taken into consideration by the use of the effective indices n_{e1} and n_{e2}.

[5] Note that the TE_{mn} channel waveguide mode is still polarized predominantly in the y direction in the y–z plane.

the result, the simple effective index analysis provides much insight about properties of channel guided-wave modes.

For the waveguide shown in Figure 5.9(a), there is no planar waveguide mode in the cladding. The x variation of the tangential field of a core planar guided wave propagating at angle δ must be matched by the summation of the continuous cladding modes at $|y| = W/2$. Since $n_{e1}\cos\delta > n_s$ and n_c, all continuous modes decay exponentially away from the $|y| = W/2$ boundary. The core planar guided wave is again totally internally reflected back and forth. The sum of all the reflected core planar guided waves would yield a non-zero solution only when the round trip phase shift of total internal reflection for specific values of δ is a multiple of 2π. These special sets of totally internally reflected core planar waveguide modes constitute the channel guided-wave modes. However, in this case, we only know the n_{e1} of the core TE planar guided-wave mode. We do not know n_{e2} outside the core. Since a combination of substrates and air modes must be used to match the x variation of the core guided wave at $y = \pm W/2$, the equivalent value of n_{e2} should be somewhere between n_c and the substrate index n_s. The best equivalent index n_{e2} to be used for the cladding region in the TM equation in y will depend on the profile of the core TE mode. For a core guided wave with a substantial evanescent tail in the x direction in the substrate, we may use the substrate index. Fortunately, for well-guided channel modes in the core, the solution of n_{eff} and the y variation is not very sensitive to the value of n_{e2} used for the calculation.

Clearly, the accuracy of the effective index method may not be very good for such a structure.

5.6.2 An example of the effective index method

Consider first a GaAs planar waveguide with $n_2 = 3.27$ and $n_s = 3.19$, and $t = 0.9$ μm in the core region operating at $\lambda = 1.5$ μm. This waveguide is exposed to air with $n_c = 1$. The GaAs layer has been partially etched away at $|y| \geq W/2$, $W = 3$ μm. In the lateral cladding region, $t = 0.6$ μm. We would like to find the effective index and the field of the lowest order TE-like channel waveguide mode.

The first step of our calculation is to find the effective index of the TE_0 planar guided wave in the core region at $W/2 \geq |y|$ and in the cladding region at $|y| > W/2$. From Eq. (5.5), we find the TE planar guided-wave modes for the core and the cladding regions, $n_{e1} = 3.223$ and $n_{e2} = 3.211$. According to Section 5.4.2, we solve the following equations to obtain the lowest-order symmetrical channel waveguide TE_{00} mode, which is polarized approximately in the y direction:

$$\tan\left[(h'_n/k)\frac{kW}{2}\right] = \frac{n_{e1}{}^2 p'_n/k}{n_{e2}{}^2 h'_n/k}, \qquad \left(\frac{h'_n}{k}\right)^2 + \left(\frac{p'_n}{k}\right)^2 = n_{e1}{}^2 - n_{e2}{}^2 \tag{5.67}$$

The solution is $(h'_0/k) = 0.1795$, which gives $n_{eff,0} = 3.218$ and $p'_0/k = 0.2121$. The field distributions are approximately

$$
\begin{aligned}
E_y &= A\sin(h_0x+\varphi_0)\cos(h'_0y)e^{-jn_{eff,0}kz} && \text{for} \quad 0<x<t,\ y\leq|W/2| \\
&= A\sin\varphi_0 e^{q_0x}\cos(h'_0y)e^{-jn_{eff,0}kz} && \text{for} \quad x\leq 0,\ y\leq|W/2| \\
&= A\sin(h_0t+\varphi_0)e^{-p_0(x-t)}\cos(h'_0y)e^{-jn_{eff,0}kz} && \text{for} \quad x\geq t,\ y\leq|W/2| \\
&= A\sin(h_0x+\varphi_0)\cos\left(\frac{h'_0W}{2}\right)e^{-p'_0\left(y-\frac{W}{2}\right)}e^{-jn_{eff,0}kz} && \text{for} \quad 0<x<t,\ y>W/2 \\
&= A\sin(h_0x+\varphi_0)\cos\left(\frac{h'_0W}{2}\right)e^{+p'_0\left(y+\frac{W}{2}\right)}e^{-jn_{eff,0}kz} && \text{for} \quad 0<x<t,\ y<-W/2 \\
&= A\sin(h_0t+\varphi_0)e^{-p_0(x-t)}\cos\left(\frac{h'_0W}{2}\right)e^{-p'_0\left(y-\frac{W}{2}\right)}e^{-jn_{eff,0}kz} && \text{for} \quad x>t, y>W/2 \\
&= A\sin(h_0t+\varphi_0)e^{q_0x}\cos\left(\frac{h'_0W}{2}\right)e^{-p'_0\left(y-\frac{W}{2}\right)}e^{-jn_{eff,0}kz} && \text{for} \quad x<0, y>W/2 \\
&= A\sin(h_0t+\varphi_0)e^{-p_0(x-t)}\cos\left(\frac{h'_0W}{2}\right)e^{p'_0\left(y+\frac{W}{2}\right)}e^{-jn_{eff,0}kz} && \text{for} \quad x>t, y<-W/2 \\
&= A\sin(h_0t+\varphi_0)e^{q_0x}\cos\left(\frac{h'_0W}{2}\right)e^{-p'_0\left(y+\frac{W}{2}\right)}e^{-jn_{eff,0}kz} && \text{for} \quad x<0, y<-W/2
\end{aligned}
\tag{5.68}
$$

Here φ_0, q_0, h_0, and p_0 are parameters of the planar guided-wave TE_0 mode in the core (given by Eq. (5.4) with β_m = 3.223k). Since we do not know in detail what the radiation modes are, we cannot find the field distributions accurately in the cladding regions ($x > t$, $|y| > W/2$) and ($x < 0$, $|y| > W/2$) from the effective index method. A reasonable estimation is that the fields near $|y| = W/2$ have an x variation similar to the field pattern of the TE_0 mode in the core in the x direction.

5.6.3 Channel waveguide modes of complex structures

An important assumption made in the effective index method is that the field variations in the x direction in the cladding and core regions near the boundaries in the y direction are similar. In an actual multi-layer channel waveguide, the material indices and thicknesses may eventually vary considerably in the y direction. Thus the pattern of the planar waveguide mode may also vary considerably at different y locations. In this case, the entire waveguide may then be approximated by sections of local waveguides in the y direction. Each local planar waveguide section has constant cross-sectional index variation that is slightly different than its neighboring section. The local planar waveguide may have multiple modes. For TE modes, the local $E_z(x)$ and $H_x(x)$ for TE_m modes in a given local section are matched at the y boundaries by the similar E_z and H_x of the TE_m mode in its neighboring sections by the effective index method. The effective index of the total composite mode is determined from the equations obtained from all the boundaries. For such structures, numerical simulation is usually employed to determine the mode patterns and the n_{eff}.

5.7 Guided-wave modes in optical fibers

Optical fibers are used for low loss transmission of optical signals, often over long distances. There are already many books that discuss the modes of various optical fibers[1]. We will not repeat those discussions here. Guided-wave modes in round, step-index optical fibers are presented here for three reasons: (1) They are the only analytical solutions of optical fibers. (2) The properties of all the optical fibers can be discussed using these solutions. (3) The presentation of the analytical solution allows us to understand the discussions of optical fiber modes in the literature. However, step-index fibers are not used in practical applications.

The cross-section of a step-index optical fiber with uniform cladding and core has already been shown in Figure 5.1(a). The core has radius a. The core index n_1 is larger than the cladding index n_2. In contrast to channel waveguides with rectangular cross-sections, there are analytical solutions of guided-wave modes in single-mode step-index fibers because of its cylindrical symmetry. Although the field distribution and the effective index (especially the dispersion) of modern graded index fibers used in communication systems are different than those of the step-index fibers, step-index fiber modes are used here to demonstrate many properties of the modes of round fibers.

5.7.1 Guided-wave solutions of Maxwell's equations

The vector wave equations obtained from Maxwell's equations in a homogeneous medium with refractive index n are [1]:

$$(\nabla^2 + n^2k^2)\underline{E} = 0 \quad \text{and} \quad (\nabla^2 + n^2k^2)\underline{H} = 0 \tag{5.69}$$

In addition, we have the curl equations relating $\underline{E}$ and $\underline{H}$. If we assume that guided-wave modes have $e^{-j\beta z}$ variation along the z direction, which is also the fiber axis, then in the cylindrical coordinates we have:

$$\nabla^2 = [\nabla_t^2] + \frac{\partial^2}{\partial z^2} = \left[\frac{1}{\rho}\frac{\partial}{\partial \rho}\left(\rho\frac{\partial}{\partial \rho}\right) + \frac{1}{\rho^2}\frac{\partial^2}{\partial \vartheta^2}\right] + \frac{\partial^2}{\partial z^2},$$

$$(\nabla_t^2 + k_t^2)E_z = 0, \qquad (\nabla_t^2 + k_t^2)H_z = 0,$$

$$k_t^2 = n^2k^2 - \beta^2 \tag{5.70}$$

The remaining transverse components of the fields are related to E_z and H_z as follows:

$$E_\rho = -\frac{j}{k_t^2}\left[\beta\frac{\partial E_z}{\partial \rho} + \frac{\omega\mu}{\rho}\frac{\partial H_z}{\partial \vartheta}\right],$$

$$E_\vartheta = -\frac{j}{k_t^2}\left[\frac{\beta}{\rho}\frac{\partial E_z}{\partial \vartheta} - \omega\mu\frac{\partial H_z}{\partial \rho}\right],$$

$$H_\rho = \frac{j}{k_t^2}\left[\frac{\omega\varepsilon_0 n^2}{\rho}\frac{\partial E_z}{\partial \vartheta} - \beta\frac{\partial H_z}{\partial \rho}\right],$$

$$H_\vartheta = -\frac{j}{k_t^2}\left[\omega\varepsilon_0 n^2\frac{\partial E_z}{\partial \rho} + \frac{\beta}{\rho}\frac{\partial H_z}{\partial \vartheta}\right] \tag{5.71}$$

The solutions of Eq. (5.70) are:

$$E_z = AJ_m(k_{t1}\rho)\cos(m\theta) \quad \text{and} \quad H_z = BJ_m(k_{t1}\rho)\sin(m\theta),$$
$$k_{t1} = \sqrt{n_1{}^2k^2 - \beta^2} \tag{5.72}$$

for $a \geq \rho$, and,

$$E_z = CH_m{}^{(2)}(jk_{t2}\rho)\cos(m\theta) \quad \text{and} \quad H_z = DH_m{}^{(2)}(jk_{t2}\rho)\sin(m\theta),$$
$$jk_{t2} = \sqrt{\beta^2 - n_2{}^2k^2} \tag{5.73}$$

for $\rho > a$. Here n_1 is the refractive index of the core, and n_2 is the refractive index of the cladding. There is a second set of solutions in which E_z has the $\sin(m\theta)$ variation and H_z has the $\cos(m\theta)$ variation. J_m is the Bessel function of the first kind and order m; $H_m^{(2)}$ is the Hankel function of the second kind of order m; and m is an integer. Similar to the guided waves in planar and channel waveguides, the Hankel function gives an exponential decay as $\rho \rightarrow \infty$ in the cladding. E_ρ, E_θ, H_ρ, and H_θ are obtained from E_z and H_z from Eq. (5.71). Continuity of E_z, H_z, E_θ, and H_θ at $\rho = a$ yields the relationship among A, B, C, and D coefficients and the characteristic equation that determines the discrete values of β for the mode. Note that since the fields decay exponentially in the radial direction in the cladding, the thickness of the cladding does not affect the solution, as long as it is sufficiently thick. The effective index, n_{eff}, of the mode is β/k. Similar to the channel waveguide modes, each mode has a cut-off condition. The higher the order of the mode, the larger the value of $ka\sqrt{n_1^2 - n_2^2}$ for the cut-off.

In many ways, the step-index fiber resembles a planar waveguide wrapped around cylindrically. However, because of the cylindrical geometry, the mathematical expressions appear to be more complicated.

5.7.2 Properties of the modes in fibers

It is interesting to note that the axially symmetric modes have $m = 0$. In that case, we have again TE (with non-zero H_z, E_θ, and H_ρ, called H_{op} modes) and TM (with non-zero E_z, H_θ, and E_ρ, called E_{op} modes) modes. However, the lowest-order mode that has the largest decay constant in the cladding, jk_{t2}, is not an axially symmetric mode. The lowest-order mode is the HE_{11} mode, which has $m = 1$ and the lowest-order radial solution of the characteristic equations. For $m \neq 0$, the modes lose their transverse character. They are known as hybrid modes. There is no cut-off for the HE_{11} mode. In HE– modes the longitudinal electric field is bigger than the longitudinal magnetic field. There are also EH– modes, in which the longitudinal magnetic field is dominant. The TM (i.e. E_{op}) modes are the axially symmetric members of the HE– family of modes. The H_{op} modes are the axially symmetric members of the EH– family of modes.

For weakly guiding modes, $\Delta = (n_1 - n_2)/n_1$ is small compared to unity. The characteristic equation for HE_{mp} modes is:

$$k_{t1}a\frac{J_m(k_{t1}a)}{J_{m-1}(k_{t1}a)} = (jk_{t2}a)\frac{H_m{}^{(2)}(jk_{t2}a)}{H_{m-1}{}^{(2)}(jk_{t2}a)} \tag{5.74}$$

where the subscript p refers to the pth root of the above equation. The characteristic equation for EH_{mp} modes is:

$$k_{t1}a\frac{J_{m+2}(k_{t1}a)}{J_{m+1}(k_{t1}a)} = (jk_{t2}a)\frac{H_{m+2}{}^{(2)}(jk_{t2}a)}{H_{m+1}{}^{(2)}(jk_{t2}a)} \tag{5.75}$$

Both the HE– and HE– modes exhibit nearly transverse field distribution. The longitudinal components have a phase shift of $\pi/2$ with respect to the transverse components; they remain small compared to the transverse field. The characteristic equation for HE_{mp} modes is the same as the characteristic equation for $EH_{m-2,p}$ modes. Therefore, for weakly guiding fibers, any $HE_{l+1,p}$ mode is degenerate with $EH_{l-1,p}$ modes (i.e. they have the same propagation constants or effective index).

When we linearly combine the degenerate $HE_{l+1,p}$ and $EH_{l-1,p}$ modes together, we obtain the linearly polarized LP_{lp} mode, which has the same effective index as the $HE_{l+1,p}$ mode. The LP_{lp} mode has only E_x and H_y in the core and cladding, it is nearly uniformly polarized over the fiber cross-section. The LP_{01} mode is just the HE_{11} mode. Each LP mode occurs in four different versions, two orthogonal directions of polarization, each with $\cos l\theta$ and $\sin l\theta$ variations. Figure 5.11 shows the phase parameter B as a function of fiber parameter $V = ka\sqrt{n_1^2 - n_2^2}$ for low-order LP_{lp} modes, taken from Unger's book [1].

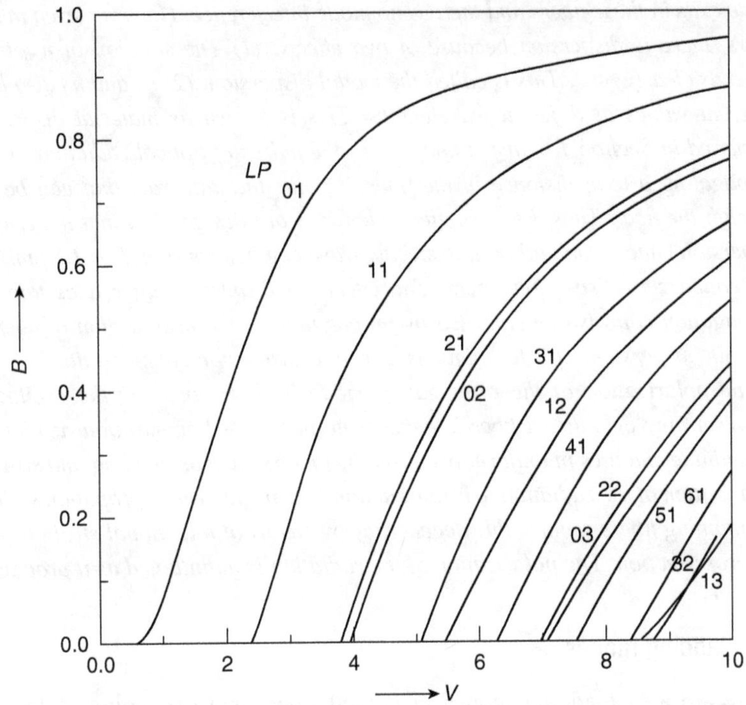

Figure 5.11 The phase parameter B of propagating modes in step-index, round fibers. The phase parameter B is related to the effective index n_{eff} of the propagating modes, $B = (n_{eff}^2 - n_2^2)/(n_1^2 - n_2^2)$. It is shown as a function of the fiber material parameter V, $V = ka\sqrt{n_1^2 - n_2^2}$, for lower-order LP modes in the weakly guiding fibers. The figure is taken from reference [1] with copyright permission from Oxford University Press.

5.7.3 Properties of optical fibers in applications

There are very few devices made from fibers. The most prominent devices are the fiber lasers, fiber optical amplifiers, directional couplers, and grating filters.[6] *Optical fibers are used primarily as transmission lines for optical pulses, often over many kilometers of distance. As the optical pulse propagates, its pulse width widens, and its polarization changes. Therefore, the properties of modes in fibers that are most important to fiber communications are the number of propagating modes, the attenuation rate of the modes, the dispersion, the polarization of the excited mode, and the change of the state of polarization as the mode propagates. In addition, there are applications of short fibers in instrumentation, where multi-mode fibers are used, and the important considerations are not the modal analysis but the physical features of the fibers.*

The wavelength dependence of the material absorption requires long-distance optical fiber networks to operate at approximately 1.3 and 1.55 μm wavelengths. In order to minimize dispersion, most fiber transmission lines use single-mode fibers. Only the HE_{11}

[6] See Chapters 8 and 9 for discussions of waveguide gratings, filters and directional couplers. The optical fiber devices operate very similarly to the waveguide devices.

mode exits in these fibers and there is no mode interference. However, even in single-mode fibers, there is dispersion because of two effects: (1) The solution of n_{eff} in Eq. (5.70) depends clearly on λ. This is called the modal dispersion. (2) n_1 and n_2 also have slightly different values at different wavelengths. This is known as material dispersion. As we discussed in Section 1.6, dispersion causes the pulses of optical radiation to spread after propagating a long distance in the fiber. It limits the data rate that can be transmitted through the fiber. Thus, some single-mode fibers are designed so that material and mode dispersion cancel each other at a specific wavelength, such as 1.3 or 1.5 μm. These fibers are called zero-dispersion fibers. The effects of dispersion after a certain distance of propagation can also be canceled by propagating in the next section of fiber which has opposite dispersion. This technique is known as dispersion compensation.

The polarization of the propagating mode is determined by the excitation source. However, any cylindrical fiber is degenerate in two orthogonal polarization directions. Any minute changes in uniformity caused by factors such as bending and stress cause the polarization of the radiation to rotate randomly in the fiber as it propagates. Polarization-maintaining fibers remove this degeneracy by means of intentional strain or ellipticity of the cross-section. The polarization of the radiation is maintained as it propagates.

5.7.4 The cladding modes

There are also cladding modes in optical fibers, corresponding to the continuous substrate and air modes in the planar and channel waveguides. They are excited whenever there is a defect, bending of the fiber, or dielectric discontinuity. Cladding modes are solutions of the boundary value equations. Their effective indices are less than n_2. These modes do not exponentially decay away from the core. A typical single-mode fiber has a core about 10 μm in diameter, while the cladding has a diameter of the order of 100 μm. Thus there are many propagating cladding modes, with the effective indices very close to each other, resembling a continuous mode distribution. In the absence of the exponential decay, cladding modes have high attenuation. Their amplitude is very small at distances far away from the discontinuity. Cladding modes are utilized in short fibers used for instrumentation, but little modal analysis is required for these applications.

Chapter summary

Modes of optical fibers and waveguides are presented in this chapter. Because of the complex mathematics, only the analytical solutions of modes in round, step-index fiber and planar waveguides have been presented in detail. These solutions demonstrate not only the mathematical techniques for finding the modes, but also important properties of the guided-wave modes, namely the effective index, the evanescent decay of the modes in the cladding, the othorgonality of modes and the dispersion. Since there are no analytical solutions for modes in channel waveguides, these modes are discussed using the effective index approximation.

The discussions in this chapter demonstrate clearly the advantages and necessity of modal analysis.

Although the methods to solve the modes of fibers and channel waveguides are similar, their applications are not. In long-distance optical-fiber transmission lines, the low attenuation and the dispersion of the optical pulses are most important. Therefore, dispersion in fibers is important to understand and to discuss. In opto-electronics, devices are short. The performance of devices depends on the interaction of the excited modes in channel waveguides. Therefore, how to understand and control the mode pattern (using the effective index and the evanescent field), the orthogonality of modes, and the excitation of modes becomes most important. These modal properties of channel waveguides are used to discuss modal interactions and device properties in Chapters 6, 7, and 8.

References

[1] H. G. Unger, *Planar Optical Waveguides and Fibers*, Oxford University Press, 1977.

[2] D. Marcuse, *Light Transmission Optics*, Van Nostrand Reinhold, 1972, Section 8.5.

[3] William S. C. Chang, *Fundamentals of Guided-Wave Opto-Electronic Devices*, Cambridge University Press, 2010.

[4] Michael C. Hamilton and Anthony E. Spezio, Spectrum analysis with integrated optics, Chapter 7, in *Guided Wave Acousto-optics*, ed. C. S. Tsai, Springer, 1990.

[5] Paul R. Ashley, Fresnel lens in a thin film waveguide, *Applied Physics Letters*, **33**, 490, 1978.

[6] P. K. Tien and R. Ulrich, Theory of prism-film couplers and thin-film light guides, *Journal of the Optical Society of America*, **60**, 1325, 1970.

[7] Donald L. Lee, *Electromagnetic Principles of Integrated Optics*, John Wiley & Sons, 1986.

[8] D. Marcuse, *Theory of Dielectric Waveguides*, Academic Press, 1974.

[9] R. Searmozzino, A. Gopinath, R. Pregla, and S. Helfert, Numerical techniques for modeling guided-wave photonic devices, *IEEE Journal of Selected Topics in Quantum Electronics*, **6**, 150, 2000.

6 Guided-wave interactions

The operation of many photonic devices is based on the interactions among optical guided waves in channel waveguides. We have already discussed the modes of individual fiber and channel waveguides in Chapter 5. From that discussion, it is clear that approximation methods need to be used to obtain the modes of channel waveguides. How to analyze the interactions of these modes is the focus of this chapter; much of this analysis is also based on approximation methods.

There are three types of guided-wave interactions that are the basis of the operation of most photonic devices: (1) The adiabatic transition of guided-wave modes in waveguides (or fiber structures). In these devices, the cross-section of the waveguide at one longitudinal position is transformed gradually to a different cross-section at another longitudinal position as the modes propagate. An example of this type of device is the symmetrical Y-branch that splits one channel waveguide into two identical channel waveguides (see Section 6.7.2). (2) The phase-matched interaction between guided-wave modes of two waveguides over a specific interaction distance. A well-known example of photonic devices based on this type of interaction is the directional coupler in waveguides (or fibers, see Sections 6.3.4 and 6.6). (3) Interaction of guided-wave modes through periodic perturbation of the optical waveguide. An example of this is the grating filter in channel waveguides (or optical fibers, see Section 6.3.3).

In this chapter, we first introduce two techniques that can be used to analyze approximately the interactions of weakly coupled guided waves. These are perturbation analysis and coupled mode analysis [1,2]. They are most accurate when the mutual interaction is moderate. Following that, analysis of coupled waveguides by super modes of the total structure is presented. The super mode analysis allows us to view the interactions among both weakly and strongly coupled waveguides from another point of view. It also allows us to understand the properties of strongly coupled waveguides such as the Y-junction and the Mach–Zehnder interferometer, even without the exact knowledge of the profile and the effective index of the modes. Much of the discussion in this chapter is taken from my earlier book [3].

In guided-wave devices, radiation modes are excited at any dielectric discontinuity. Rigorous modal analysis of propagation in a waveguide with varying cross-section in the direction of propagation should involve, in principle, all these modes. However, radiation modes usually fade away at some reasonable distance from the discontinuity. They are important only when radiation loss must be accounted for. Thus in the discussion of guided-wave interactions in this chapter, radiation modes such as the substrate and air modes in waveguides (and the cladding modes in fibers) are not discussed.

6.1 Review of properties of the modes in a waveguide

In order to simplify our discussion, the formalism of modal equations and the properties of the guided-wave modes are summarized first.

In any waveguide (or fiber) that has a transverse index variation independent of z (i.e. independent of its longitudinal position), the electric and magnetic fields, $\underline{E}(x,y,z)$ and $\underline{H}(x,y,z)$, can be explicitly expressed in terms of the longitudinal ($\underline{E_z}$, $\underline{H_z}$) and transverse ($\underline{E_t}$,$\underline{H_t}$) fields as follows:

$$\underline{E} = [E_x \underline{i}_x + E_y \underline{i}_y] + E_z \underline{i}_z = \underline{E}_t + E_z \underline{i}_z = \underline{E}(x,y) e^{-j\beta z} e^{j\omega t},$$

$$\underline{H} = [H_x \underline{i}_x + H_y \underline{i}_y] + H_z \underline{i}_z = \underline{H}_t + H_z \underline{i}_z = \underline{H}(x,y) e^{-j\beta z} e^{j\omega t},$$

$$\nabla = \left[\frac{\partial}{\partial x}\underline{i}_x + \frac{\partial}{\partial y}\underline{i}_y\right] + \frac{\partial}{\partial z}\underline{i}_z = \nabla_t + \frac{\partial}{\partial z}\underline{i}_z, \tag{6.1}$$

$$\nabla_t \times \underline{E}_t = -j\omega\mu H_z \underline{i}_z, \qquad \nabla_t \times \underline{H}_t = j\omega\varepsilon(x,y) E_z \underline{i}_z,$$

$$\nabla_t \times E_z \underline{i}_z - j\beta \underline{i}_z \times \underline{E}_t = -j\omega\mu \underline{H}_t,$$

$$\nabla_t \times H_z \underline{i}_z - j\beta \underline{i}_z \times \underline{H}_t = j\omega\varepsilon(x,y) \underline{E}_t$$

Equation (6.1) implies that the transverse fields can be obtained directly from the longitudinal fields, or vice versa. One only needs to use either set of them to specify the field.

The nth guided-wave mode, given by $\underline{e_n}$ and $\underline{h_n}$, is the nth discrete eigenvalue solution of $\underline{E}$ and $\underline{H}$ in the above vector wave equation that satisfies the condition of the continuity of tangential electric and magnetic fields across all boundaries. In view of the properties of the modes discussed in Chapter 5, we expect the following properties for the $\underline{e_n}$ and $\underline{h_n}$ modes of any general waveguide with constant cross-section in z.

(1) The magnitude of the fields outside the higher-index core or channel region decays exponentially away from the high-index region in lateral directions.
(2) The higher the order of the mode, the slower is the exponential decay rate of the evanescent tail.
(3) The effective index $n_{eff,n}$ ($n_{eff,n} = \beta_n/k$) is less than the highest index that is in the core index and larger than the index of the cladding or the substrate. n_{eff} is larger for a lower-order mode.
(4) Most importantly, it can be shown from the theory of differential equations, that the guided-wave modes of lossless waveguides are orthogonal to each other and to the

substrate or cladding modes. Mathematically, this is expressed for guided-wave modes as,

$$\iint_S (\underline{e}_{t,m} \times \underline{h}_{t,n}^*) \cdot \underline{i}_z \mathrm{d}s = \int_{-\infty}^{\infty}\int_{-\infty}^{\infty} (\underline{e}_{t,m} \times \underline{h}_{t,n}^*) \cdot \underline{i}_z \mathrm{d}x\mathrm{d}y = 0 \qquad \text{for} \quad n \neq m \tag{6.2}$$

where the surface integral is carried out over the entire transverse cross-section with integration limits extending to $\pm\infty$. $\underline{e}_{t,m}$ is the transverse component of $\underline{e}_m$. Guided-wave modes plus all the radiation modes constitute a complete orthogonal set of modes so that any field can be represented as a superposition of the modes. Moreover, the channel guided-wave modes are normalized, i.e.:

$$\frac{1}{2}\mathrm{Re}\left[\iint_S (\underline{e}_{t,n} \times \underline{h}_{t,n}^*) \cdot \underline{i}_z \mathrm{d}S\right] = 1. \tag{6.3}$$

For planar guided-wave modes, the modes are also orthogonal and normalized in x variation, as shown in Eq. (5.6) of Section 5.3.1 and Eq. (5.33) of Section 5.4.1. However, integration with respect to the y coordinate is absent. The normalization means that the power carried by the mth normalized planar guided-wave mode is 1 W per unit distance (i.e. meter) in the y direction.

6.2 Perturbation analysis

It is difficult to calculate the modes of a complex waveguide structure. However, when an original waveguide is perturbed by another object nearby, the perturbation analysis allows us to calculate approximately the change in $\underline{E}$ and $\underline{H}$ of the guided-wave modes in the original waveguide without solving Maxwell's equations for the total waveguide structure. Perturbation analysis is applicable as long as the perturbing object is either small or at a position reasonably far away from the waveguide core so that the evanescent tail of the mode for the original waveguides has decayed.

6.2.1 Derivation of perturbation analysis

Consider two waveguide structures that have the cross-sectional ε variation of the core shown in Figure 6.1(a) and (b). The original waveguide core is shown in Figure 6.1(a). The original waveguide core plus a perturbation waveguide core are shown in Figure 6.1(b). The two structures differ in the dielectric perturbation $\Delta\varepsilon$ shown in Figure 6.1(c), where $\Delta\varepsilon(x,y) = \varepsilon'(x,y) - \varepsilon(x,y)$. Let $\underline{E}$ and $\underline{H}$ be solutions of Eq. (6.1) of the previous section for the original waveguide with index profile $\varepsilon(x,y)$ shown in Figure 6.1(a). Let $\underline{E}'$ and $\underline{H}'$ be solutions of Eq. (6.1) for the waveguide structure with index profile $\varepsilon'(x,y)$ shown in Figure 6.1(b). Let us assume that $\underline{E}$, $\underline{H}$, and the guided-wave modes of the structure in Figure 6.1 (a) are already known. The guided-wave modes of the original waveguide in Figure 6.1(b) are the perturbation of the guided-wave modes of the structure in Figure 6.1(a) due to $\Delta\varepsilon$.

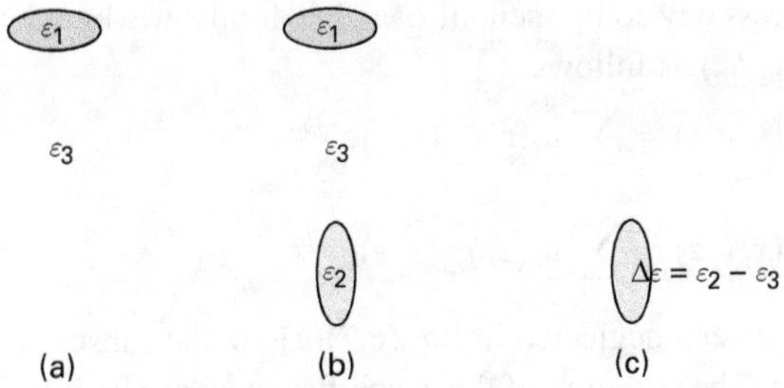

Figure 6.1 The index profile of a waveguide perturbed by $\Delta\varepsilon$. (a) The permittivity variation, $\varepsilon(x,y)$, of the original unperturbed waveguide structure. (b) The permittivity variation, $\varepsilon'(x,y)$, of the perturbed waveguide. (c) The permittivity perturbation from the additional material, $\Delta\varepsilon$, to the original waveguide structure.

Mathematically, from vector calculus and Eq. (6.1), we know,

$$\nabla \cdot [\underline{E}^* \times \underline{H}' + \underline{E}' \times \underline{H}^*] = -j\omega\, \Delta\varepsilon\, \underline{E}^* \cdot \underline{E}' \tag{6.4}$$

Let us apply volume integration to both sides of this equation over a cylindrical volume, V.

$$\iiint_V \nabla \cdot [\underline{E}^* \times \underline{H}' + \underline{E}' \times \underline{H}^*]\mathrm{d}x\mathrm{d}y\mathrm{d}z = -j\omega \iiint_V \Delta\varepsilon \underline{E}' \cdot \underline{E}^* \mathrm{d}x\mathrm{d}y\mathrm{d}z \tag{6.5}$$

The cylinder has flat circular ends parallel to the x–y plane. It has an infinitely large radius for the circular ends and a short length dz along the z-axis. According to advanced calculus, the volume integration on the left-hand side of this equation can be replaced by the surface integration of $[\underline{E}^* \times \underline{H}' + \underline{E}' \times \underline{H}^*]$ on the cylinder. The contribution of the surface integration over the cylindrical surface is zero because the guided-wave fields $\underline{E}$ and $\underline{E}'$ have already decayed to zero at the surface. For a sufficiently small dz, $\underline{E}^* \cdot \underline{E}'$ is approximately a constant from z to z + dz. Therefore, we obtain:

$$\iint_S \{[\underline{E}^* \times \underline{H}' + \underline{E}' \times \underline{H}^*]|_{z+\mathrm{d}z} - [\underline{E}^* \times \underline{H}' + \underline{E}' \times \underline{H}^*]|_z\} \cdot \underline{i}_z \mathrm{d}S$$

$$= -j\omega \left[\iint_S \Delta\varepsilon\, \underline{E}' \cdot \underline{E}^* \mathrm{d}S \right] \mathrm{d}z$$

Here S is the flat end surface of the cylinder oriented in the $+z$ direction. In other words,

$$\iint_S \frac{\partial}{\partial z} [\underline{E}_t{}^* \times \underline{H}_t{}' + \underline{E}_t{}' \times \underline{H}_t{}^*] \cdot \underline{i}_z \mathrm{d}S = -j\omega \iint_S \Delta\varepsilon(x,y) \underline{E}' \cdot \underline{E}^* \mathrm{d}S \tag{6.6}$$

Mathematically, $\underline{E}'$ and $\underline{H}'$ can be represented by summation of any set of modes. They can be either the modes of the structure shown in Figure 6.1(a) or the modes of the structure shown in Figure 6.1(b). Both sets of modes, $(\underline{e}_{t,j}, \underline{h}_{t,j})$ and $(\underline{e}_{tk}', \underline{h}_{tk}')$, form a complete orthogonal set. From the perturbation analysis point of view, we are not interested in the exact fields or modes of the structure shown in Figure 6.1 (b). We only want to know how the fields for the waveguide in Figure 6.1(a) are affected by $\Delta\varepsilon$.

In Eq. (6.6), let us express any component of $\underline{E}'$ and $\underline{H}'$ for the waveguide with core ε_1 in terms of the modes ($\underline{e}_{ti}$, $\underline{h}_{ti}$) as follows:

$$\underline{E}_t'(x,y,z) = \sum_j a_j(z)\underline{e}_{t,j}(x,y)\mathrm{e}^{-j\beta_j z},$$

$$\underline{H}_t'(x,y,z) = \sum_j a_j(z)\underline{h}_{t,j}(x,y)\mathrm{e}^{-j\beta_j z} \tag{6.7}$$

The radiation modes have been neglected in Eq. (6.7). Here, the subscript t designates the transverse component. The variation of the a_j coefficient shows how $\underline{E}_t'$ and $\underline{H}_t'$ vary as a function of z. Substituting Eq. (6.7) into Eq. (6.6), letting $\underline{E}_t = \underline{e}_{t,n}$ and $\underline{H}_t = \underline{h}_{t,n}$, and utilizing the orthogonality and normalization relations,[1] we obtain:

$$\frac{\mathrm{d}a_n}{\mathrm{d}z} = -j\sum_m a_m C_{m,n}\mathrm{e}^{+j(\beta_n-\beta_m)z}$$

$$C_{m,n} = \frac{\omega}{4}\iint_S \Delta\varepsilon\,(\underline{e}_m\cdot\underline{e}_n{}^*)\mathrm{d}S \tag{6.8}$$

This is the basic result of the perturbation analysis [3]. It tells us how to find the a_j coefficients. Once we know the a_j coefficients, we know $\underline{E}_t'(x,y,z)$ and $\underline{H}_t'(x,y,z)$ for the waveguide with core ε_1, perturbed by $\Delta\varepsilon$. Please note that the results shown in Eq. (6.8) do not tell us about the field around the waveguide with core ε_2 in Figure 6.1(b). It only shows us how to evaluate the change in the fields of the waveguide with core ε_1 as a function of z in terms of its own modes.

6.2.2 A simple application of perturbation analysis: perturbation by a nearby dielectric

In order to demonstrate the power of the results shown in Eq. (6.8), let us find the change in the propagation constant β_0 of a forward propagating guided-wave mode caused by the addition of another dielectric material with index ε' in the vicinity of the original waveguide.

Let the original waveguide be located at $x = 0$ and $y = 0$. The waveguide is surrounded by medium with permittivity ε_1. The dielectric material with ε' is located at $\infty > x \geq L$ and $\infty > y > -\infty$. L is located reasonably far away from the waveguide. Let us apply this $\Delta\varepsilon$ to Eq. (6.8). If the original waveguide has only a single mode, $\underline{e}_0$, then we do not need to carry out the summation in Eq. (6.8). We obtain,

$$\frac{\mathrm{d}\,a_0}{\mathrm{d}\,z} = -ja_0\left[\frac{\omega}{4}\int_L^\infty\int_{-\infty}^\infty(\varepsilon'-\varepsilon_1)\underline{e}_0\cdot\underline{e}_0{}^*\mathrm{d}x\mathrm{d}y\right] = -j\Delta\beta\,a_0$$

or

$$a_0 = A\,\mathrm{e}^{-j\Delta\beta z},\quad \Delta\beta = \frac{\omega}{4}(\varepsilon'-\varepsilon)\int_t^{+\infty}\int_{-\infty}^{+\infty}\underline{e}_0\cdot\underline{e}_0{}^*\mathrm{d}x\mathrm{d}y.$$

$$\underline{E}_t' = A\underline{e}_0(x,y)\mathrm{e}^{j(\beta+\Delta\beta)z} \tag{6.9}$$

[1] The orthogonality relation has been proven only for modes in lossless waveguides. However, the modes are often considered orthogonal in the literature, even when the modes are lossy.

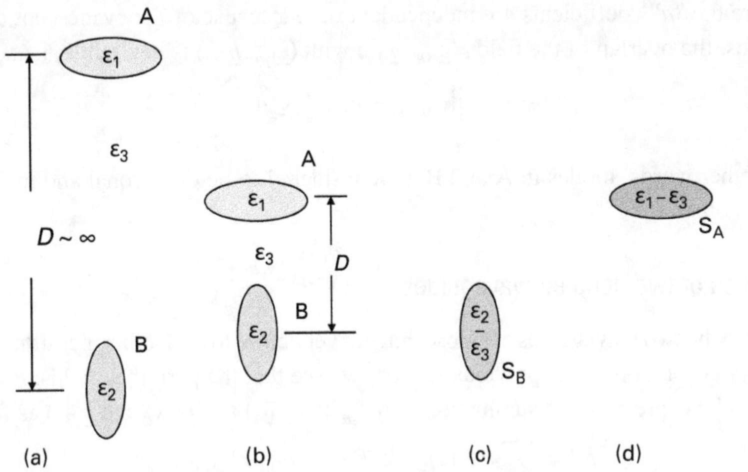

Figure 6.2 Mutual perturbation of two waveguides. (a) The permittivity profile of two well-separated waveguides. (b) The permittivity of two waveguides, A and B, with core dielectric constants ε_1 and ε_2 separated by a modest distance D. (c) The perturbation of ε_3 by ε_2 of waveguide B for modes in waveguide A. (d) The perturbation of ε_3 by ε_1 of waveguide A for the modes in waveguide B.

Clearly β_0 for the guided mode $\underline{e}_0$ is changed by $\Delta\beta$. Notice that the perturbation analysis does not address the field distribution in the region $x > L$. The perturbation analysis allows us to calculate $\Delta\beta$ for the original waveguide mode without solving the differential equation of the waveguide with perturbation.

6.3 Coupled mode analysis

When there are two waveguides located close to each other within the evanescent tail distance, coupled mode analysis allows us to calculate the amplitudes and effective index of the modes in both waveguides due to the mutual coupling. Since it is a perturbation analysis, it is good when the coupling among the waveguides is moderate.

6.3.1 Modes of two uncoupled parallel waveguides

Consider the two waveguides shown in Figure 6.2(b). Let the distance of separation D between the two waveguides, A and B, be very large at first in Figure 6.2(a). In that case, the modes of A and B will not be affected by each other. The modes of the total structure, $\underline{e}_{tn}$ and $\underline{h}_{tn}$, are just the modes of individual waveguides, ($\underline{e}_{An}$, $\underline{h}_{An}$) and ($\underline{e}_{Bn}$, $\underline{h}_{Bn}$), or a linear combination of them. The fields of the total structure can be expressed as the summation of all the modes of the waveguides A and B.

$$\underline{E} = \sum_n a_{An}\underline{e}_{An}e^{-j\beta_{An}z} + a_{Bn}\underline{e}_{Bn}e^{-j\beta_{Bn}z}$$

$$\underline{H} = \sum_n a_{An}\underline{h}_{An}e^{-j\beta_{An}z} + a_{Bn}\underline{h}_{Bn}e^{-j\beta_{Bn}z} \tag{6.10}$$

Here the "a" coefficients are independent of z. Because of the evanescent decay of the fields, the overlap of the fields ($\underline{e}_{An}$, $\underline{h}_{An}$) with ($\underline{e}_{Bn}$, $\underline{h}_{Bn}$) is negligible, i.e.:

$$\iint_S (\underline{e}_{t,An} \times \underline{h}_{t,Bm}{}^*) \cdot \underline{i}_z \mathrm{d}S = 0 \tag{6.11}$$

In other words, modes in A and B are considered to be orthogonal and independent of each other.

6.3.2 Modes of two coupled waveguides

When the two waveguides are close, but not very close to each other, perturbation analysis is applicable. From the previous section, we see that the perturbed fields, $\underline{E}'$ and $\underline{H}'$, can again be expressed as a summation of ($\underline{e}_{An}$ and $\underline{e}_{Bn}$) and ($\underline{h}_{An}$ and $\underline{h}_{Bn}$) as follows:

$$\begin{aligned} \underline{E}' &= \sum_n a_{An}(z)\underline{e}_{An} \mathrm{e}^{-j\beta_{An}z} + a_{Bn}(z)\underline{e}_{Bn} \mathrm{e}^{-j\beta_{Bn}z} \\ \underline{H}' &= \sum_n a_{An}(z)\underline{h}_{An} \mathrm{e}^{-j\beta_{An}z} + a_{Bn}(z)\underline{h}_{Bn} \mathrm{e}^{-j\beta_{Bn}z} \end{aligned} \tag{6.12}$$

Here the "a" coefficients are functions of z. However, the effect of the perturbation created by the finite separation distance D will be different for A modes and for B modes, as shown below.

Let the two waveguides, A and B, be separated by a finite distance D, as shown in Figure 6.2(b). For modes of waveguide A, the significant change from waveguide A shown in Figure 6.2(a) is the increase of permittivity from ε_3 to ε_2 at the position of the B waveguide. This perturbation is shown in Figure 6.2(c). For modes of waveguide B, the perturbation of the waveguide B is the increase of permittivity from ε_3 to ε_1 at the position of waveguide A. This perturbation is shown in Figure 6.2(d). Applying the result in Eq. (6.8) to waveguide A and B separately, we obtain:

$$\frac{\mathrm{d}a_{An}}{\mathrm{d}z} = -j\left[C_{An,An}a_{An} + \sum_m C_{Bm,An}\mathrm{e}^{j(\beta_{An}-\beta_{Bm})z}a_{Bm}\right]$$

$$\frac{\mathrm{d}a_{Bn}}{\mathrm{d}z} = -j\left[C_{Bn,Bn}a_{Bn} + \sum_m C_{Am,Bn}\mathrm{e}^{j(\beta_{Bn}-\beta_{Am})z}a_{Am}\right]$$

where

$$\begin{aligned} C_{An,An} &= \frac{\omega}{4}\iint_{S_B}(\varepsilon_2 - \varepsilon_3)[\underline{e}_{An} \cdot \underline{e}_{An}{}^*]\mathrm{d}S \\ C_{Bm,An} &= \frac{\omega}{4}\iint_{S_B}(\varepsilon_2 - \varepsilon_3)[\underline{e}_{Bm} \cdot \underline{e}_{An}{}^*]\mathrm{d}S \\ C_{Bn,Bn} &= \frac{\omega}{4}\iint_{S_A}(\varepsilon_1 - \varepsilon_3)[\underline{e}_{Bn} \cdot \underline{e}_{Bn}{}^*]\mathrm{d}S \\ C_{Am,Bn} &= \frac{\omega}{4}\iint_{S_A}(\varepsilon_1 - \varepsilon_3)[\underline{e}_{Am} \cdot \underline{e}_{Bn}{}^*]\mathrm{d}S \end{aligned} \tag{6.13}$$

Eq. (6.13) is the well-known coupled-mode equation [1]. It is used extensively to analyze waveguide devices.

The physical meaning of the results expressed in Eq. (6.13) is that the amplitude of mode propagation in A is affected by the presence of B and its modes. Conversely, the amplitudes of modes propagating in B are affected by the presence of A and its modes. $C_{An,An}$ *(or* $C_{Bn,Bn}$*) represents the effect on the propagation wave number (or effective index) of the* $\underline{e}_{An}$ *(or* $\underline{e}_{Bn}$*) mode due to the presence of the second waveguide.* $C_{Bm,An}$ *represents the effect on the amplitude* a_{An} *of A by the mode* a_{Bn} *excited in the second waveguide (or* a_{Bn} *by* a_{Am} *of B by* $C_{Am,Bn}$*).*

The change in amplitudes a_{An} *and* a_{Bn} *and propagation wave numbers is obtained without knowing the modes of the total composite waveguide A and B together. The C coefficients are a measure of the coupling strength among modes.*

There are number of ways in which Eq. (6.13) may be simplified: (1) If one is only interested in the mutual interaction among the modes excited in the waveguides, the effect of the change in phase velocity (i.e. the effective index) may not be important. Then $C_{An,An}$ or $C_{Bn,Bn}$ may not need to be calculated. (2) The example given in Section 6.2.2 illustrates the case when $C_{An,An}$ cannot be neglected because a_{Bm} is zero. Here, $C_{An,An}$ (or $C_{Bn,Bn}$) is used to calculate the change of the propagation wave number (i.e. effective index) of the modes just due to the presence of the second waveguide. (3) When there are higher-order modes in waveguides A and B, there should also be more terms, such as $C_{An,Aj}$, $C_{Bn,Bj}$, $C_{Bj,An}$, and $C_{Aj,Bn}$, in a more precise analysis. However, these C coefficients are even smaller than $C_{An,An}$, $C_{Bn,Bn}$, $C_{Bm,An}$, and $C_{Am,Bn}$, because of the orthogonality properties and the faster evanescent decay of the higher-order modes. Therefore, those terms have not been included in Eq. (6.13).

6.3.3 An example of coupled mode analysis: the grating reflection filter

Grating filters are very important devices in wavelength division multiplexed (WDM) optical fiber communication networks. In such networks, signals are transmitted via optical carriers that have slightly different wavelengths. The purpose of a filter is to select an optical carrier at a specific wavelength (or a group of optical carrier wavelengths within a specific band of wavelengths) to direct it (or them) to a different direction of propagation (e.g. reflection) [4]. The desired characteristics of a grating filter are: (1) High and uniform reflection of incident waves in a waveguide within the selected wavelength band. (2) Sharp reduction of reflectivity immediately outside the band. (3) High contrast ratio of the intensity of reflected optical carriers inside and outside the band. In distributed feedback lasers, gratings are fabricated on channel waveguides so that the forward and backward waves will be coupled and reflected to form a resonator.

A grating reflection filter utilizes a perturbation of the waveguide by a periodic $\Delta\varepsilon$. The $\Delta\varepsilon$ couples the forward-propagating guided-wave mode to the reflected guided-wave mode. Let us consider a grating layer which has a cosine variation of the dielectric constant along the z direction, i.e. $\Delta\varepsilon(z)$, within a thickness d in the x direction and width W in the y direction. It is placed on top of a channel waveguide that has thickness, t.

Alternatively, a periodic variation $\Delta\varepsilon(z)$ can also be obtained in the cladding of an optical fiber by photo-refractive methods.

Let us assume that the waveguide has only a single mode. Mathematically, the waveguide is perturbed by $\Delta\varepsilon$ where

$$\Delta\varepsilon(z) = \Delta\varepsilon_0 \cos(Kz)\,\mathrm{rect}\left(\frac{2(x-H)}{d}\right)\mathrm{rect}\left(\frac{2y}{W}\right) \tag{6.14}$$

It has a periodicity $T = 2\pi/K$ in the z direction and a maximum change of dielectric constant $\Delta\varepsilon_0$. The $\Delta\varepsilon$ perturbation layer is centered at $x = H$ in the cladding, where $H \geq t + (d/2)$. This is a change in the cladding refractive index n_c of the waveguide or fiber. The above mathematical expression for $\Delta\varepsilon$ is a simplification from a realistic grating. For example, an ideal etched grating may have $\Delta\varepsilon$ described by a rectangular function of both x and z.[2]

Let the complex amplitude of the forward-propagating guided-wave mode be a_f and the amplitude of the backward-propagating mode at the same wavelength a_b. Then the application of Eq. (6.13) to the fields in the waveguide within the grating section from $z = 0$ to $z = L$ yields:

$$\underline{E}_t'(x,y,z) = [a_f(z)\mathrm{e}^{-j\beta_0 z} + a_b(z)\mathrm{e}^{+j\beta_0 z}]\underline{e}_{t,0}(x,y)$$

$$\frac{\mathrm{d}a_f}{\mathrm{d}z} = -jC_{ff}a_f - jC_{bf}a_b\mathrm{e}^{-j2\beta_0 z}$$

$$\frac{\mathrm{d}a_b}{\mathrm{d}z} = -jC_{bb}a_b - jC_{fb}a_f\mathrm{e}^{j2\beta_0 z} \tag{6.15}$$

$$C_{ff} = -C_{bb} = -C_{fb} = C_{bf} = \frac{\omega}{4}\left[\int_{H-\frac{d}{2}}^{H+\frac{d}{2}}\int_{-\frac{W}{2}}^{\frac{W}{2}} \Delta\varepsilon_0 |\underline{e}_0 \cdot \underline{e}_0{}^*|\mathrm{d}x\mathrm{d}y\right]\left[\frac{1}{2}(\mathrm{e}^{jKz} + \mathrm{e}^{-jKz})\right]$$

There are minus signs on C_{bb} and C_{fb}, because, in the normalization of the modes shown in Eq. (6.2), $\underline{i}_z$ is pointed toward the $+z$ direction. $\underline{i}_z$ for the backward wave is pointing toward the $-z$ direction. β_0 is the propagation wave number of the incident mode, $\beta_0 = n_{eff,m}k$.

Clearly a_f and a_b will only affect each other significantly along the z direction when the driving terms on the right-hand side of Eq. (6.15) have a slow z variation. Since the perturbation has a coz(Kz) variation, the maximum coupling between a_f and a_b will take place when $K = 2\beta_0$. This is known as the phase matching (or the Bragg) condition of the forward- and backward-propagating waves. When the Bragg condition is satisfied, the

[2] The mathematical expression of $\Delta\varepsilon$ for rectangular grating grooves has been given in Eqs. (1.103) to (1.105). Any periodic $\Delta\varepsilon$ can be expressed as a summation of Fourier components of sinusoidal gratings. Thus the $\Delta\varepsilon$ in Eq. (6.14) is just the fundamental component of all the Fourier components.

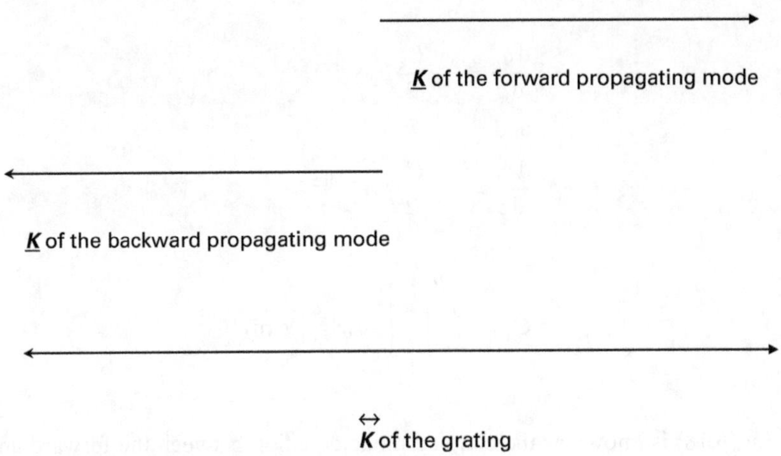

Figure 6.3 Propagation wave vectors for forward and backward waves and the grating. The propagation wave vectors of the forward and backward guided waves are shown as $\underline{K}$ vectors in the $+z$ and $-z$ directions. The $\overleftrightarrow{K}$ of the grating is shown as a bi-directional vector. Phase matching is achieved when the magnitude of $|\overleftrightarrow{K}|$ is the sum of all the $|\underline{K}|$.

relationship between β and K is illustrated in Figure 6.3, where β_0 of the forward- and backward-propagating modes with $e^{\pm j\beta_0 z}$ variations are represented by vectors with magnitude β_0 in the $\pm z$ directions. Since a cosine function is the sum of two exponential functions, $\underline{K}$ is represented as a bi-directional vector of magnitude K.

If we designate λ_o as the free-space wavelength in which the maximum coupling takes place, then the phase-matching condition is satisfied when K is given by:

$$K = \frac{4\pi n_{eff}}{\lambda_o}. \tag{6.16}$$

Here n_{eff} is the effective index of the guided-wave mode. When $K \cong 2\beta_0$, the terms involving C_{ff} and C_{bb} in Eq. (6.16) will have negligible effect on the magnitude of $\mathrm{d}a_f/\mathrm{d}z$ and $\mathrm{d}a_b/\mathrm{d}z$ because of its sinusoidal variation in z. In comparison, the terms involving C_{fb} and C_{bf} will have the dominant effect on da_f/dz and da_b/dz because the z variation in the $e^{\pm j2\beta_o z}$ term is cancelled by one of the $e^{\pm jKz}$ terms.

Since β is inversely proportional to λ, Eq. (6.16) will not be satisfied simultaneously for all β within the desired wavelength band. In order to analyze the grating properties as a function of wavelength for a given K, we need to consider the solution of Eq. (6.16) under approximate phase-matching conditions. Let

$$2\beta_0 - K = \delta_K \tag{6.17}$$

Under this condition, we obtain from Eq. (6.16),

$$\frac{\mathrm{d}a_f}{\mathrm{d}z} = -j\frac{C_g}{2}a_b e^{j\delta_K z}$$

and

$$\frac{\mathrm{d}a_b}{\mathrm{d}z} = +j\frac{C_g}{2}a_f e^{-j\delta_K z}$$

where

$$C_g = \frac{\omega}{4}\int_{H-\frac{d}{2}}^{H+\frac{d}{2}}\int_{-\frac{W}{2}}^{\frac{W}{2}} \Delta\varepsilon_0 \left|\underline{e}_o\right|^2 \mathrm{d}x\mathrm{d}y \tag{6.18}$$

Eq. (6.18) is known as the coupled mode equation between the forward and the backward propagating modes. We know the solutions for such a differential equation are the familiar exponential functions, $e^{\gamma^+ z}$and $e^{\gamma^- z}$. Specifically, the solutions of Eq. (6.18) for the forward- and backward-propagating waves are:

$$\begin{aligned} a_b(z) &= A_1 e^{\gamma^+ z} + A_2 e^{\gamma^- z} \\ a_f(z) &= -j\frac{2}{C_g}\left[A_1\gamma^+ e^{-\gamma^- z} + A_2\gamma^- e^{-\gamma^+ z}\right] \\ \gamma^+ &= -j\frac{\delta_K}{2} + Q,\ \ \gamma^- = -j\frac{\delta_K}{2} - Q \\ Q &= \sqrt{\left(\frac{C_g}{2}\right)^2 - \left(\frac{\delta_K}{2}\right)^2} \end{aligned} \tag{6.19}$$

The A_1 and A_2 coefficients will be determined from initial conditions at $z = 0$ and $z = L$.

For a grating that begins at $z = 0$ and terminates at $z = L$, the amplitude of $a_f(z = 0)$ is the same as the incident wave. The reflected wave in the input waveguide is equal to $a_b(z = 0)$. The amplitude of the output wave at $z > L$ is equal to $a_f(z = L)$, and a_b must be zero at $z \geq L$. Thus:

$$\begin{aligned} A_2 &= -A_1 e^{2QL} \\ a_b &= -A_1 2e^{QL-j\left(\frac{\delta_K}{2}z\right)} \sinh[Q(L-z)] \\ a_f &= -jA_1\frac{4}{C_g}e^{QL+j\left(\frac{\delta_K}{2}z\right)}\left[j\frac{\delta_K}{2}\sinh\left(Q(L-z)\right) + Q\cosh\left(Q(L-z)\right)\right] \end{aligned} \tag{6.20}$$

At $z = 0$, the ratio of the reflected power to the incident power (for $\delta_k < C_g$) is:

$$\frac{|a_b(z=0)|^2}{|a_f(z=0)|^2} = \frac{\left(\frac{C_g}{2}\right)^2 \sin\text{h}^2 QL}{Q^2 \cos\text{h}^2 QL + (\delta_K/2)^2 \sin\text{h}^2 QL} \tag{6.21}$$

At $z = L$, the ratio of the transmitted power in the forward propagating mode to the incident power of the forward mode at $z = 0$ is:

$$\frac{|a_f(z=L)|^2}{|a_f(z=0)|^2} = \frac{Q^2}{Q^2 \cos\text{h}^2 QL + (\delta_k/2)^2 \sin\text{h}^2 QL} \tag{6.22}$$

Since $|a_f(z=L)|^2 + |a_b(z=0)|^2 = |a_f(z=0)|^2$, the conservation of power of the incident, transmitted, and reflected waves is verified.

For a reflection filter, we want $|a_b(z=0)/a_f(z=0)|^2$ large within a desired band of wavelength, and small outside this band. We also like to control the width of the desired wavelength band. These features can be controlled by L, δ_K, and C_g, of the grating. Notice that $|a_b\,(z=0)|$ is larger for larger L and smaller δ_K/C_g. At $\lambda = \lambda_g$, δ_K is 0, and the grating reflection is a maximum. The maximum possible value of $|a_b(z=0)/a_f(z=0)|^2$ is 1. At $\delta_K = C_g$, Q is 0. When $\delta_k > C_g$, Q becomes imaginary. When Q is imaginary, $\sin hQL = j\sin|Q|L$ and $\cos hQL = \cos|Q|L$. So $|a_b(z=0)/a_f(z=0)|^2$ becomes oscillatory and decreases in peak values as δ_k is increased. Let $\Delta\lambda_g$ be the wavelength deviation form λ_g such that, when $\lambda = \lambda_g \pm \Delta\lambda_g$, $Q = 0$. Then $2\Delta\lambda_g$ is the pass band of the filter,

$$\Delta\lambda_g = \pm\frac{4\pi\, C_g\, n_{eff}}{K^2} \tag{6.23}$$

In summary, K of the grating can be used to control the center wavelength λ_g at which the transmission of the forward-propagating wave is blocked. C_g of the grating is used to control the wavelength width $\Delta\lambda_g$ within which effective reflection occurs. The smaller the C_g, the narrower the range of transmission wavelengths. For a given transmission range, L is used to control the magnitudes of the reflected and transmitted waves. These are useful parameters for designing grating reflection filters. Since the maximum reflection takes place at $K = 4\pi n_{eff}/\lambda_o$, which is known in the literature as the Bragg condition, such a reflector is also called a distributed Bragg reflector, DBR. If the $\Delta\varepsilon$ variation of the grating groves is not sinusoidal, as shown in Eq. (6.14), *any periodic $\Delta\varepsilon$ can be written as a summation of Fourier components. The higher-order Fourier components will provide phase matching at $mK = 4\pi n_{eff}/\lambda_o$, $m = 2, 3, 4, \ldots$*

Therefore higher-order Bragg reflection can take place at λ_o/m. However, the higher the order, the weaker the diffraction, because the Fourier component is usually smaller.[3]

[3] Sometimes the shape of the grating grooves is blazed to enhance diffraction of a specific order by increasing the Fourier component.

6.3.4 Another example of coupled mode analysis: the directional coupler

A directional coupler has an interaction region between two parallel waveguides (or fibers). A prescribed fraction of power in waveguide A is transferred into waveguide B within the interaction region and vice versa.

A top view of a channel waveguide directional coupler is illustrated in Figure 6.4(a). Two well-separated waveguides are brought together into the interaction region by transition waveguides. After the interaction region, the two waveguides are separated again via transition waveguides. Within the interaction region, the waveguides are separated from each other by a distance D, which is usually of the order of the evanescent decay length. Directional couplers could

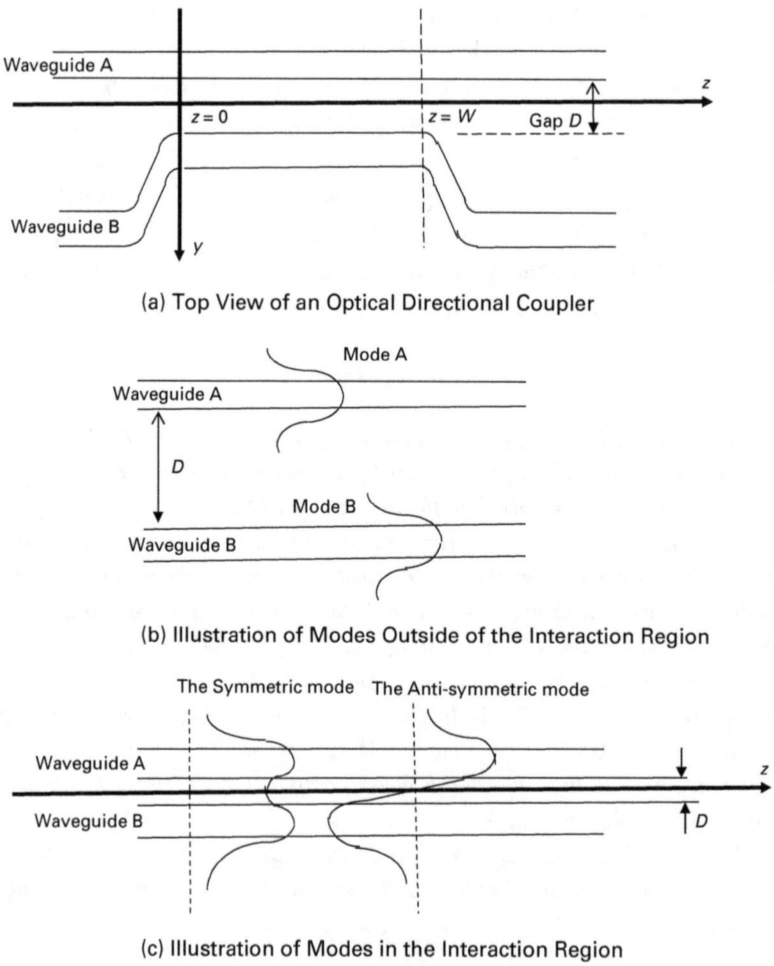

Figure 6.4 Top view of a directional coupler and illustration of its modes. (a) Top view of a channel waveguide directional coupler. The interaction region begins at $z = 0$ and ends at $z = W$. (b) Illustration of modes outside of the interaction region where the waveguides are well separated. (c) Illustration of modes in the interaction region.

also be obtained using two optical fibers or planar waveguides, provided there is a similar coupling region.

Let the length of the interaction section be W. Clearly, Eq. (6.13) is directly applicable to the modes of the individual waveguides in the interaction region. Outside the interaction region, the waveguides are well separated from each other without any further interaction.

Let $\underline{e}_A$ and $\underline{e}_B$ be the modes of the two waveguides (or fibers) that are interacting with each other through their evanescent field in the interaction section. They have complex amplitudes, a_A and a_B. Let the two waveguides have cores with cross-sections, S_A and S_B, and dielectric constants, ε_A and ε_B. The cores are surrounded by a medium that has dielectric constant ε_3. Let the coupling region begin at $z = 0$ and ends at $z = W$ as shown in Figure 6.4(a). For mathematical convenience, the coupling is assumed to be uniform within this distance. Application of Eq. (6.13) yields

$$\begin{aligned}
\frac{da_A}{dz} &= -jC_{BA}e^{j\Delta\beta z}a_B(z) \\
\frac{da_B}{dz} &= -jC_{AB}e^{-j\Delta\beta z}a_A(z) \\
C_{AB} &= \frac{\omega}{4}\iint_{S_A}(\varepsilon_A - \varepsilon_3)[\underline{e}_A \cdot \underline{e}_B{}^*]dS \\
C_{BA} &= \frac{\omega}{4}\iint_{S_B}(\varepsilon_B - \varepsilon_3)[\underline{e}_B \cdot \underline{e}_A{}^*]dS \\
\Delta\beta &= \beta_A - \beta_B
\end{aligned} \tag{6.24}$$

Here, C_{AA} and C_{BB} have been neglected because we are only interested in the change in the amplitude of a_A and a_B produced by C_{AB} and C_{BA}. Solution of a_A and a_B will depend again on initial conditions.

Let the initial condition be $a_A = \mathrm{A}$ and $a_B = 0$ at $z = 0$. Then, we obtain

$$\begin{aligned}
a_A &= Ae^{j\frac{\Delta\beta}{2}z}\left[\cos\left(\sqrt{C_{BA}C_{AB} + \left(\frac{\Delta\beta}{2}\right)^2}\,z\right) - j\frac{\left(\frac{\Delta\beta}{2}\right)}{\sqrt{C_{BA}C_{AB} + \left(\frac{\Delta\beta}{2}\right)^2}}\sin\left(\sqrt{C_{BA}C_{AB} + \left(\frac{\Delta\beta}{2}\right)^2}\,z\right)\right] \\
a_B &= \frac{-jC_{AB}A}{\sqrt{C_{BA}C_{AB} + \left(\frac{\Delta\beta}{2}\right)^2}}e^{-j\frac{\Delta\beta}{2}z}\sin\left(\sqrt{C_{BA}C_{AB} + \left(\frac{\Delta\beta}{2}\right)^2}\,z\right)
\end{aligned}$$

for $0 \le z \le W$

$$\tag{6.25}$$

Similarly, if the boundary condition is $a_B = B$ and $a_A = 0$ at $z = 0$, we obtain:

$$a_A = \frac{-jC_{BA}B}{\sqrt{C_{BA}C_{AB} + \left(\frac{\Delta\beta}{2}\right)^2}} e^{+j\frac{\Delta\beta}{2}z} \sin\left(\sqrt{C_{BA}C_{AB} + \left(\frac{\Delta\beta}{2}\right)^2}\, z\right)$$

$$a_B = Be^{-j\frac{\Delta\beta}{2}z}\left[\cos\left(\sqrt{C_{BA}C_{AB} + \left(\frac{\Delta\beta}{2}\right)^2}\, z\right) + j\frac{\left(\frac{\Delta\beta}{2}\right)}{\sqrt{C_{BA}C_{AB} + \left(\frac{\Delta\beta}{2}\right)^2}} \sin\left(\sqrt{C_{BA}C_{AB} + \left(\frac{\Delta\beta}{2}\right)^2}\, z\right)\right]$$

$$\text{for} \quad 0 \leq z \leq W \tag{6.26}$$

At $z = W$, the power transmitted from one waveguide to another and the power remained in the original waveguide are obtained from a_B and a_A. Note that, unless $\Delta\beta = 0$, there cannot be full transfer of power from A to B. Substantial transfer of power from A to B (or vice versa) at $z = W$ can take place only when $\Delta\beta$ is small. Thus $\beta_A = \beta_B$ is the phase-matching condition for maximum transfer of power. The C coefficients, W, and $\Delta\beta$, control the power transfer from A to B and from B to A. When W is increased beyond the length for maximum transfer of power, then a_A and a_B will exhibit oscillatory variation as W is increased.

Usually, the directional coupler has two identical waveguides. In that case, $C_{BA} = C_{AB} = C$, and the ratio of $|a_B|^2/|a_A|^2$ is the power distribution among the two waveguides. At $z = 0$, let there be an input power I_{in} in waveguide A, no input power in waveguide B. Then the output power I_{out} in waveguide B after an interaction distance W is given directly by Eq. (6.25). It is:

$$I_{out}/I_{in} = \frac{1}{C^2 + \left(\frac{\Delta\beta}{2}\right)^2} \sin^2\left(\sqrt{C^2 + \left(\frac{\Delta\beta}{2}\right)^2}\, W\right) \tag{6.27}$$

Clearly the ratio I_{out}/I_{in} can be controlled by W, C, and $\Delta\beta$. C is determined by the separation distance D and the evanescent decay of the modes involved. The maximum I_{out} is obtained whenever $\sqrt{C^2 + (\Delta\beta/2)^2}\, W = \pi/2, 3\pi/2, 5\pi/2, \ldots$. If for any reason W becomes too long or too short, I_{out} will oscillate between zero and the maximum. It means also that, for a given W, the maximum I_{out} is obtained only at specific wavelengths. The value of W that equals $\pi/2C$ is known as the coupling length of the directional coupler. The bandwidth of I_{out} is determined by $\Delta\beta$ within the wavelength *range between the wavelengths at which* $\sqrt{C^2 + (\Delta\beta/2)^2}W = \pi, 2\pi, 3\pi$ *(i.e.* $I_{out} = 0$*) and the wavelengths at which* I_{out} *is maximum.*

6.4 Super mode analysis

The operation of a number of devices such as directional and grating couplers was analyzed in the previous sections by perturbation and coupled mode analysis. However, the perturbation approach breaks down when the interaction is too strong. The alternative analysis of the operation of such a device is the super mode analysis of the total waveguide structure.

What is a super mode? For infinitely long parallel waveguides with uniform cross-section and distance of separation, the modes of the total structure are the super modes. These super modes can be calculated only in limited cases. Quite often, we cannot calculate the mode profiles and effective index of the super modes because the total waveguide configuration is too complex. However, even in that case, we can still make important conclusions about the properties of the device without knowing the effective index and mode profile of the modes.

In short, super mode analysis is an analysis of waveguide devices based on the interference of the modes of the total structure. It is different from coupled mode analysis because it does not assume that the modes of the individual waveguides are just perturbed by its neighbor. Therefore the super mode analysis is accurate when the separation between waveguides is very small, or even zero. Viewing the devices from the super mode point of view also sheds a different light on their operation than the coupled mode analysis.

In the following sections, we will present first how to find the super modes of two coupled waveguides. This will be compared with the modes obtained from the coupled mode analysis so the differences and the similarities of the two approaches can be clearly demonstrated. After that, three sample devices, the directional coupler, the Y-branch coupler, and the Mach–Zehnder interferometer, are presented to demonstrate the super mode analysis. Among these examples, the directional coupler has already been presented in Section 6.3.4 using the coupled mode analysis. Thus we can also compare the results of two different approaches. The Y-branch coupler is an example that cannot be analyzed by coupled mode analysis. The Mach–Zehnder interferometer is an example in which the simplicity of super mode analysis is clearly demonstrated.

6.5 Super modes of two parallel waveguides

In order to understand the inter-relation between the super modes and individual modes clearly, we first present a general discussion of super modes in the following two subsections.

6.5.1 Super modes of two well-separated waveguides

Consider the two waveguides shown in Figure 6.4(a) outside of the interaction region. The distance of separation D between the two waveguides, A and B, is very large. In this case, the fields of the total structure can be expressed as the summation of all the modes of the waveguides A and B.

$$\underline{E} = \sum_n a_{An}\underline{e}_{An}e^{-j\beta_{An}z} + a_{Bn}\underline{e}_{Bn}e^{-j\beta_{Bn}z}$$

$$\underline{H} = \sum_n a_{An}\underline{h}_{An}e^{-j\beta_{An}z} + a_{Bn}\underline{h}_{Bn}e^{-j\beta_{Bn}z} \tag{6.28}$$

Here the "a" coefficients are independent of z. Since there is evanescent decay of the fields, the overlap of the fields ($\underline{e}_{An}$, $\underline{h}_{An}$) with ($\underline{e}_{Bn}$, $\underline{h}_{Bn}$) is negligible, i.e.:

$$\iint_S (\underline{e}_{t,An} \times \underline{h}_{t,Bm}{}^*)\cdot\underline{i}_z \mathrm{d}S = 0$$

In other words, modes of A and B are considered to be orthogonal to each other. The super-modes of the total structure, ($\underline{e}_{sn}$,$\underline{h}_{sn}$) and ($\underline{e}_{an}$,$\underline{h}_{an}$) are just linear combinations of the modes of individual waveguides, ($\underline{e}_{An}$,$\underline{h}_{An}$) and ($\underline{e}_{Bn}$,$\underline{h}_{Bn}$), such that

$$\underline{e}_{sn} = \frac{1}{\sqrt{2}}(\underline{e}_{An} + \underline{e}_{Bn}) \quad \text{and} \quad \underline{e}_{an} = \frac{1}{\sqrt{2}}(\underline{e}_{An} - \underline{e}_{Bn}) \tag{6.29}$$

Note that for such uncoupled waveguides, the magnitudes of the modes, $|a_{An}|$ and $|a_{Bn}|$, do not change as the modes propagate. This is the same conclusion that we have reached in Section 6.3.1.

6.5.2 Super modes of two coupled waveguides

When the distance of separation between the two waveguides is close, as shown in Figure 6.2(b), we can use the effective index approximation or numerical methods to find the super modes.

Consider a two-channel waveguide, as depicted in Figure 6.5. It is just a waveguide A and a waveguide B coupled through a gap. Figure 6.5(a) shows the cross-sectional view in the x–y plane, while Figure 6.5(b) shows the top view in the y–z plane. In this illustration, channel A (or waveguide A) has core thickness t_A and width W_A, while channel B (or waveguide B) has core thickness t_B and width W_B. The width of the gap between two channels is G. The thickness of the waveguide core in the cladding region and in the gap is t_c. The substrate index is n_{sub}, while the index of the core of the waveguide is n_{wg}.

According to the effective index method presented in Section 5.6.1, we first find the effective indices of the planar waveguide modes for the channel A and channel B waveguides separately, as we did in Section 5.2. For simplicity, let us assume that there is only a single TE_0 mode in the x direction. Let the effective index for planar

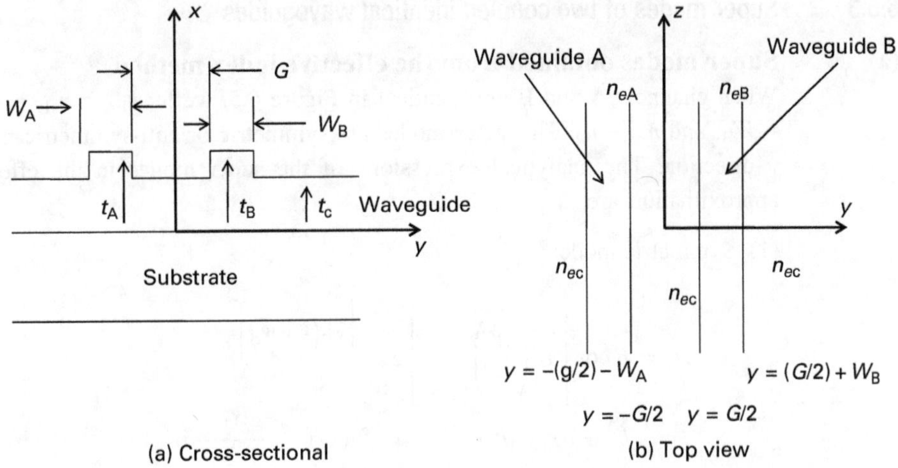

Figure 6.5 The two-channel waveguide. (a) The cross-sectional view. The two parallel ridge waveguides, with W_A and W_B wide ridges, are separated by a gap G. (b) The top view.

mode in waveguide A be n_{eA}, the effective index for the planar mode in waveguide B be n_{eB}, and the effective index for planar mode in the gap and cladding regions be n_{ec}. The lateral variation of the super mode in the y direction is then found by solving the equivalent TM planar waveguide mode for H_x:

$$\left[\frac{\partial^2}{\partial z^2}+\frac{\partial^2}{\partial y^2}+\omega^2\varepsilon(y)\mu\right]H_x(y,z)=0$$
$$\varepsilon(y)=\varepsilon_o n_{ej}^{\;2} \qquad j=A,\ B \text{ or } c \tag{6.30}$$

$$\frac{\partial}{\partial x}\equiv 0 \quad \text{for planar TM mode approximation} \tag{6.31}$$

$$E_y=\frac{j}{\omega\varepsilon(y)}\frac{\partial H_x}{\partial z} \tag{6.32}$$

$$E_z=\frac{-j}{\omega\varepsilon(y)}\frac{\partial H_x}{\partial y} \tag{6.33}$$

where the boundary conditions are the continuity of H_x and E_z at $y = \pm|G/2|$ and $y = \pm|W+(G/2)|$; $W = W_A$ or W_B.

The lowest-order super modes are illustrated in Figure 6.4(c). In the following subsection, solutions of the super modes of two identical waveguides are obtained explicitly. Super modes of two identical coupled waveguides have already been obtained by the coupled mode analysis. It is instructional to compare the two results side by side.

6.5.3 Super modes of two coupled identical waveguides

(a) Super modes obtained from the effective index method

When channels A and B are identical in Figure 6.5, we let $n_{eA} = n_{eB} = n_1$, $W = W_A = W_B$, and $n_{ec} = n_2$. The super modes are symmetric or anti-symmetric modes in the y direction. The analytical expressions of the super modes in the effective index approximation are:

(1) Symmetric mode:

$$H_x = B\cos\left[h_1\left(\frac{W}{2}\right)+\varphi\right]e^{+q_2\left[y+\left(\frac{G}{2}+W\right)\right]}$$
$$\beta^2 - {q_2}^2 = k^2{n_2}^2 \qquad \text{for} \quad y \leq -\frac{G}{2} - W \tag{6.34a}$$

$$H_x = B\cos\left[h_1\left(y+\frac{G+W}{2}\right)-\varphi\right]$$
$$\beta^2 + {h_1}^2 = k^2{n_1}^2 \qquad \text{for} \quad -\left(\frac{G}{2}+W\right) \leq y \leq -\frac{G}{2} \tag{6.34b}$$

$$H_x = B'[e^{-q_2 y} + e^{+q_2 y}]$$
$$\beta^2 - {q_2}^2 = k^2{n_2}^2 \qquad \text{for} \quad -\frac{G}{2} \leq y \leq +\frac{G}{2} \tag{6.34c}$$

$$H_x = B\cos\left[h_1\left(y-\frac{G+W}{2}\right)+\varphi\right]$$
$$\text{for} \quad \frac{G}{2} \leq y \leq \frac{G}{2} + W \tag{6.34d}$$

$$H_x = B\cos\left[h_1\frac{W}{2}+\varphi\right]e^{-q_2\left[y-\left(\frac{G}{2}+W\right)\right]}$$
$$\text{for} \quad \frac{G}{2} + W \leq y \tag{6.34e}$$

where B and B' are related by

$$B'\left[e^{-q_2\frac{G}{2}} + e^{+q_2\frac{G}{2}}\right] = B\cos\left[h_1\frac{W}{2} - \varphi\right] \tag{6.35}$$

φ, q_2, and h_1 of the symmetric mode are obtained from the following transcendental equations derived from the boundary conditions at $y = \pm |G/2|$ and at $y = \pm |W + (G/2)|$:

$$\frac{h_1}{n_1{}^2}\sin\left(h_1\frac{W}{2}+\varphi\right) = \frac{q_2}{n_2{}^2}\cos\left(h_1\frac{W}{2}+\varphi\right) \tag{6.36a}$$

$$B\frac{h_1}{n_1{}^2}\sin\left(h_1\frac{W}{2}-\varphi\right) = B'\frac{q_2}{n_2{}^2}\left[\mathrm{e}^{+q_2\frac{G}{2}} - \mathrm{e}^{-q_2\frac{G}{2}}\right] \tag{6.36b}$$

$$h_1{}^2 + q_2{}^2 = (n_1{}^2 - n_2{}^2)k^2 \tag{6.36c}$$

(2) Anti-symmetric mode

$$\begin{aligned} &H_x = -B\cos\left[h_1\left(\frac{W}{2}\right)+\varphi\right]\mathrm{e}^{+q_2\left[y+\left(\frac{G}{2}+W\right)\right]} \\ &\beta^2 - q_2{}^2 = k^2 n_2{}^2 \qquad \text{for} \quad y \le -\frac{G}{2} - W \end{aligned} \tag{6.37a}$$

$$\begin{aligned} &H_x = -B\cos\left[h_1\left(y+\frac{G+W}{2}\right)-\varphi\right] \\ &\beta^2 + h_1{}^2 = n_1{}^2k^2 \qquad \text{for} \quad -\left(\frac{G}{2}+W\right) \le y \le -\frac{G}{2} \end{aligned} \tag{6.37b}$$

$$\begin{aligned} &H_x = B'[\mathrm{e}^{+q_2 y} - \mathrm{e}^{-q_2 y}] \\ &\qquad \text{for} \quad -\frac{G}{2} \le y \le +\frac{G}{2} \end{aligned} \tag{6.37c}$$

$$\begin{aligned} &H_x = B\cos\left[h_1\left(y-\frac{G+W}{2}\right)+\varphi\right] \\ &\qquad \text{for} \quad \frac{G}{2} \le y \le \frac{G}{2} + W \end{aligned} \tag{6.37d}$$

$$\begin{aligned} &H_x = B\cos\left[h_1\frac{W}{2}+\varphi\right]\mathrm{e}^{-q_2\left[y-\left(\frac{G}{2}+W\right)\right]} \\ &\qquad \text{for} \quad \frac{G}{2} + W \le y \end{aligned} \tag{6.37e}$$

Where B and B' are related by

$$B'\left[e^{-q_2\frac{G}{2}} - e^{+q_2\frac{G}{2}}\right] = -B\cos\left[h_1\frac{W}{2} - \varphi\right] \tag{6.38}$$

φ, h_1, and q_2 of the anti-symmetric mode are solutions of the following transcendental equations obtained from the boundary conditions:

$$\frac{h_1}{n_1^2}\sin\left(\frac{h_1 W}{2} + \varphi\right) = \frac{q_2}{n_2^2}\cos\left(\frac{h_1 W}{2} + \varphi\right) \tag{6.39a}$$

$$B'\frac{q_2}{n_2^2}\left[e^{-q_2\frac{G}{2}} + e^{+q_2\frac{G}{2}}\right] = B\frac{h_1}{n_1^2}\sin\left(h_1\frac{W}{2} - \varphi\right) \tag{6.39b}$$

$$h_1^2 + q_2^2 = (n_1^2 - n_2^2)k^2 \tag{6.39c}$$

For both symmetric and anti-symmetric modes, the dominant electric field E_y is related to H_x by

$$E_y \frac{-\beta}{n^2\omega\varepsilon_o} H_x$$

(b) Super modes obtained from coupled mode analysis

It is interesting to note that, in the coupled mode equations (Eq. (6.13)) the modes of the total structure are just linear combinations of the modes of the uncoupled waveguides, e_A and e_B, which are illustrated in Figure 6.4(b).

This is the classical example of a pair of coupled identical waveguides. Mathematically, in terms of Eq. (6.13), we have $\beta_A = \beta_B$, $\varepsilon_A = \varepsilon_B$, and $C_{AB} = C_{BA} = C$. Then, the solution of Eq. (6.24) is:

$$a_A(z) = \frac{1}{2}(A - B)e^{+jCz} + \frac{1}{2}(A + B)e^{-jCz}$$

$$a_B(z) = \frac{1}{2}(B - A)e^{+jCz} + \frac{1}{2}(A + B)e^{-jCz}$$

$$C = \frac{\omega}{4}\iint_{S_B}(\varepsilon_A - \varepsilon_B)[\underline{e}_B \cdot \underline{e}_A]dS \tag{6.40}$$

A and B are determined from the initial condition at $z = 0$. Substituting this result into Eq. (6.12), we obtain:

$$\underline{E}' = \frac{1}{\sqrt{2}}(A - B)\left[\frac{1}{\sqrt{2}}(\underline{e}_A - \underline{e}_B)\right]e^{-j(\beta - C)z} + \frac{1}{\sqrt{2}}(A + B)\left[\frac{1}{\sqrt{2}}(\underline{e}_A + \underline{e}_B)\right]e^{-j(\beta + C)z} \tag{6.41}$$

The symmetric combination, $\underline{e}_s = 1/\sqrt{2}(\underline{e}_A + \underline{e}_B)$, is a normalized symmetric eigenmode with $\beta_s = \beta + C$. The anti-symmetric combination, $\underline{e}_a = 1/\sqrt{2}(\underline{e}_A - \underline{e}_B)$, is a normallized anti-symmetric eigenmode with $\beta_a = \beta - C$. The total electric field of two identical waveguides is a superposition of two super modes, e_s and e_a. In other words,[4]

$$\underline{E}' = \frac{1}{\sqrt{2}}(A - B)\underline{e}_a e^{-j\beta_a z} + \frac{1}{\sqrt{2}}(A + B)\underline{e}_s e^{-j\beta_s z}$$
$$= A_a \underline{e}_a e^{-j\beta_a z} + A_s \underline{e}_s e^{-j\beta_s z} \tag{6.42}$$

When we compare the super mode result in (b) with the result in (a), we note the similarity between two results. When the coupling is weak, the symmetric mode obtained in (a) is approximately the same as $e_A + e_B$ obtained in (b), while the anti-symmetric mode obtained in (a) is approximately the same as $e_A - e_B$ obtained in (b). However, it is much easier to obtain the results by the coupled mode analysis. On the other hand, the mode profile and β_a and β_s obtained from the coupled analysis are inaccurate when the separation, G, of the two waveguides is small. The correct answer is given only by the super mode analysis.

6.6 Directional coupling of two identical waveguides viewed as super modes

The modes of individual isolated waveguides at $z < 0$ in the uncoupled region have been illustrated in Figure 6.4(b). The symmetric mode e_s is $1/\sqrt{2}(e_A + e_B)$, the anti-symmetric mode e_a is $1/\sqrt{2}(e_A - e_B)$. e_s, e_a, e_A, and e_B all have the same propagation wave number (or effective index). There is no transfer of power from one mode to another. The total field at any position z (e.g. $W > z > 0$) in the coupled waveguide depends on the excitation. When the excitation field at $z = 0$ is symmetric in A and B, only the symmetric modes exist. When the excitation field is anti-symmetric at $z = 0$, only the anti-symmetric mode exists. The lowest order e_s and e_a in the coupled region are illustrated in Figure 6.4(c). When there is only incident radiation to waveguide A and there is no incident radiation to waveguide B at $z = 0$, both the symmetric and the anti-symmetric modes exist with equal amplitude. The power carried in A and B for $W > z > 0$ depends on the interference of the symmetric and anti-symmetric modes. At $z > W$, the waveguides again become uncoupled, and powers in A and B remain the same as the powers in A and B at $z = W$.

Let the difference in effective index for these two modes be Δn_{eff} at $0 < z < W$. As an example, let the excitation be A = 1 and B = 0 at $z = 0$. When $\Delta n_{eff} kW = \pi$, the sum of the symmetric and anti-symmetric modes will have no power in channel A at $z = W$; all the power is in channel B. The minimum length at which complete transfer of power takes place is called the characteristic length, which is $W_c = \pi/\Delta n_{eff} k$.[5] For $z > W$, the two waveguides are well separated from each other. The power in waveguide A and B in

[4] It has also been shown by coupled mode analysis that when waveguides A and B are not identical, there are also two modes. Their propagation wave numbers are: $\beta = \frac{\beta_A + \beta_B}{2} \pm \sqrt{C_{BA} C_{AB} + (\Delta\beta/2)^2}$.

[5] From Eq. (6.41), it is clear that Δn_{eff} of the symmetric and anti-symmetric modes is equivalent to $2C$ in the coupled mode analysis for weakly or moderately coupled directional coupler.

those regions is independent of z. Therefore all the optical power P will remain in B for $z > W$. The power in channel B oscillates as a sinusoidal function of $\Delta n_{eff}kW$. The maxima are at $\Delta n_{eff}kW = (2n + 1)\pi$; the minima are at $\Delta n_{eff}kW = (2n + 2)\pi$.

Compare this result with the results obtained in Section 6.3.4: the results are the same. However, when the coupling is strong, the mode profiles and the C coefficient obtained by the coupled mode analysis will be inaccurate.

6.7 Super mode analysis of the adiabatic Y-branch and Mach–Zehnder interferometer

In the following subsections, a new concept, the adiabatic transition, will be introduced first. The operation of the Y-branch splitter depends on an adiabatic transition and will be discussed next. Lastly, the Mach–Zehnder interferometer, which consists of two parallel waveguides connected to input and output waveguides by two Y-branch splitters, will be discussed.

6.7.1 The adiabatic horn

Consider the transition for a guided-wave mode propagating from waveguide C into waveguide D, as shown in Figure 6.6(a), known commonly also as a waveguide horn. Let waveguide C be a single-mode waveguide and waveguide D be a multi-mode waveguide. As the waveguide cross-section expands, the second mode emerges at $z = z_1$ (i.e. there exists a second mode in the electromagnetic solution of an infinitely long waveguide that has the greater width at $z = z_1$). The third mode emerges at $z = z_2$, etc. The transition section can be approximated by many steps of local waveguides that have constant local cross-section within each step, as shown in Figure 6.6(b). At each junction of two adjacent steps, modal analysis can be used to calculate the excitation of the modes in the new step by the modes in the previous step. For adiabatic transition in the forward direction, the steps are so small that only the same-order mode is excited in the next section by the mode in the previous section. In other words, the overlap integral of the lowest-order mode in the transmitted section to the same-order mode in the incident section is approximately one, while the overlap integrals to other orders of modes in the transmitted section to the incident order of mode in the incident section are approximately 0. In other words, a negligible amount of power is coupled from the input lowest-order mode into higher-order modes and radiation modes. Similarly, if the waveguide C can support multi-modes, only those modes excited in C will be transmitted into D.

Let us now consider a reverse transition from $z > z_3$ to $z = 0$, where the incident field excites several modes at D. D is a multi-mode waveguide; C is a single-mode waveguide. Whenever a higher-order mode propagating in the $-z$ direction is excited at D, it will not be transmitted to C. The power in this higher-order mode will only be transferred into the radiation modes at the z position where this mode is cut off. Only the power in the lowest-order mode at D will be transmitted to the lowest-order mode at C. An important practical application of this result is that when an LED is used to excite a

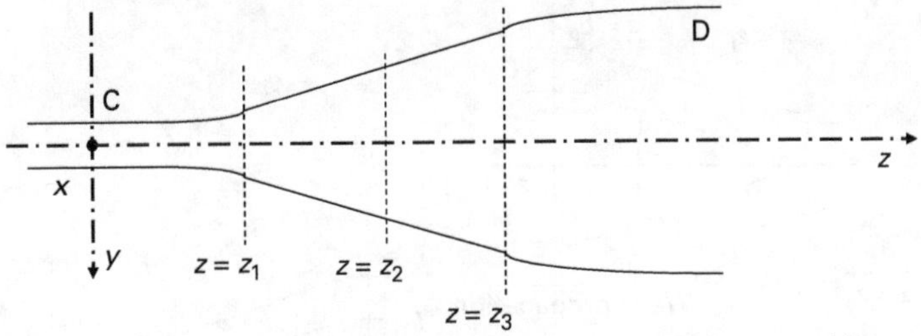

(a) The transition from a single-mode channel waveguide to a multi-mode channel waveguide

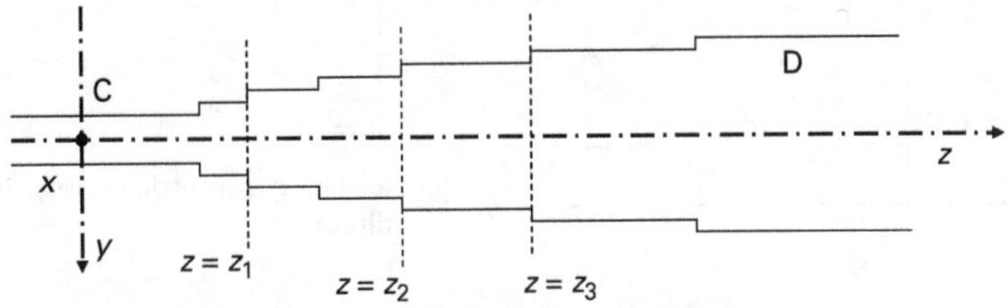

(b) The step approximation of the transition

Figure 6.6 Top view of an adiabatic transition and its step approximation. (a) The transition from a single-mode channel waveguide to a multi-mode channel waveguide (i.e. a waveguide horn). (b) The step approximation of the transition. Within each local section of the waveguide the dielectric constant profile is independent of z. The second mode exists for $z > z_1$, the third mode for $z > z_2$, and the fourth mode for $z > z_3$.

single-mode waveguide via a waveguide horn, the efficiency will be very low because only the fraction of the power in the lowest-order mode will be transmitted. Note that if C is also a multi-mode waveguide that supports the first m modes, only these m modes excited in D will be transmitted into C without any loss of amplitude.

In an expanding adiabatic transition, only the lowest-order mode is excited in the multi-mode output waveguide by the lowest-order mode in the input section, and there is no power loss. Conversion of power into higher-order modes will occur when the tapering is not sufficiently adiabatic or when there is scattering. The same conclusion can be drawn for propagation of the lowest-order mode in the reverse direction, i.e. from D to C. Power in higher-order modes will be dissipated before it reaches C.

6.7.2 Super mode analysis of a symmetric Y-branch

How a Y-branch functions depends on the modes that are supported by the waveguides that make up the Y-branch. Two examples are presented here to demonstrate this effect.

(a) A single-mode Y-branch

A guided-wave component used frequently in fiber and channel waveguide devices is a single-mode waveguide symmetric Y-branch. Its top view in the y–z plane is illustrated in Figure 6.7(a). A single-mode channel waveguide is connected to two single-mode

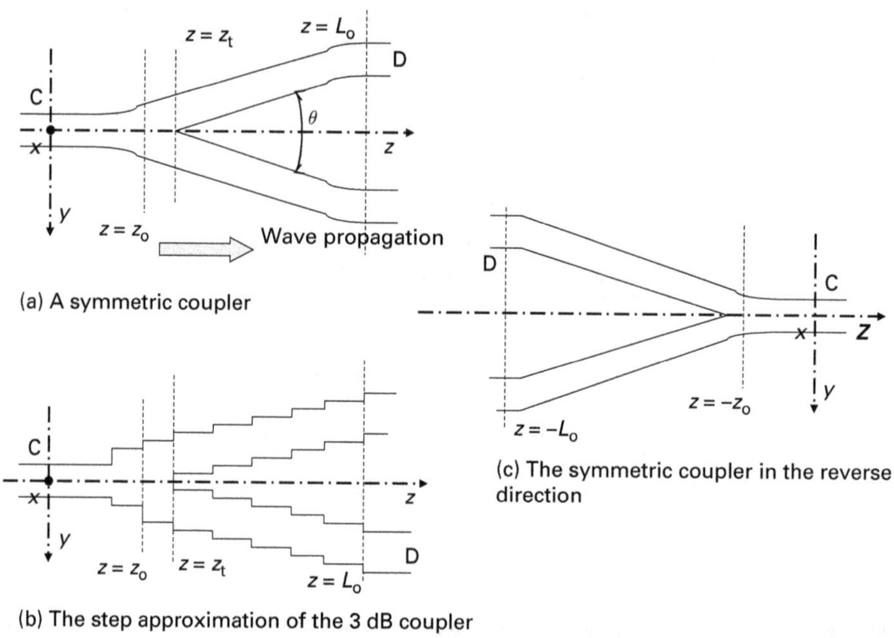

Figure 6.7 Top view of a symmetric Y-branch coupler. (a) A symmetric 3 dB coupler that splits the power in the input channel waveguide at C into two identical channel waveguides at D. (b) The step approximation of the Y-branch 3 dB coupler. (c) The reverse symmetric coupler that combines the field from two input waveguides into a single-mode output waveguide at C.

channel waveguides at $z = 0$. It is symmetric in the y direction with respect to the x–z plane. At $z > L_0$, the two waveguides are well separated from each other. All waveguides have identical cross-sectional index profiles in the x and y directions.

The practical application of such a device is to split the forward-propagating power in the original waveguide at C equally into two waveguides at D. The symmetric Y-branch functions like a 3 dB coupler from the input waveguide to the two output waveguides. In an ideal adiabatic transition, the angle of the branching, θ, is sufficiently small that the scattering and conversion loss from $z = 0$ to $z = L_0$ are negligible. Thus, losses are not included in the following analysis.

The forward propagation of guided waves in an ideal Y-branch coupler can be analyzed as follows. The input wave is a single TE_0 mode with amplitude a_{in} at $z < 0$. Only the TE_0 mode is excited by any incident radiation. The waveguide width in the y direction begins to broaden at $z > 0$. After $z > z_0$, the waveguide (or the split waveguides) has two modes. At $z > z_t$, there are two waveguides. From $z = z_0$ to $z \cong L_0$, the two super modes are the symmetric mode, $1/\sqrt{2}(e_A + e_B)$, and the anti-symmetric mode, $1/\sqrt{2}(e_A - e_B)$. As the guided wave propagates from $z < 0$ to $z = L_0$, only the symmetric mode is excited at each successive junction in Figure 6.7(b). No anti-symmetric mode is excited in such an adiabatic transition. At $z > L_0$, the coupling between the two waveguides is zero. Thus the optical power in the input waveguide is split equally into waveguides A and B. The amplitude of the modes in A and B are a_A and a_B. In a lossless Y-branch, the power is conserved. Thus $|a_A| = |a_B| = \frac{1}{\sqrt{2}}|a_{in}|$

The reverse situation is shown in Figure 6.7(c). A radiation is incident backwards on the Y-branch at z = D. If the incident field at z = L is symmetric, it will excite only the lowest-order symmetric mode of the double waveguides. This symmetric mode is transmitted without loss to the output waveguide at C as the TE_0 mode. However, if the incident mode is an anti-symmetric mode, it will continue to propagate as the anti-symmetric mode from $z = -L_0$ to its cut-off point. Just before the cut-off point, the anti-symmetric mode has a very long evanescent tail, and its n_{eff} is very close to the effective index of cladding or substrate modes. The anti-symmetric mode begins to transfer its energy into the radiation mode in the cladding or the substrate. After the cut-off, there is no anti-symmetric mode propagating in the device. Therefore, when the incident radiation excites both the symmetric and anti-symmetric mode at $z = L_0$, the lowest-order symmetric mode is transmitted to the waveguide at $z < 0$, and the anti-symmetric mode any high-order symmetric mode is blocked.

(b) A double-mode Y-branch

In order to demonstrate the characteristics of Y-branch when there is a change in the waveguide structure, let the input waveguide at $z \leq 0$ in Figure 6.7(a) have two modes. The waveguides after $z > z_t$ are still single-mode waveguides. Let the two modes at $z \leq 0$ be the lowest-order symmetric and anti-symmetric modes. In the case of a forward-propagating Y-branch, if the incident radiation is just in the symmetric mode, it will be transmitted as the symmetric mode at $z = L_0$ as discussed in the preceding paragraph. If the incident radiation is just in the anti-symmetric mode, it will also be transmitted as the anti-symmetric mode at $z = L_0$. If both the symmetric and the anti-symmetric modes are excited at $z = 0$ they will all be transmitted without any change in magnitude to $z = L_0$. However, symmetric and anti-symmetric modes have different phase velocities, i.e. n_{eff}. n_{eff} is also a function of the separation of the two waveguides, which is a function of z. As the modes propagate, their relative phase will change. The profile of the total field at $z = L_0$ will depend on the relative phase and amplitude of symmetric and anti-symmetric modes. Consequently, the power-splitting ratio will depend on the design of the horn and the excitation.

In the reverse coupler shown in Figure 6.7(c), when the waveguide at $z < 0$ has two modes, radiation in both the symmetric and anti-symmetric modes will be transmitted without loss to z = 0. However, the total field pattern at z = 0 will be very different, depending on the relative phase between them, which is the total cumulative phase difference between the two modes from $z = -L_0$ to $z = 0$.

6.7.3 Super mode analysis of the Mach–Zehnder interferometer

The Mach–Zehnder interferometer, illustrated in Figure 6.8, consists of two symmetric single-mode Y-branches (one is a forward-expanding Y-branch, the second a reverse Y-branch) connected by two parallel propagating single-mode channel waveguides that are well separated from each other. The objective is to control a specific fraction of the input

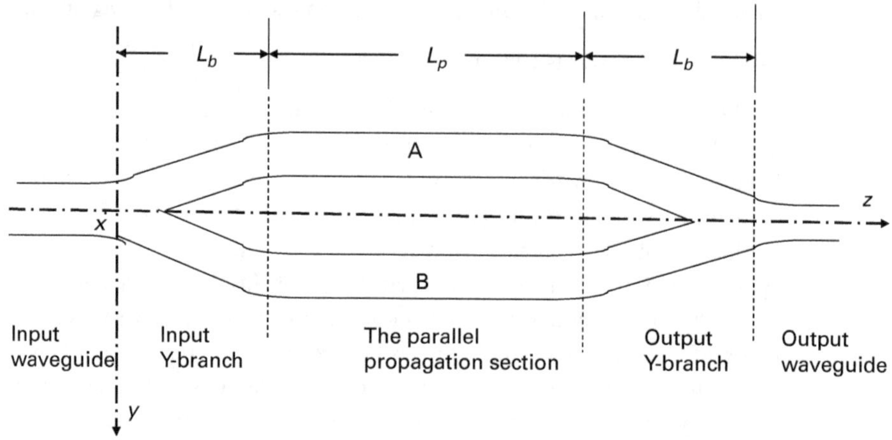

Figure 6.8 Top view of a channel waveguide Mach–Zehnder interferometer. Two isolated waveguides, A and B, connect an input single-mode symmetrical Y-branch 3 dB coupler to an output reversed single-mode symmetric Y-branch coupler. Waveguides A and B are well separated from each other. Only the power in the symmetric super mode at the input of the output Y-branch coupler will be transmitted to the single-mode output waveguide.

optical power to the output by controlling the propagation constant of one (or both) of the connecting propagating waveguides.

Let the waveguides A and B in the parallel propagation section have modes, e_A and e_B, with amplitude (including phase), a_A and a_B. The incident radiation excites the fundamental symmetric mode in the input waveguide at $z = 0$. The incident radiation excites the modes of individual identical waveguides in the propagating section, A and B, equally in amplitude and phase, $|a_A| = |a_B| = 1/\sqrt{2}|a_{in}|$. Since the connecting waveguides, A and B (located from $z = L_b$ to $z = L_b + L_p$), are well separated, modes will propagate in them without any interaction. If the refractive indices of the materials in A and B differ, the magnitude of a_A and a_B will not change, but the phase of a_A and a_B will differ at $z = L_b + L_p$. If the relative phase between a_A and a_B at $z = L_b + L_p$ is π, the total incident mode to the output Y-branch is anti-symmetric: it excites only the anti-symmetric mode. The anti-symmetric mode is dissipated into the substrate in the reverse Y-branch. Thus the output power is zero. If the relative phase is 2π, the symmetric mode is excited at the output Y-branch. All the optical power is transmitted to the output. For other relative phase differences between a_A and a_B, there will be a mixture of symmetric and anti-symmetric modes excited. Only the symmetric mode excited at the reverse output coupler is transmitted as the output. Therefore the amount of power transmitted to the output waveguide can be varied from 0% to 100% by controlling the relative phases of a_A and a_B. This is the principle of the Mach–Zehnder interferometer.

Beside differences in phase, there can also be other differences between waveguides. For example, waveguide B could have absorption such that $a_B = 0$ at $z = L_b + L_p$. In that case, $|a_s| = |a_a| = 1/\sqrt{2}|a_A| = \frac{1}{2}|a_{in}|$. Therefore only ¼ of the input power is transmitted to the output.

Chapter summary

Three different analytical techniques that can be used to analyze interactions of guided waves in devices have been presented. It is important to recognize the differences, the similarities, and the limitations of these techniques. Four commonly used applications, the directional coupler, the waveguide Y-branch, the grating filter, and the Mach–Zehnder interferometer have been used as examples to demonstrate these techniques. It is interesting to note that how a waveguide Y-branch and a Mach-Zehnder interferometer operate can by analyzed by the super mode analysis without even knowing the effective index and the profile of the modes. These techniques will be used again in Chapters 7 and 8 to analyze various opto-electronic devices.

References

[1] A. Yariv, *Optical Electronics in Modern Communication*, Oxford University Press, 1997.

[2] D. L. Lee, *Electromagnetic Principles of Integrated Optics*, John Wiley & Sons, 1986, Chapter 8.

[3] William C. C. Chang, *Fundamentals of Guided-Wave Opto-Electronic Devices*, Cambridge University Press, 2010.

[4] D. C. Flanders, H, Kogelnik, R. V. Schmidt, and C. V. Shank, Grating filters for thin film optical waveguides, *Applied Physics Letters*, **24**, 195, 1974.

7 Passive waveguide devices

The passive waveguide (and fiber) devices used in optical communication are mode transformers, power dividers, wavelength filters, resonators, frequency multiplexers, and couplers. All these are optical devices. How these devices work is the focus of this chapter. The guided-wave modes presented in Chapter 5 and the analytical techniques presented in Chapter 6 will be used to analyze these devices. The performance of each device will be evaluated in terms of its application.

In the next chapter, active waveguide devices will be discussed. Active devices utilize other physical mechanisms, such as the electro-optical effect, the acousto-optical effect, or the electro-absorption effect, to achieve their function. When we discuss active devices it is also necessary to discuss these mechanisms. The electrical properties of the devices using these mechanisms are as important as their optical properties.

7.1 Waveguide and fiber tapers

Waveguide and fiber tapers are used to match the mode of one waveguide (or fiber) to the mode of another waveguide (or fiber) that has a different configuration.

In a taper, the cross-section of a waveguide or fiber is adiabatically tapered to a new dimension to transform the profile of the mode. The ideal taper has already been illustrated and analyzed in Section 6.7.1. In realistic tapers, there will be conversion losses into radiation or other guided-wave modes caused by fabrication defects. These losses need to be minimized. The performance of a realistic taper is measured by how efficiently the mode can be transformed.

7.2 Power dividers

In guided-wave and fiber optical systems, power dividers are used to distribute specific fractions of input power into different output channels. The input and output waveguides (or fibers) are often interconnected to other waveguides (or fibers). The performance of power dividers is measured by their desired output power distribution, wavelength variation of the power distribution, physical size, and insertion losses, which include the coupling loss to other input and output waveguides (or fibers). The commonly used power dividers are the Y-branch equal-power splitter, the directional coupler, the multi-mode interference coupler, and the Star coupler.

7.2.1 The Y-branch equal-power splitter

An adiabatic symmetric single-mode Y-branch waveguide splits the optical power in the input waveguide equally into two output waveguides. It is a 3 dB power splitter. In this device, a single-mode waveguide is interconnected with two identical single-mode waveguides in a Y-branch configuration. In most applications, the input and output waveguides are also coupled into other waveguide devices or optical fibers. The ideal device has already been analyzed and discussed in Section 6.7.2 and illustrated in Figure 6.7. In any realistic Y-branch coupler, there is an excess insertion loss caused by the power scattered into the substrate or the cladding at the intersection region by the defects produced in fabrication processes. There are also coupling losses to and from other waveguides or fibers. The performance of any splitter is characterized by its total insertion loss and by the evenness of its distribution of input power into the output waveguides (or fibers). Since the power splitting ratio is independent of wavelength, and since the excess and propagation losses vary slowly with wavelength, the characteristics of a Y-branch equal-power splitter are only mildly dependent on wavelength. Conversely, a Y-branch coupler in the reverse direction can also be used as a power combiner. Note that, in order to function as a 3 dB coupler for all incident radiation, the input waveguide must be a single-mode waveguide. The efficiency of the combiner in the inverse direction depends on the phase of excited modes.

7.2.2 The directional coupler

A waveguide directional coupler consists of two parallel waveguides, A and B, coupled to each other in a coupling section W long. Within the coupling region, the guided waves in the two waveguides interact with each other via the evanescent fields. The coupling section is connected to waveguides in the input and output sections through transition waveguides. Outside the coupling region, the input and output waveguides are well isolated from each other, and the waveguides function as individual isolated waveguides.

An ideal channel waveguide directional coupler was discussed and analyzed in Section 6.3.4 by coupled mode analysis and in Section 6.6 by super mode analysis; it was illustrated in Figure 6.4. In most directional couplers, A and B are identical waveguides. There is input to only one of the waveguides. Let the input power be in the A waveguide. From the coupled mode analysis point of view, $\Delta\beta = 0$ between A and B. When propagation and coupling losses are neglected and when there is no input power in B, 100% of the input power to waveguide A is transferred into waveguide B for $CW = (n + 1/2)\pi$, and all power is retained in waveguide A for $CW = n\pi$. At appropriate value of CW within $n\pi < CW < (n + 1/2)\pi$, any desired distribution of power in A and B can be obtained. Conversely, if the input is in B, there is 100% transfer of power from waveguide B to waveguide A when $CW = (n + 1/2)\pi$, and all power is retained in waveguide B when $CW = n\pi$. In reality, there will be an insertion loss caused by the propagation loss and the excess scattering loss produced by the defects created in the fabrication processes.

In a more sophisticated operation, the directional coupler can also combine two inputs into two outputs. If there are inputs to both A and B, the amplitudes of the modes transferred to the outputs are superimposed. However, there will be interferences of the modes in the output waveguides that depend on the relative phase of the inputs.

From the point of view of super mode analysis, any input radiation to the total waveguide structure that includes both A and B excites the symmetric and the anti-symmetric modes. As the excited modes propagate through the coupling region, they interfere with each other because anti-symmetric and symmetric modes have different effective indices. The output powers in A and B are determined by the sum of the amplitudes (including phase) of symmetric and anti-symmetric modes at the exit of the coupling section.

The directional coupler is a reciprocal device. Reflected optical power in the output waveguides will also be distributed in the same ratio back to the input waveguides. Most commonly, the fibers (or waveguides) at the input and output ends are match terminated, meaning that no reflected power at the outputs will be reflected back to the input (or inputs).

Since the coupling coefficient C and $\Delta\beta$ (or the n_{eff} of the super modes) are dependent on the wavelength, the distribution of power will depend on the wavelength. Furthermore, when the coupling coefficient C or the $\Delta\beta$ (or Δn_{eff}) is controlled electro-optically, the outputs are a function of the applied electrical signal. Then the directional coupler becomes a switch (or modulator), which will be discussed in Chapter 8.

Although the directional coupler discussed here uses two channel waveguides coupled to each other side by side, there are also directional couplers using two waveguides that are coupled vertically. In that case, the coupling region consists of two waveguides fabricated on top of each other and separated by an isolation layer between them.

In optical fibers, the directional coupler can also be made when the cladding is partially removed to provide the coupling via the evanescent field. The length of this interaction region and the proximity of fiber cores control the power splitting ratio between the two fibers. The advantage of a fiber directional coupler is that there is no need to couple the power in the input fiber into the output fibers, which may have insertion loss.

Comparing the directional coupler with the Y-branch power splitter, it is clear that the directional coupler is a more flexible device. The Y-branch is a 3 dB power splitter. What fraction of the input power is split into the output waveguides in a directional coupler can be easily varied. On the other hand, the Y-branch is easy to make. The operation of the single-mode Y-branch power splitter is independent of wavelength, while the power splitting in directional couplers is wavelength dependent.

7.2.3 The multi-mode interference coupler

A multi-mode interference coupler consists of a section of multi-mode channel wave-guide, L long, abruptly terminated at both ends. A number of access channel waveguides (usually single-mode waveguides) are connected to it at the beginning and at the end.

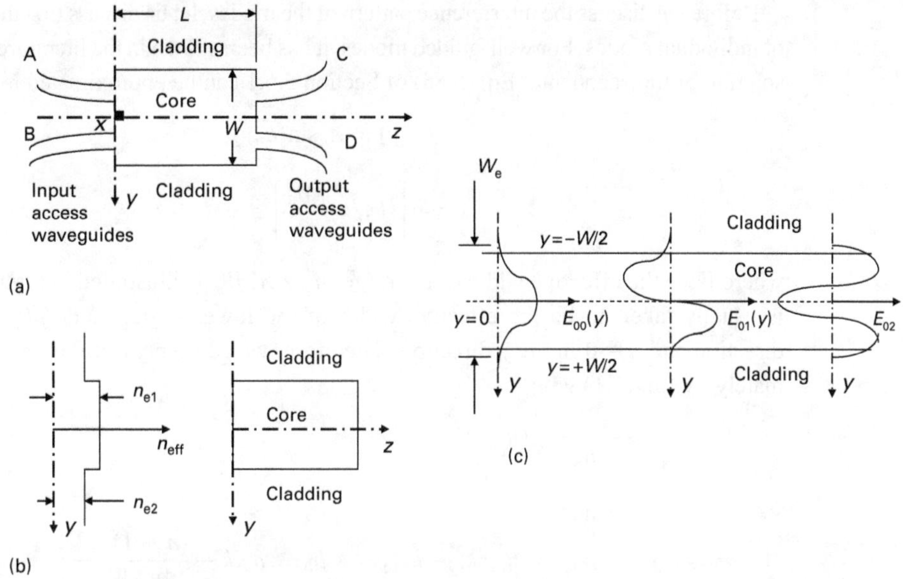

Figure 7.1 A multi-mode interference coupler. (a) The top view of a 2 × 2 multi-mode interference coupler. The multi-mode waveguide is L long and W wide. (b) The effective index profile of the multi-mode waveguide. (c) The field patterns (as a function of y) of the lowest-order modes of the multi-mode section.

The modes in the multi-mode channel are excited by the radiation from the input access waveguides. As they propagate to the end of the multi-mode channel, their relative phases are shifted because of the difference of effective indices. The amplitudes of the modes excited in the output access waveguides are determined by the interference pattern of the modes after propagating L distance in the multi-mode channel. Such devices are generally referred to as $N \times M$ multi-mode interference (MMI) couplers, where N and M are the number of input and output waveguides, respectively [1].

Figure 7.1(a) illustrates a multi-mode interference coupler with two input and two output access waveguides. The multi-mode section is shown here as a step-index ridge waveguide with width, W, and length, L. It is single mode in the depth direction x and multi-mode ($n \geq 3$) in the lateral direction y. The objective of such a multi-mode coupler is to redistribute the powers in the input access waveguides transferred into the output access waveguides.

Let the multi-mode waveguide be a ridge waveguide, as shown in Figure 5.9(b). The profile of the effective index of the planar TE_0 modes in the y direction is illustrated in Figure 7.1(b). For the planar waveguide mode in the core (i.e. within the ridge), there is just a single TE_0 mode in the x direction with effective index n_{e1}. There is also a TE_0 mode in the cladding with an effective index n_{e2}; $n_{e1} > n_{e2}$. The channel guided-wave modes in the core can be found by the effective index method discussed in Section 5.6.1 or by numerical methods. The field variation of the first few modes in the y direction is illustrated in Figure 7.1(c).

Before we discuss the interference pattern of the modes, let us discuss first the properties of individual modes. For well-guided modes, it has been shown in the literature [2] that the solution of transcendental Eq. (5.66) of Section 5.6.1 can be approximated by

$$E_n(y) = A_n \sin(h_n y)$$

$$\tan\left[(h_n/k)\frac{kW_e}{2}\right] \cong \infty \tag{7.1}$$

where W_e is the effective width of the ridge, $W_e > W$. W_e is illustrated in Figure 7.1(c). It is usually taken to be the effective width of the lowest-order mode $m = 0$ in the x direction and $n = 0$ in the y direction. Therefore, we can represent the modes approximately by Eq. (7.1) with

$$h_n = \frac{(n+1)\pi}{W_e}$$

and

$$\beta_{0n}{}^2 = n_{e1}{}^2 k^2 - h_n{}^2, \qquad \beta_{0n} \cong n_{e1}k - \frac{(n+1)^2 \pi\lambda}{4n_{e1}W_e{}^2} \tag{7.2}$$

In Eq. (7.2), the propagation constants of the various-order modes have a quadratic dependence on n. By defining L_π as the beat length (i.e. the propagation length in which the phase difference of the two modes is π) between the $n = 0$ and $n = 1$ modes, we obtain

$$L_\pi = \frac{\pi}{\beta_0 - \beta_1}, \qquad \beta_0 - \beta_n = \frac{n(n+2)\pi}{3L_\pi} \tag{7.3}$$

Let us examine the total field of all the modes. Let there be N modes in the multi-mode channel. The y variation of any input field at $z = 0$, $E_{in}(y, z = 0)$, can be expressed as the summation of the E_n modes. Thus,

$$E_{in}(y,0) = \sum_{n=0}^{n=N-1} C_n E_n(y)$$

$$E_{in}(y,z) = \sum_{n=0}^{n=N-1} \left\{ C_n E_n(y) e^{\left[j\frac{n(n+2)\pi}{3L_\pi}z\right]} \right\} e^{-j\beta_0 z} \tag{7.4}$$

$$E_n(y) = A_n \sin(h_n y)$$

Any input field at $z = 0$ will be repeated or mirrored at $z = L$, whenever

$$e^{j\frac{n(n+2)\pi}{3L_\pi}L} = 1 \tag{7.5}$$

or

$$e^{j\frac{n(n+2)\pi}{3L_\pi}L} = (-1)^n \tag{7.6}$$

When the condition in Eq. (7.5) is satisfied, the field at $z = L$ is a direct replica of the input field. When the condition in Eq.(7.6) is satisfied, the even modes will have the same phase as the input, but the odd modes will have a negative phase, producing a mirrored image of the input field. For the 2 × 2 coupler shown in Figure 7.1(a), this means that power in input A will be transferred to output C when Eq. (7.5) is satisfied and to output D when Eq. (7.6) with odd n is satisfied.[1]

More general analysis of the mode interference pattern can be obtained as follows. Eqs. (7.1) and (7.2) show that the y variation of the field of a well-guided multi-mode channel-waveguide mode resembles the lowest-order sine terms of a Fourier series in y within the period from $y = -W_e/2$ to $y = +W_e/2$. There are only a finite number of sine Fourier series terms representing the modes. In order to analyze the more complex interference patterns, let us now extend the expression for the modes to outside of the range $-W_e/2$ to $W_e/2$ in a periodic manner so that we can take advantage of our knowledge of the Fourier series. Since these modes have a half-cycle sine variation within $-W_e/2 < y < W_e/2$, the extended mode in $-3W_e/2 < y < -W_e/2$ and in $W_e/2 < y < 3W_e/2$ should be anti-symmetric with respect to the mode in $-W_e/2 < y < W_e/2$. A similar extension can be made beyond $y > |3W_e/2|$. Consider now the total extended field over all y coordinates, including the periodic extension of the fields outside the multi-mode waveguide region. The extended input field from all the input access waveguides (periodically repeated outside the region from $y = -W_e/2$ to $W_e/2$) can then be expressed as a summation of these Fourier terms. Eq. (7.4) shows that the relative phase of the Fourier terms dependent on L. Different multi-fold images in the y direction at the end of the multi-imode section can be formed by summing these phase terms with different L. As an example, let us consider $L = 3pL_\pi/2$, where p is an odd integer. Then

$$\begin{aligned} E_{in}\left(y, \frac{3pL_\pi}{2}\right) &= \sum_{n\,\text{even}} C_n E_n(y) + \sum_{n\,\text{odd}} (-j)^p C_n E_n(y) \\ &= \frac{1 + (-j)^p}{2} E_0(y, 0) + \frac{1 - (-j)^p}{2} E_0(-y, 0) \end{aligned} \tag{7.7}$$

The last equation represents a pair of images of E_{in} in quadrature, with amplitudes $1/\sqrt{2}$, at distances $z = 3L_\pi/2,\ 9L_\pi/2,\ \ldots$. The replicated, the mirrored, and the double images of E_0 at various z distances are illustrated in Figure 7.2. Clearly, for a 2 × 2 coupler, we have a 3 dB power splitter from input B into output waveguides, C and D, at $z = 3L_\pi/2$ and at $z = 9L_\pi/2$. We transfer the power from B to C (called the cross-state) when $z = 3L_\pi$, and from B to D (called the through-state) when $z = 6L_\pi$.

The preceding discussion is for an ideal $M \times N$ interference coupler. A realistic 2 × 2 InGaAsP MMI cross-coupler has been made with $W = 8$ μm and

[1] Note that a two-mode interference coupler is identical to a two-waveguide directional coupler with zero gap of separation. Therefore it can also be analyzed by super mode analysis. This concept can also be extended to the MMI coupler.

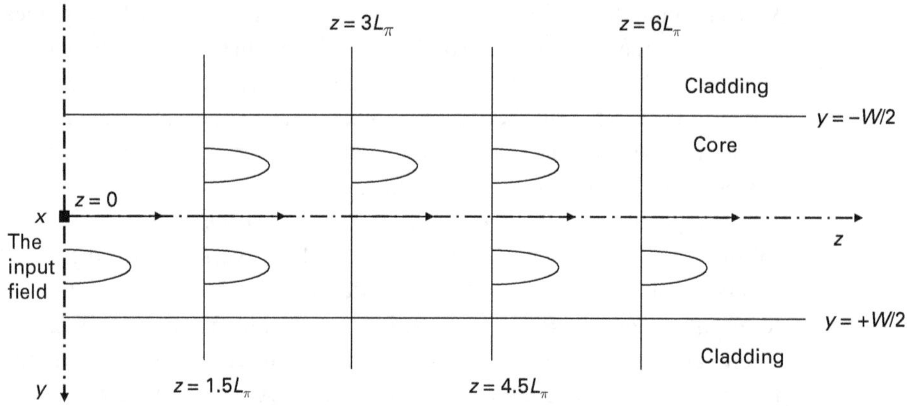

Figure 7.2 Images of the input field at various distances in a multi-mode interference coupler. The input field is shown at $z = 0$. It can be decomposed into a summation of modes. Each mode has a different phase velocity. The total profile of the summation of these modes will yield a two-fold image of the input at $z = 1.5\,L_\pi$ and at $z = 4.5L_\pi$, a mirror single image at $z = 3L_\pi$, and a direct image at $z = 6L_\pi$.

L = 500 μm, which gives excess loss of 0.4 to 0.7dB and an extinction ratio of 28 dB, and a 3 dB splitter with L = 250 μm and imbalances between C and D well below 0.1 dB [1].

The actual design of an MMI coupler must take into account the number of input and output access waveguides, the number of modes in the multi-mode waveguide, the relative phase and amplitude of the incident modes in the input access waveguides, and the position and width of access waveguides. Compared with directional couplers and Y-branch couplers, the MMI coupler is very compact. It allows $M \times N$ coupling. However, it can only redistribute the power at specific ratios. An MMI coupler is likely to have more insertion loss, and the division of power into the output waveguides is not flexible.

7.2.4 The Star coupler

The diffraction of the radiation from an input channel waveguide mode into a planar waveguide produces a broadened beam in the planar waveguide as the planar guided-wave propagates. When there are N output channel waveguides placed at the end of the planar waveguide section, the input power is distributed into the output waveguides. This is the basic principle of a Star coupler.

If there are N such input ports and N output ports, the Star coupler is an $N \times N$ power distributor. It is used in the wavelength division multiplexed (WDM) fiber optical systems. An example of a Star coupler is illustrated in Figure 7.3 [3].

The Star coupler consists of two arrays of N uniformly spaced identical ports fed into the planar waveguide in the horizontal direction. Each port is a TE_{00} mode channel waveguide which has a width "a." The planar waveguide also supports a TE_0 mode that matches the TE_{00} channel waveguide mode in the vertical direction. Ports (i.e. ends of

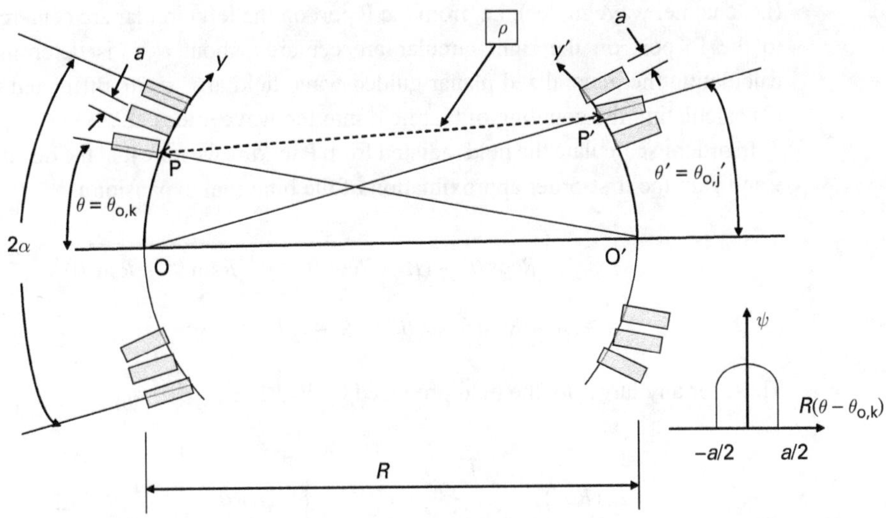

Figure 7.3 The Star coupler. (Taken with permission from Ref. [3] with permission from IEEE.)

channel waveguides) in each array are located on a circular arc with radius R. There are two circular arcs facing each other. One for the input ports, and one for the output ports. The center of the circle of the array on the left is at O′, which is also the middle of the circular arc for the array on the right. Conversely, the center of the circle on the right is at O, which is also the middle of the circular arc or the array on the left. In other words, these are confocal circular arcs. The center position of the kth port on the left arc is given by $R\theta_{o,k}$, and the center position of the jth port on the right arc is given by $R\theta'_{o,j}$. The power entering the single-mode planar waveguide region (from any one of the $2N$ waveguides) will be diffracted and propagated in the horizontal plane of the planar waveguide. Waveguides on the opposite circular arc are excited by this planar guided wave.

The objective of the Star coupler is to maximize the power transfer from any one of the channel waveguides in the left array to the waveguides in the right array. Ideally, there is no power loss and the power from any input waveguide is divided uniformly into the N output channels. In that case the transfer efficiency will be $1/N$. However, this is difficult to achieve in practice. In this section we will analyze the diffraction of the planar TE_0 guided-wave mode. In particular we will calculate the field at the output array produced by the radiation from a given channel waveguide in the input array. We will calculate the excitation of the mode of the channel waveguide in the output array by this field, thereby determining the power transfer from the input channel to the output channels.

The incident field at each port is the mode of the input channel waveguide. Let us assume here that the E_y of the guided-wave mode for all input and output channels in the horizontal plane is $\psi(y)$ or $\psi(y')$, where y (or y') is the coordinate along the left (or right) circular arc, as shown in Figure 7.3. The transmission between any two ports (i.e. channel waveguides), i.e. from the P port on the left circular arc centered about $\theta_{o,k}$ to the P′ port on the right circular arc centered about $\theta_{o,j}'$, is determined by: (1) calculating the generalized planar guided-wave field at $y' = R\theta'$ diffracted from P, and (2) calculating the coupling of that field into the waveguides at P′.

In order to calculate the field radiated from P to $R\theta'$, we note that the distance between y and y' in the first-order approximation of the binomial expansion is

$$\rho = \sqrt{[R\cos\theta' - (R - R\cos\theta)]^2 + (R\sin\theta' - R\sin\theta)^2}$$
$$\cong R - R\sin\theta'\sin\theta \cong R - R\theta\theta'. \tag{7.8}$$

Thus, for any large $\beta\rho$, the field produced by P at P′ is;

$$E_y(R\theta') \cong \sqrt{\frac{n_{eff}k}{j2\pi R}}\,e^{-jn_{eff}kR}\int_{\theta_{0,k}-\frac{a}{2R}}^{\theta_{0,k}+\frac{a}{2R}} \psi(R\theta)e^{+j2\pi\left(\frac{n_{eff}}{\lambda}\vartheta'\right)R\vartheta}R\,d\theta \tag{7.9}$$

Here, we have assumed that the field for the kth port is confined approximately within the waveguide, as shown in the inset of Figure 7.3. Note that the phase factor, $-jn_{eff}kR$, is now a constant on the circular arc on the right. The positioning of the ports on confocal circular arcs serves the function of creating a constant phase factor, $-jn_{eff}kR$, similar to the spherical reflectors in a confocal resonator in three dimensions.[2] The relation between $E_y(R\theta')$ and $\psi(R\theta)$ is related by an integral resembling Fourier transform as follows.

Using a change of variable, $u = 2R/a(\theta - \theta_{0,k})$, we obtain:

$$E_y(R\theta') \cong a\sqrt{\frac{n_{eff}}{j\lambda R}}\,e^{-jn_{eff}kR}\,e^{+j2\pi\frac{n_{eff}R\theta_{0,k}\theta'}{\lambda}}\varphi(R\theta')$$

where,

$$\varphi(R\theta') = \frac{1}{2}\int_{-1}^{+1}\psi\left(\frac{au}{2}\right)e^{+j2\pi\left(\frac{n_{eff}a\theta'}{2\lambda}\right)u}du \tag{7.10}$$

Since $\psi(au/2)$ is identical for all the waveguides, the φ factor is independent of $\theta_{0,k}$. E_y is only dependent on the center position $R\theta_{0,k}$ of the input channel through the factor $e^{j2\pi\frac{n_{eff}R\theta_{0,k}\theta'}{\lambda}}$.

Let the total E_y at $R\theta'$ be expressed as a summation of the fields of all the channel guides, $\psi_i(R\theta')$, on the right circular arc array plus the stray guided-wave fields in the

[2] The confocal mirrors in the confocal resonator of laser cavities in Chapter 4 also allowed us to simplify the diffraction integral equations.

gaps between channel guides, $\zeta(R\theta')$. Let us assume, as an approximation, that there is negligible overlap among all ψ_i and ζ. Then

$$E_y(R\theta') = \sum_i b_i\psi_i(R\theta') + \zeta(R\theta'). \tag{7.11}$$

Here, $\psi_i(R\theta')$ is the ψ centered about $\theta_{o,i}$. Multiplying both sides by $\psi_j^*(R\theta')$ and integrating with respect to $R\theta'$ from $-\infty$ to $+\infty$, we obtain:

$$\int_{\vartheta_{o,j}-\frac{a}{2R}}^{\vartheta_{o,j}+\frac{a}{2R}} E_y(R\theta')\psi(R\theta')R\mathrm{d}\theta' \cong b_j \int_{\theta_{o,j}-\frac{a}{2R}}^{\theta_{o,j}+\frac{a}{2R}} |\psi(R\theta')|^2 R\mathrm{d}\theta' \tag{7.12}$$

Utilizing once more the change of variable, $u' = 2R/a(\theta' - \theta_o')$, we obtain,

$$|b_j|^2 \left[\frac{a}{2}\int_{-1}^{+1}\left|\psi\left(\frac{a}{2}u' + R\theta_{o,j}\right)\right|^2 \mathrm{d}u'\right]^2 = \left[\frac{n_{eff}a^4}{\lambda R}\left|\varphi(R\theta_{o,k})\right|^2\left|\varphi(R\theta_{o,j})\right|^2\right], \tag{7.13}$$

or

$$|b_j|^2 = \frac{4n_{eff}a^2}{\lambda R}\frac{|\varphi(R\theta_{o,k})|^2|\varphi(R\theta_{o,j})|^2.}{\left[\int_{-1}^{+1}\left|\psi\left(\frac{a}{2}u + R\theta_{o,k}\right)\right|^2 \mathrm{d}u\right]^2} \tag{7.14}$$

Since the power contained in the total E_y is proportional to $\int |E_y|^2 R\mathrm{d}\theta$, which is approximately equal to $\sum_i |b_i|^2 \int |\psi|^2 R\mathrm{d}\theta$, $|b_j|^2$ is the power transfer from the channel waveguide centered at $\theta_{o,k}$ to the channel waveguide centered at $\theta_{o,j}$.

In an actual Star coupler, R, N and "a" are designed to optimize the power transfer. C. Dragone and his colleagues have optimized the design, which gives $0.34(1/N)$ to $0.55(1/N)$ of the input power to any one of the output channels [3].

In summary, comparing the Star coupler with the Y-branch coupler, the directional coupler, and the MMI coupler, it is clear that the power in any one of the N input ports in the Star coupler is always distributed as evenly as possible to all the N output ports with equal phase. It works in both the forward and the backward directions. The insertion loss and the uniformity of the output power distribution are the major issues in its performance. On the other hand, the Y-branch single-mode coupler achieves even distribution of power for one input waveguide coupled to two output waveguides, i.e. it is only a 1 × 2 coupler. Its insertion loss can be very small, and the balance of output power is very good. Although the Y-branch coupler can be repeated to obtain 1 × 2 N coupling in the forward direction, then the total device will be long. The directional coupler is a much more flexible 2 × 2 coupler that can split the input power at any ratio into the output channels in both the forward and backward directions. It can also be fabricated in fibers. However, it is wavelength dependent. The MMI coupler is the smallest N × M coupler. It can transfer the input power into the output waveguides only at fixed ratios, dictated by the various image patterns. Its insertion loss is an issue. It is also wavelength dependent.

7.3 The phased array channel waveguide frequency demultiplexer

In wavelength division multiplexed (WDM) optical fiber systems, optical radiations at different wavelengths from an input waveguide need to be directed into different output waveguides.

Let us consider a component called a PHASAR demultiplexer [4]. In this device, two Star couplers discussed in Section 7.2.4 are interconnected by an array of identical channel waveguides, each with length L_j, as shown in Figure 7.4(a). On the input side of the first Star coupler, only the kth waveguide (i.e. the transmitting waveguide) on the input side is excited. All other input waveguides have zero power. The electric field of the input transmitting channel waveguide at the kth position will create a field distribution $E_y(R\theta')$ at the output circle in the first Star coupler. If all the interconnecting waveguides have equal length, and if the stray fields, ζ, in the gap between channel guides are small, a field distribution identical to $E_y(R\theta)$ in the first Star coupler will be created on the input side of the second Star coupler. By reciprocity, this field distribution on the input side of the second Star coupler will create a field distribution on the output side, which excites only ψ_k at the position of the kth output waveguide and not elsewhere. In other words, the power in the kth transmitting waveguide of the first Star coupler will now be transmitted exclusively to the kth output channel of the second Star coupler. The situation does not change if the lengths of the interconnecting waveguides between the two Star couplers differ from each other so that the phase shift between adjacent interconnecting waveguide is 2π, i.e.

$$\frac{2\pi\, n_{eff,c}}{\lambda}(L_j - L_{j-1}) = \frac{2\pi n_{eff,c}}{\lambda}\Delta L = 2\pi \tag{7.15}$$

Here, $n_{eff,c}$ is the effective index of the channel waveguide. The physical ΔL required to meet this condition will depend on λ.

Let the spacing between adjacent channel waveguides be d_a ($d_a = R\Delta\alpha$) in the first Star coupler. Then, according to Eq. (7.10) of Section 7.2.4, E_y (created by the field of the kth channel waveguide in the input array) at the center of the mth waveguide in its output circular array, has the phase

$$e^{j2\pi\frac{n_{eff}R}{\lambda}(k\Delta\alpha)(m\Delta\alpha)} \tag{7.16}$$

$kR\Delta\alpha$ and $mR\Delta\alpha$ are the center angular positions of the kth and mth channel waveguide in the input and output array of the Star coupler, as shown in Figures 7.3 and 7.4(b). k and m are integers, ranging from $-(N-1)/2$ to $(N-1)/2$. n_{eff} is the effective index of the planar

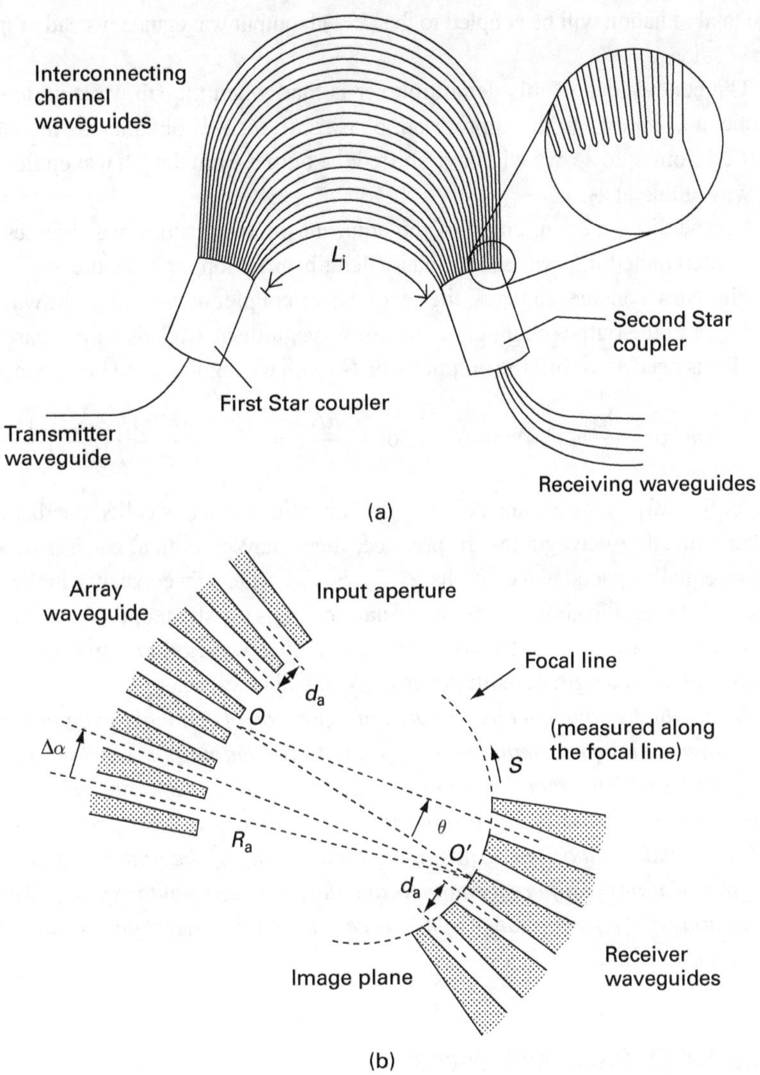

Figure 7.4 The PHASAR demultiplexer. (a) The layout. (b) Geometry of the Star coupler on the receiver side. This figure is taken from Ref. [4] with permission from IEEE. The two Star couplers are connected by an array of interconnecting channel waveguides that have different lengths. Optical radiation from the input waveguide is transmitted to the interconnecting waveguides by the input Star coupler. The input radiation to the output Star coupler will have phase shifts controlled by the wavelength, as well as by the length increments of the interconnecting waveguides. The objective is to create an appropriate phase shift so that radiation at different wavelengths is transmitted to different output waveguides of the output receiver Star coupler.

waveguide mode in the Star coupler. If the excitation changed from the kth waveguide to the $(k+1)$th waveguide, the difference in E_y caused by this change is just a phase difference, $m\Delta\varphi = 2\pi(R\Delta\alpha)(n_{eff}/\lambda)(m\Delta\alpha)$, at the center of the mth waveguide. Conversely, when the radiation in the array of input waveguides in the second Star coupler has a total E_y field that contains this extra phase factor $m\Delta\varphi$ for each input waveguide, $m = -(N-1)/2$ to $(N-1)/2$, the total radiation will be coupled to the $(k+1)$th output waveguide instead of the kth output waveguide.

The central idea of this demultiplexer is that, when the kth waveguide is the output guide at λ_1 and when the appropriate phase shift $m\Delta\varphi$ is obtained as the wavelength is shifted from λ_1 to λ_2, we will have shifted the output from the kth waveguide to the $(k+1)$th waveguide at λ_2.

Let the difference in length of the adjacent interconnecting waveguides be ΔL. The mth interconnecting waveguide has a length $m\Delta L$ longer than the waveguide at the origin. Now consider in detail the second Star coupler at two different wavelengths, λ_1 and λ_2. Let the output channel be the kth waveguide at λ_1. This extra phase factor $m\Delta\varphi$ (which is needed to shift the output to the $(k+1)$th waveguide) will be obtained at λ_2 when

$$m\Delta\varphi = \frac{2\pi}{c} n_{eff,c}(\Delta f) m\Delta L, \quad \text{or} \quad \frac{R\Delta\alpha}{\Delta f} = \frac{d_a}{\Delta f} = \left(\frac{n_{eff,c}}{n_{eff}}\right)\left(\frac{\Delta L}{\Delta\alpha}\right)\frac{1}{f_2} \tag{7.17}$$

Here, $f_1 = c/\lambda_1, f_2 = c/\lambda_2$, and $\Delta f = f_1 - f_2$. The ratio of $d_a/\Delta f$ is called the dispersion of the interconnecting waveguides. In practice, there may be optical carriers at a number of close, equally spaced wavelengths, $\lambda_1, \lambda_2, \lambda_3 \ldots\ldots$ (i.e. Δf = constant), in the transmitting channel. When the above dispersion relationship is satisfied, optical carriers at different wavelengths are transmitted to a different output waveguide. This device is called a PHASAR wavelength demultiplexer in WDM fiber systems.

The use of the confocal circular arc configuration of channel waveguides to distribute the power with equal amplitude and phase to the output ports and the use of channel waveguides with unequal length to control the phase distribution are clever uses of the waveguide properties to achieve frequency demultiplexing. The properties of the channel waveguides important to this application are $n_{eff,c}$, the uniformity of $n_{eff,c}$, and the length variations of different channels and the attenuation of the waveguides. The major limitations of the performance of a Star coupler are the insertion loss and the degree of the uniformity that can be achieved.

7.4 Wavelength filters and resonators

7.4.1 Grating filters

The most commonly used wavelength filter or deflector is a periodic grating. There are two types:

(1) Gratings with periodic variation of dielectric constant or reflectivity transverse to the direction of propagation of the incident optical radiation. Within this category, grating diffraction of plane waves was discussed in Section 1.5. Deflection of an incoming planar guided wave into different directions was discussed in Section 5.5.6. They all operate in a similar manner: the incident radiation is diffracted into different directions that correspond to different orders of diffraction. The direction of a given order of diffracted radiation depends on the periodicity of

the grating grooves and the wavelength of the radiation, known as the dispersion of the grating. Large dispersion of these transverse gratings yields reasonably high sensitivity to wavelength variation. The distinct features of this type of grating are: they operate within fairly large range of optical wavelength. The transverse gratings are used commonly in instruments such as grating spectrometers, beam scanners, etc. There is no resonance effect.

(2) Gratings with periodic variation of dielectric constant in the direction of propagation of incident optical radiation. Longitudinal grating diffraction in channel waveguides was discussed in Section 6.3.3. In this case, the grating functions like a reflection filter, known as a distributed Bragg reflector, DBR. The center frequency of reflection is determined by the periodicity of the grating shown in Eq. (6.16). The reflection coefficient is high within a narrow band of frequency that is controlled by the magnitude of the periodic grating perturbation C_g and the length of the grating L. Longitudinal gratings can also be made in optical fibers [5].

Although the analyses of both types of gratings are similar, how to utilize the diffraction effect is very different. One is for deflection of radiation into different directions, utilizing the dispersion properties. The other is to obtain very narrow band filtering. Very narrower band wavelength filter DBR grating reflectors can be obtained by using small C_g and long L.

7.4.2 DBR resonators

When two identical longitudinal grating reflectors are placed consecutively on the same waveguide separated by a distance, d, the guided-wave mode reflected back and forth between the DBR reflectors behaves like a Fabry–Perot resonator at optical wavelengths close to the wavelength that satisfies the Bragg condition. In order to analyze such a resonator, we note that the reflection coefficient Γ of a DBR is given in Eq. (6.21). It has a phase φ_Γ. Since there are two identical DBRs, the resonance condition is

$$2\beta_o d + 2\varphi_\Gamma = 2n\pi \tag{7.18}$$

Like any Fabry–Perot resonator, if d and Γ are large, the resonance will have a much sharper frequency response than the DBR reflector itself. The frequency response of all Fabry–Perot resonators is similar for the same Γ and d. Fabry–Perot resonance filtering of the plane waves is presented in Sections 2.2.1 and 2.2.2. Properties of Fabry–Perot resonators presented in Section 2.2.2 are applicable to DBR resonators.[3]

7.4.3 The ring resonator wavelength filter

Channel waveguides can also be made into a ring (or loop), as illustrated in Figure 7.5(a). Resonances in an isolated waveguide ring occur at frequencies, ω, when the phase shift of a guided-wave mode after one round of propagation is a multiple of 2π,

[3] A single grating with appropriate $\beta_0 d$ can also resonate. When such a resonator is used in semiconductor lasers it is known as the distributed feedback (DFB) lasers.

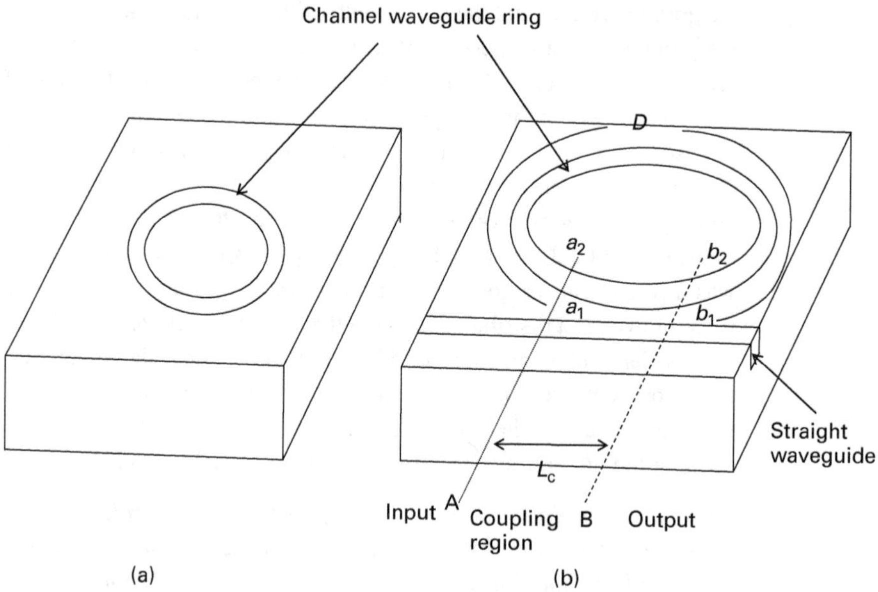

Figure 7.5 Ring resonators. (a) An isolated ring resonator. (b) A ring resonator coupled to a straight input–output waveguide. The coupling region is L_c long.

$$\frac{\omega n_{eff}}{c}(2\pi R) = 2n\pi \tag{7.19}$$

where R is the radius of the ring. The Q-factor and the finesse of the resonator are affected by the propagation loss of the guided wave. However, the ring resonance cannot be utilized in applications unless another waveguide is coupled to the ring. A ring resonator coupled to a straight waveguide via a variable-gap directional coupling interaction is illustrated in Figure 7.5(b). The coupling region is L_c long. The coupling changes the losses of the resonator, as well as providing input and output to the ring.

In order to analyze the resonator shown in Figure 7.5(b), how the ring resonator and the straight waveguide are coupled needs to be discussed first. Then the resonance condition and the finesse and Q of the resonator, which are affected by the coupling, can be analyzed.

(a) Variable-gap directional coupling

Directional coupling between two adjacent waveguides was discussed in Section 6.3.4 for a constant coupling gap. Directional coupling between two waveguides with a variable coupling gap can be approximated as a cascade of short, local directional couplers that has a constant coupling gap within each local section.

Results obtained in Eqs. (6.25) and (6.26) for two coupled waveguides could be rewritten in matrix form for the jth local section with constant coupling gap as

$$\begin{vmatrix} b_{1j} \\ b_{2j} \end{vmatrix} = \begin{vmatrix} t_j & \kappa_j \\ -\kappa_j^* & t_j^* \end{vmatrix} \begin{vmatrix} a_{1j} \\ a_{2j} \end{vmatrix} \tag{7.20}$$

where a_{1j} and b_{1j} are the complex amplitude of the guided wave at the input and output of the local straight waveguide, while a_{2j} and b_{2j} are for the local ring waveguide. t_j and κ_j are the abbreviated expressions of the results given in these two equations.

At the junction of the next section $a_{1,j+1} = b_{1j}$ and $a_{2,j+1} = b_{2j}$. Therefore, the effect of the variable coupling for the total coupling region L_c can be expressed as a matrix,

$$\begin{vmatrix} t & k \\ -k^* & t^* \end{vmatrix}$$

which is the product of all these $[t_j\ \kappa_j]$ matrices.[4]

In Figure 7.5 (b), A marks the beginning of the coupling region and B marks the end of the coupling region. The distance between A and B of the coupling region is L_c. The length of the isolated waveguide in the ring is D. The incident optical guided wave in the straight waveguide is shown to have complex amplitude a_1 at position A before the coupling. The exit optical wave in the straight waveguide is shown to have complex amplitude b_1 at position B. The complex amplitude of the guided wave in the ring resonator is a_2 at A and b_2 at B. a_1, a_2, b_1, and b_2 are related by the variable coupler as

$$\begin{vmatrix} b_1 \\ b_2 \end{vmatrix} = \begin{vmatrix} t & \kappa \\ -\kappa^* & t^* \end{vmatrix} \begin{vmatrix} a_1 \\ a_2 \end{vmatrix} \tag{7.21}$$

Where

$$\kappa\kappa^* + tt^* = 1 \tag{7.22}$$

(b) The resonance condition of the coupled ring

For the guided wave propagating from B to A in the ring, the distance of propagation is D in the isolated waveguide of the ring. Therefore

$$a_2 = b_2 e^{-\alpha D} e^{-jn_{eff}kD} = b_2 e^{-\alpha D} e^{-j\theta} \tag{7.23}$$

Where $\theta = n_{eff}kD$. α is the attenuation coefficient of the guided-wave mode and n_{eff} is its effective index. Note that, from Eq. (7.21), $b_2 = -\kappa^* a_1 + t^* a_2$ where $t = |t|e^{j\varphi_t}$. Thus the phase shift for one round of propagation is $\theta + \varphi_t$. Similar to the discussion of the resonance condition expressed in Eq. (7.19), resonance for CW radiation at a single free-space wavelength λ_o now occurs when $\theta + \varphi_t = 2n\pi$, which is:

[4] Alternatively, the coupled mode equation can be solved with a variable coupling coefficient C.

$$\theta_o + \varphi_t \cong \frac{2\pi n_{eff}(D + L_c)}{\lambda_o} = 2n\pi \tag{7.24}$$

(c) Power transfer

The power transmitted from the input guided-wave a_1 to the output is $|b_1/a_1|^2$, and the power transmitted from the input guided-wave a_1 to the recirculating guidedwave in the ring is $|a_2/a_1|^2$ They have been calculated from Eq. (7.22) by Yariv [6] to be:

$$\left|\frac{b_1}{a_1}\right|^2 = \frac{e^{-2\alpha D} + |t|^2 - 2e^{-\alpha D}|t|\cos(\theta + \varphi_t)}{1 + e^{-2\alpha D}|t|^2 - 2e^{-\alpha D}|t|\cos(\theta + \varphi_t)} \tag{7.25}$$

$$\left|\frac{a_2}{a_1}\right|^2 = \frac{e^{-2\alpha D}(1 - |t|^2)}{1 - 2e^{-\alpha D}|t|\cos(\theta + \varphi_t) + e^{-2\alpha D}|t|^2} \tag{7.26}$$

At resonance, $\cos(\theta + \varphi_t) = 1$. $|b_1/a_1|^2$ drops to zero when $e^{-\alpha D} = |t|$, known as the critical coupling condition. At critical coupling (i.e. $e^{-\alpha D} = |t|$), there is perfect destructive interference between the guided wave in the output waveguide coupled from the ring and from the input. The amplitude $|a_2/a_1|^2$also soars rapidly to a high value near resonance. Its maximum value at resonance is

$$\left|\frac{a_2}{a_1}\right|^2_{\max} = \frac{|t|^2}{1 - |t|^2} \tag{7.27}$$

(d) The free spectral range and the Q-factor

The free spectral range FSR of adjacent resonances at wavelengths, λ_n and λ_{n+1}, is

$$FSR = \omega_{n+1} - \omega_n = \frac{2c\pi}{n_{eff}(D + L_c)} \tag{7.28}$$

As the wavelength changes, "$\theta + \varphi_t$" deviates from its resonance condition by

$$\Delta(\theta + \varphi_t) \cong \frac{n_{eff}(D + L_c)(\omega - \omega_o)}{c} \tag{7.29}$$

As $\Delta(\theta + \varphi_t)$ increases, $|b_1/a_1|$ will increase and $|a_2/a_1|$ will decrease. If there is critical coupling, $|a_2/a_1|^2$ drops to half of its maximum value when

$$\Delta\omega = \frac{(1 - |t|^2)c}{n_{eff}D|t|} = \frac{|\kappa|^2 c}{n_{eff}D|t|} \tag{7.30}$$

Here $\Delta\omega$ is defined as half the linewidth when the intensity drops to ½ of its maximum. Again, we can calculate the Q factor of the resonator to be

$$Q = \frac{\omega_o}{2\Delta\omega} = \frac{\omega_o n_{eff} D|t|}{2c|\kappa|^2} \tag{7.31}$$

Assuming that since $L_c << D$, the finesse of the ring resonator is approximately

$$F = \frac{\omega_{n+1} - \omega_n}{2\Delta\omega} \cong \frac{\pi|t|}{(1 - |t|^2)} \tag{7.32}$$

High Q or high finesse depends on the ability to achieve small α in the ring. There are usually two major factors that may contribute to the propagation loss: (1) the scattering and absorption loss of the channel waveguide and (2) the radiation loss of a curved waveguide.

The radiation loss of a curved waveguide

In a straight channel waveguide, the guided-wave mode can be considered as planar guided waves totally reflected at the lateral boundaries of the core. There is an evanescent field in all the cladding regions because β_z (i.e. $n_{eff}kz$) in the direction of propagation is so large that the propagation wavenumbers of the fields in the lateral directions in the cladding are imaginary, as they are required by the continuity of the fields in the longitudinal direction. Total internal reflection has zero propagation loss in the cladding regions, as long as the propagation wavenumbers in the lateral directions, β_x and β_y, are imaginary. When the waveguide has curvature ρ, the lateral region outside of the curved waveguide fans out. The electromagnetic field in the expanded lateral region propagates with a new expanded coordinate in the z direction, which increases as the distance from the waveguide increases. At some distance away from the waveguide, β of the fields in the lateral direction outside the curve no longer needs to be imaginary in order to meet the continuity condition of $n_{eff}kz$. At this point, the fields become propagating waves and energy will radiate away. The total internal reflection will now have a radiation loss. The smaller the curvature, ρ, the larger the radiation loss. Unger has presented clearly an analysis of the radiation loss in a curved planar waveguide [7]. His analysis shows that the radiation loss from a curved planar waveguide increases exponentially as $k\rho$ is decreased. Kominato *et al.* have shown experimentally that the radiation loss increases dramatically in their waveguides for a bending radius if less than 4 mm [8].

The propagation loss

There are two kinds of propagation loss in waveguides: absorption loss and scattering loss. Volume scattering is usually caused by defects in materials, while surface scattering is caused by defects on the interface created during processing. Low-loss straight channel waveguides have been made in $LiNbO_3$ waveguides by diffusion. However, the propagation loss of curved $LiNbO_3$ waveguides is generally unknown. Absorption loss occurs in semiconductors due to dopands and free carriers. Although absorption in intrinsic semiconductors can be kept very low, substantial surface scattering loss occurs quite often in channel waveguides in high refractive index crystalline medium because of the defects produced in the fabrication processes. For this reason, low-loss semiconductor waveguides are usually ridged waveguides, discussed in Section 5.6. Surface scattering loss is especially high in curved semiconductor waveguides because etching tends to follow crystalline orientation, thereby creating large defects along the curved boundary. Low-loss ring resonators have been made primarily with doped silica waveguides on Si substrates.

There are many similarities between the ring resonator and the Fabry–Perot resonator. Assuming that the reflectivity, R, in the Fabry–Perot resonator and the coupling, κ, in ring resonators are independent of frequency, both Fabry–Perot and ring resonatos have many sharp resonances that are equally spaced in frequency. However, there are four important considerations: (1) In order to get the same Q, D of the ring resonator can be much smaller than the L of the Fabry–Perot resonator. It is hard to get very long low-loss waveguides for Fabry–Perot resonators. It is easy to get a low-loss silica waveguide ring within a ring. Extremely high Q has been obtained in silica ring resonators. (2) The radiation loss of channel waveguide ring resonators increases with decrease in the radius of curvature. Therefore, large ring size needs to be used to obtain very high Q. (3) The FSR, i.e. the separation of the adjacent resonances, is usually much larger in Fabry–Perot resonators than in ring resonators. Kominato et al. have shown that a finesse, F, larger than 30 has been obtained in ring resonators made from GeO_2-doped silica waveguides with a ring radius of 6.5 mm at $\lambda = 1.55\ \mu m$ [9]. However, the FSR of their resonator is only 5 GHz. (4) Techniques such as double ring resonators need to be employed to achieve a wide FSR [10]. A double ring resonator with 100 GHz of FSR and $F > 138$ was demonstrated by S. Suzuki et al. [11].

7.4.4 The ring resonator delay line

When a pulsed optical signal is injected into the input waveguide of a ring resonator, it is coupled into and recirculated in the ring. The optical signal pulse is transmitted periodically to the output, whenever it reaches the output port. Therefore, there are delayed output optical signals at multiple-delay time intervals of $(D + L_c)/v_g$ in ring resonators). For low-loss resonators, the output pulses will be repeated many times. If there are n_g output pulses and if the last pulse is used for signal processing, then the total available time delay of this pulse from the input pulse is n_g times the single time delay interval of the resonator.

Note that the time response of a resonator is related directly to the frequency response of the resonator (e.g. FSR) discussed in the previous section. It is well known that when there are N outputs at discrete frequencies separated at equal frequency intervals $\delta\omega$ around a center frequency ω_o, we obtain mathematically,

$$E = \sum_{-(N-1)/2}^{+(N-1)/2} A e^{j(\omega_o + n\delta\omega + \varphi)t} = A e^{j\omega_o t} e^{j\varphi} \frac{\sin(N\delta\omega t/2)}{\sin(\delta\omega t/2)} \tag{7.33}$$

where A and φ are amplitudes and phases of all the outputs. n identifies the individual field at frequency, $\omega_o + n\delta\omega$ and varies from $-(N-1)/2$ to $+(N-1)/2$ for odd N. E is now periodic in t with period $T = 2\pi/\delta\omega$.

Chapter summary

Passive optical waveguide devices are used to transform the mode profile from one location to another, to divide optical power from various input ports into output ports, to split optical power into different outputs at specified distribution ratio, to redirect optical energy according to its wavelength, to filter optical signals according to wavelength, to create resonances that have very narrow bandwidth, and to provide time delays of signals. Although resonance, beam splitting, and wavelength filtering functions can also be obtained by plane wave and TEM wave devices, the performances of the waveguide devices are much superior. The analyses presented in this chapter let us understand how the performances of these devices are controlled by various design factors. The analyses are equally applicable to channel waveguides and to optical fibers. Note that waveguide devices can only be analyzed by the modal analyses presented in Chapters 5 and 6.

References

[1] L. B. Soldano and Erik C. M. Pennings, Optical multi-mode interference devices based on self-imaging: principles and applications, *Journal of Lightwave Technology*, **13**, 615, 1995.

[2] N. S. Kapany and J. J. Burke, *Optical Waveguides*, Academic Press, 1972.

[3] C. Dragone, Efficient N×N coupler using Fourier optics, *Journal of Lightwave Technology*, **7**, 479, 1989.

[4] M. K. Smit and C. van Dam, Phasar based WDM: devices, principles, design, and applications, *IEEE Journal of Selected Topics in Quantum Electronics*, **2**, 236, 1996.

[5] G. Meltz, W. W. Morey, and W. H. Glen, Formation of Bragg gratings in optical fibers by a transverse holographic method, *Optics Letters*, **14**, 823, 1989.

[6] A. Yariv, Universal relations for coupling of optical power between microresonators and dielectric waveguides, *Electronics Letters*, **36**, 321, 2000.

[7] H. G. Unger, *Planar Optical Waveguides and Fibers*, Oxford University Press, 1977, Section 2.8.

[8] T. Kominato, Y. Ohmori, N. Takato, H. Okazaki, and M. Yasu, Ring resonators composed of GeO_2-doped silica waveguides, *Journal of Lightwave Technology*, **10**, 1781, 1992.

[9] T. Kominato, Y. Ohmori, N. Takato, H. Okazaki, and M. Yasu, Ring resonators composed of GeO_2-doped silica waveguide, *Journal of Lightwave Technology*, **10**, 1781, 1992.

[10] Kauhiro Oda, Norio Takato, and Hiroma Toba, A wide FSR waveguide double-ring resonator for optical FDM transmission systems, *Journal of Lightwave Technology*, **9**, 728, 1991.

[11] S. Suzuki, K. Oda, and Y. Hibino, Integrated-optic double-ring resonators with a wide free spectral range of 100 GHz, *Journal of Lightwave Technology*, **13**, 1766, 1995.

8 Active opto-electronic guided-wave components

Opto-electronic components are optical devices driven by electric signals or converters of optical signals into electrical signals. The operation of these devices depends on various electro-optical effects. In order to discuss these devices, it requires not only optical analysis, but also the analysis of the electro-optical processes and their rf circuit response. Performances of these components should also be evaluated from both optical and electrical points of view.

The most well-known electro-optical effect is probably the amplification of optical radiation by stimulated emission of radiation. When the amplification exceeds the losses and the outputs in a cavity, laser oscillation is obtained [1]. Any discussion of laser oscillators can be divided into two parts: the amplification process and the laser cavity. In gas and solid-state lasers, amplification is obtained by optical pumping or electrical discharge. In semiconductor lasers, the amplification of the guided wave is obtained via current injection in a forward biased p-n junction. When end reflections (or feedback) are absent and when there is a net gain, a laser amplifier is obtained [2]. Cavity analyses are optical; the analysis of optical cavities for solid-state and gaseous lasers was presented in Chapter 4. The waveguide cavity using DBR reflectors was discussed in Chapter 6. In a waveguide laser oscillator, distributed feedback resonance of a grating is also used to form the cavity. However, the discussion of the amplification processes involves extensive knowledge of physics.

The second well-known electro-optical effect is the detection of optical radiation by photo-generation of carriers. When optical radiation is incident on a semiconductor with photon energy greater than the semiconductor bandgap, electrical carriers are generated by absorption of the radiation. Photo-generated carriers in a reverse biased p-i-n junction are then collected and transmitted to the external circuit [3]. In the surface normal photo configuration, the optical radiation is absorbed in the absorbing layer of a reverse-biased diode. The optical analysis of the detector is simple. It consists of the plane wave propagating through a p-i-n diode that has absorption layer. In waveguide photo detectors, the optical radiation is incident onto and absorbed by a waveguide that is also a reverse-biased diode. The absorption takes place over the length of the waveguide [4]. Optical waveguides were discussed in Chapters 5 and 6. The analysis of the transport of photo-generated carriers and their transit times in p-i-n structures requires analysis of semiconductor devices.

In semiconductor lasers and photo-diodes, discussion of carrier injection, stimulated emission, recombination, and carrier transport in semiconductor junctions requires extensive review of semiconductor-device physics. Such a discussion is beyond the scope of this book and there are already many books on lasers and detectors [5,6,7]. Therefore, semiconductor lasers and photo detectors will not be discussed here.

Besides lasers and detectors, the most common electro-optical effects used in active opto-electronic guided-wave devices are the change of the absorptive or refractive properties of materials created by an applied electrical or acoustic signal. For example, in Mach–Zehnder interferometers or directional couplers, the electro-optical change in susceptibility is utilized in order to operate them as modulators or switches. How we analyze modulators and switches is the focus of the discussion in this chapter.

8.1 The effect of electro-optical χ

Let us first understand the effect of $\Delta\chi$.

Propagation of any guided wave is affected by the susceptibility, χ, of the material. In general, χ is complex,

$$\chi = \chi' - j\chi'' \tag{8.1}$$

In a homogeneous lossless isotropic material without any electro-optical effect, we have assumed in previous chapters that

$$\varepsilon = \chi_o\varepsilon_o = n^2\varepsilon_o, \quad \text{or} \quad \chi_o' = n^2 \quad \text{and} \quad \chi_o'' = 0 \tag{8.2}$$

When there is an electro-optical effect, $\Delta\chi$ is produced by the applied electric signal. $\Delta\chi$ has a real part $\Delta\chi'$ and an imaginary part $\Delta\chi''$:

$$\Delta\chi = \Delta\chi' - j\Delta\chi'' \tag{8.3}$$

If there is $\Delta\chi$, the susceptibility is changed from χ_o to χ_{eo},

$$\varepsilon = \chi_{eo}\varepsilon_o = (\chi_o + \Delta\chi)\varepsilon_o = n^2\varepsilon_o + (\Delta\chi)\varepsilon_o \tag{8.4}$$

In general, $\chi_{eo} = \chi'_{eo} - j\chi''_{eo}$, thus the real and the imaginary part of χ_{eo} are

$$\chi'_{eo} = n^2 + \Delta\chi' \quad \text{and} \quad \chi''_{eo} = \Delta\chi'' \tag{8.5}$$

8.1.1 Electro-optic effects in plane waves

If $\Delta\chi$ is created by the electric field of a DC or low-frequency rf signal whose spatial field variation is much slower than the dimensions of the opto-electronic device, then $\Delta\chi$ is considered to be uniform at any instant of time within the device. For a plane wave propagating in the z direction in a material that has a uniform susceptibility χ_{eo},

$$E(z,t) = Ee^{j(\omega t - k_{eo}z)} \tag{8.6}$$

$$k_{eo} = \omega\sqrt{\mu_o \varepsilon_{eo}} \cong \omega\sqrt{\mu_o n^2 \varepsilon_o}\left[\left(1+\frac{\Delta\chi'}{2n^2}\right)-j\left(\frac{\Delta\chi''}{2n^2}\right)\right] \tag{8.7}$$

Therefore,

$$E(z,t) = \left[E e^{j\omega t} e^{-j\omega\sqrt{\mu_o n^2 \varepsilon_o} z}\right] e^{-j\omega\sqrt{\mu_o n^2 \varepsilon_o}\frac{\Delta\chi'}{2n^2} z} e^{-\frac{\omega\sqrt{\mu_o n^2 \varepsilon_o}\Delta\chi''}{2n^2} z} \tag{8.8}$$

and

$$I(z) = \frac{2}{\sqrt{\mu_o/\varepsilon_{eo}}}|E(z,t)|^2 \cong \left[\frac{2E^2}{\sqrt{\mu_o/n^2\varepsilon_o}}\right] e^{-\frac{\omega\sqrt{\mu_o n^2 \varepsilon_o}\Delta\chi''}{n^2} z} \tag{8.9}$$

$$\frac{dI}{dz} = -\alpha I, \quad \text{with} \quad \alpha = \frac{\omega\sqrt{\mu_o \varepsilon_o}\Delta\chi''}{n} \tag{8.10}$$

Therefore, if there is $\Delta\chi'$, a plane wave exhibits an additional electro-optical phase shift, $\omega\sqrt{\mu_o\varepsilon_o}(\Delta\chi'/2n)z$. The effect of $\Delta\chi''$ is different. When $\Delta\chi''$ is positive, the intensity, I, of the plane wave is attenuated by $e^{-\frac{\omega\sqrt{\mu_o\varepsilon_o}\Delta\chi''}{n}z}$. When $\Delta\chi''$ is negative, I is amplified.

8.1.2 Electro-optic effects in waveguides at low frequencies

Similarly, the electro-optical $\Delta\chi'$ and $\Delta\chi''$ in the material creates a change in the effective index and an attenuation of the guided-wave mode. At low frequencies, $\Delta\chi$ varies uniformly in time across the entire device. In order to calculate rigorously the effect of the electro-optic $\Delta\chi$ on guided-wave modes, n^2 in Eq. (5.1) of Section 5.1 of guided-wave modes and other equations need to be replaced by $n^2 + \Delta\chi' - j\Delta\chi''$. However, to find the modes and n_{eff} with the modified n is a major undertaking. If the modes of the waveguide without the electro-optic effects are already known, the effect of $\Delta\chi'$ and $\Delta\chi''$ can be calculated much more easily as perturbations of the original guided-wave modes by $\Delta\chi'$ and $\Delta\chi''$. The perturbation analysis will be used in this chapter.

(a) Effect of $\Delta\chi'$

For analysis of modes involving $\Delta\chi'$, let us consider that there is a change in index from n to $n + \Delta n$, then

$$(n+\Delta n)^2 = n^2 + \Delta\chi' \quad \text{or} \quad \Delta n \cong \frac{1}{2n}\Delta\chi' \tag{8.11}$$

If the change in Δn covers the entire profile of the mode, then the result given in Eq. (8.8) applies directly to guided-wave modes.

However, the rf or DC electric field that creates $\Delta\chi$ may only exist within a region smaller than the size of the guided-wave mode. Then, we describe Δn by:

$$\Delta n(x,y) = \Delta n_o\, g(x,y) \tag{8.12}$$

The guided wave will now have a z variation,

$$A e^{-jn_{eff}\beta_o z}\, e^{-j\Delta n_{eff}\beta_o z} e^{j\omega t} \tag{8.13}$$

From the perturbation analysis in Eq. (6.8), we obtain

$$\beta_o \Delta n_{eff} = \frac{\omega n \varepsilon_o}{2}\Delta n_o \int_{\text{electro-optic region}} g(x,y)(\underline{e_m}\bullet\underline{e_m}^*)\mathrm{d}x\mathrm{d}y \tag{8.14}$$

Since $\underline{e_m}$ is normalized,

$$\frac{n_{eff,m}\beta_o}{2\omega\mu}\iint_{\infty} \underline{e_m}\bullet\underline{e_m}^*\mathrm{d}x\mathrm{d}y = 1$$

$$\Delta n_{eff,m} = \frac{n\Delta n_o}{n_{eff,m}}\Gamma_m \quad \Gamma_m = \frac{\iint_{\text{electro-optic region}} g(x,y)\underline{e_m}\bullet\underline{e_m}^*\mathrm{d}x\mathrm{d}y}{\int_{-\infty}^{\infty}\int_{-\infty}^{\infty} \underline{e_m}\bullet\underline{e_m}^*\mathrm{d}x\mathrm{d}y} \tag{8.15}$$

Γ_m is known as the overlap integral (or filling factor) of the electro-optic effect. Note again that the primary effect of $\Delta\chi'$ is to create a phase shift of the guided wave after propagating a distance z.

(b) **Effect of $\Delta\chi''$**

In the absence of $\Delta\chi''$, the *n*th mode of the waveguide propagating in the *z* direction was expressed as

$$a_n e^{-jn_{eff,n}\beta_o z} e^{j\omega t} \tag{8.16}$$

If $\Delta\chi''$ is uniform over the entire profile of the guided-wave mode, then the result in Eq. (8.8) again applies.

If $\Delta\chi''(x,y)$ exists only in a portion of the waveguide, we will have

$$\Delta\chi''(x,y) = \Delta\chi''_o\, g(x,y)$$

The effect of the change in susceptibility on the guided-wave mode can again be calculated by perturbation analysis. According to Eq. (6.8), we have

$$\frac{\mathrm{d}a_n}{\mathrm{d}z} = -\frac{\Delta a_n}{2}a_n, \qquad \frac{\Delta a_n}{2} = \frac{\omega}{4}\varepsilon_o\Delta\chi''_o \iint_{\text{active region}} g(x,y)(\underline{e_n}\bullet\underline{e_n}^*)\mathrm{d}x\mathrm{d}y \tag{8.17}$$

$$a_n = A e^{-\frac{\Delta a_n}{2}z} \tag{8.18}$$

In view of the normalization of $\underline{e_n}$ we can rewrite the expressions into the form:

$$\frac{\Delta\alpha_n}{2} = \frac{\Gamma\Delta\alpha_o}{2} = \frac{\beta_o}{2n_{eff}}\Gamma_n\Delta\chi''_o, \qquad \Gamma_n = \frac{\iint\limits_{\text{active region}} g(x,y(\underline{e_n}\bullet\underline{e_n}^*)ds}{\iint\limits_{\infty} \underline{e_n}\bullet\underline{e_n}^*ds} \tag{8.19}$$

Γ_n is the overlap integral (or filling factor) of $\Delta\chi''$ in the active region for the nth mode. Note that according to Eq. (8.19), the transmission of the guided wave after a distance L is $T = e^{-\alpha_o L}e^{-\Gamma\Delta\alpha_{av}L}$. If there is residual propagation loss, α_o includes all the residual attenuation that already existed in the absence of the modulation electric field.

8.2 The physical mechanisms to create $\Delta\chi$

$\Delta\chi$ is produced by the electro-optical effects of the rf electric field. Several physical mechanisms that create $\Delta\chi$ are presented here.

8.2.1 $\Delta\chi'$

The most commonly used electro-optical effect to obtain $\Delta\chi'$ is the linear Pockel's effect. An electric field $\underline{F}$ is applied to the electro-optically active material by the electrodes fabricated on the waveguide. $\Delta\chi'$ is proportional $\underline{F}$. The specific relation between $\underline{F}$ and $\Delta\chi'$ depends on material and the configuration of the device. The most common waveguide materials that have been used include $LiNbO_3$, polymers, and III–V compound semiconductors.

Electro-optic materials such as $LiNbO_3$ are often birefringent. In birefringent materials, the optical displacement $\underline{D}$ is no longer parallel to the optical electric field $\underline{E}$. Any birefringent material has principle crystalline axes, $\underline{x}$, $\underline{y}$, and $\underline{z}$. Along the principle axes, $\underline{D}$ and $\underline{E}$ have a matrix relationship:

$$\begin{vmatrix} D_x \\ D_y \\ D_z \end{vmatrix} = \varepsilon_0 \begin{Vmatrix} n_x^2 & 0 & 0 \\ 0 & n_y^2 & 0 \\ 0 & 0 & n_z^2 \end{Vmatrix} \begin{vmatrix} E_x \\ E_y \\ E_z \end{vmatrix} \tag{8.20}$$

For each plane wave propagating along a given direction of propagation $\underline{s}$ ($\underline{s} = s_x\underline{i_x} + s_y\underline{i_y} + s_z\underline{i_z}$), there are two independent solutions in which $\underline{D}$ and $\underline{E}$ are parallel. One is an extraordinary wave that has $\underline{D_e} = n_e\underline{E_e}$. The second is an ordinary wave that has $\underline{D_o} = n_o\underline{E_o}$. Both "$\underline{D_o}$ and $\underline{E_o}$" and "$\underline{D_e}$ and $\underline{E_e}$" are perpendicular to $\underline{s}$. The solutions of $\underline{D}$, $\underline{E}$, e_o, and n_e, for the ordinary and extraordinary plane waves are obtained from the following equation for any specific direction of $\underline{s}$ [8, 9]:

$$\frac{1}{\underline{E}\bullet\underline{D}}\left[\frac{D_x^2}{n_x^2} + \frac{D_y^2}{n_y^2} + \frac{D_z^2}{n_z^2}\right] = 1 \tag{8.21}$$

If we let $\frac{D_x^2}{\underline{E}\bullet\underline{D}} = X^2$, $\frac{D_y^2}{\underline{E}\bullet\underline{D}} = Y^2$, and $\frac{D_z^2}{\underline{E}\bullet\underline{D}} = Z^2$, Eq. (8.21) is simplified to:

$$\frac{X^2}{{n_x}^2}+\frac{Y^2}{{n_y}^2}+\frac{Z^2}{{n_z}^2}=1 \tag{8.22}$$

Eq. (8.22) is referred to in the literature as the index ellipsoid because it has the form of an ellipsoid that has axial lengths, n_x, n_y, and n_z in the (x, y, and z) coordinates. As an example, for a plane wave propagating in the x direction and polarized in the y direction, $D_y = n_y E_y$. For a plane wave in the x direction and polarized in the z direction, $D_z = n_z E_z$.

When an external field $\underline{F}$ is applied to the material, its effect can be expressed as a change of the index ellipsoid. Since $\underline{F}$ may not be parallel to the crystalline axes x, y, and z, the change of index ellipsoid is expressed in general as:

$$\left[\frac{1}{{n_x}^2}+\Delta\left(\frac{1}{n^2}\right)_1\right]X^2+\left[\frac{1}{{n_y}^2}+\Delta\left(\frac{1}{n^2}\right)_2\right]Y^2+\left[\frac{1}{{n_z}^2}+\Delta\left(\frac{1}{n^2}\right)_3\right]Z^2$$

$$+2\Delta\left(\frac{1}{n^2}\right)_4 YZ+2\Delta\left(\frac{1}{n^2}\right)_5 ZX+2\Delta\left(\frac{1}{n^2}\right)_6 XY=1 \tag{8.23}$$

$\Delta(1/n^2)_i$are related to $\underline{F}$ through an electro-optic tensor of the material,

$$\begin{vmatrix}\Delta\left(\frac{1}{n^2}\right)_1\\ \Delta\left(\frac{1}{n^2}\right)_2\\ \Delta\left(\frac{1}{n^2}\right)_3\\ \Delta\left(\frac{1}{n^2}\right)_4\\ \Delta\left(\frac{1}{n^2}\right)_5\\ \Delta\left(\frac{1}{n^2}\right)_6\end{vmatrix}=\begin{Vmatrix}r_{11} & r_{12} & r_{13}\\ r_{21} & r_{22} & r_{23}\\ r_{31} & r_{32} & r_{33}\\ r_{41} & r_{42} & r_{43}\\ r_{51} & r_{52} & r_{53}\\ r_{61} & r_{62} & r_{63}\end{Vmatrix}\bullet\begin{vmatrix}F_x\\ F_y\\ F_z\end{vmatrix} \tag{8.24}$$

The analysis of plane wave propagation in anisotropic media has been presented in a number of references [8,9]. The r_{ij} coefficients of different materials are also given in these references. In general, calculation of the electro-optic effect in waveguides due to $\underline{F}$ is very complex. Fortunately, the optical waveguides and $\underline{F}$ in commonly used devices are oriented along only specific directions of crystalline axes in $LiNbO_3$, polymers, and III–V semiconductors. In these devices, calculation of $\Delta\chi'$ as a function of F is not difficult. In order not to side track from the main objectives of the chapter, only a discussion of Pockel's effect along these special directions is presented here.

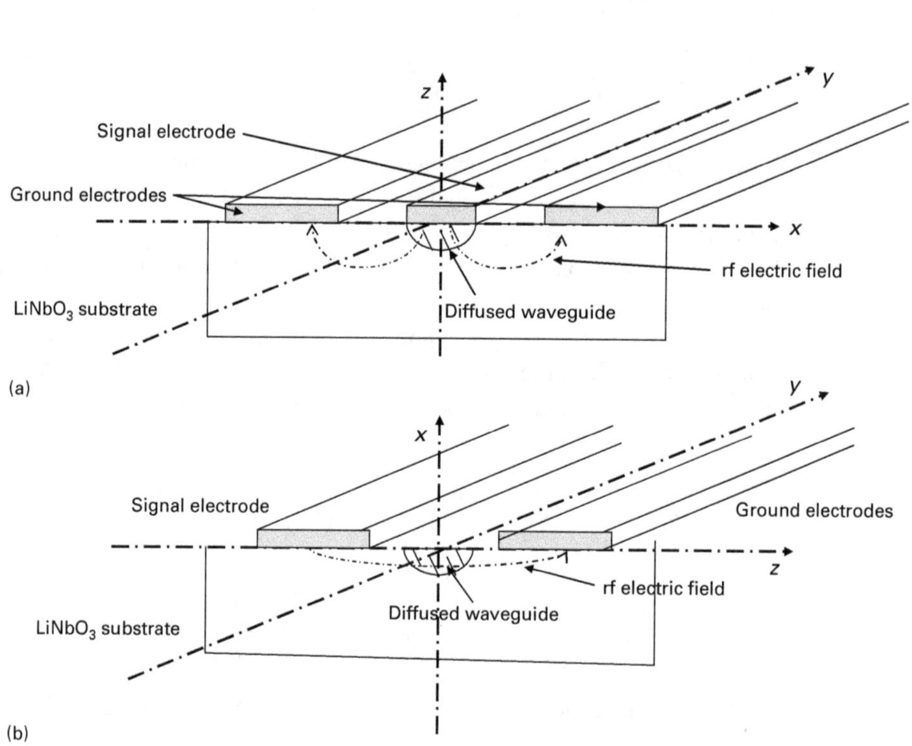

Figure 8.1 Commonly used waveguide and electrode configurations in $LiNbO_3$. (a) A diffused waveguide on z-cut substrate. (b) A diffused waveguide on x-cut substrate. The direction of propagation is in the y direction. The rf field produced by the electrode is oriented in the z direction in the core of the waveguide.

(a) The $LiNbO_3$ waveguide

$LiNbO_3$ waveguides are usually fabricated on x-cut or z-cut substrates with propagation in the y direction of the crystal [10]. Figure 8.1 illustrates these two types of waveguide. Figure 8.1(a) shows a diffused waveguide on z-cut $LiNbO_3$. In order to take advantage of the large electro-optic coefficient, r_{33}, $\underline{F}$ in the waveguide applied by the electrodes is oriented predominantly in the z direction. Let us assume for this discussion that $\underline{F}$ is uniform in the region occupied by the guided-wave mode. Therefore, $F_x = F_y \cong 0$, and the crystalline x, y, and z axes are still the axes of the index ellipsoid with the applied $\underline{F}$. Plane waves propagating in the y direction with both $\underline{D}$ and $\underline{E}$ polarized in the z direction will have $n = n_e'$, while plane waves with x polarization of $\underline{D}$ and $\underline{E}$ will have $n = n_o'$. In these two polarizations, $\underline{D}$ is parallel and proportional to $\underline{E}$. From Eqs. (8.23) and (8.24), we obtain the following n_e' and n_o':

$$n_e' = \left(\frac{1}{n_e^2} + r_{33}F_z\right)^{-1/2} \cong n_e - \frac{1}{2}n_e^3 r_{33}F_z,\ n_o' = \left(\frac{1}{n_o{}^2} + r_{13}F_z\right)^{-1/2} \cong n_o - \frac{1}{2}n_o^3 r_{13}F_z \tag{8.25}$$

Since, like plane waves, modes in waveguides also have a dominant electric and magnetic field polarized perpendicular to the direction of propagation, the n_e' and n_o'

obtained for the plane waves can be used to represent approximately the electro-optical change in material index for the guided waves in the same polarization.

For waveguides on z-cut substrates shown in Figure 8.1(a), guided-wave modes propagate in the y direction. The TE modes have a dominant optical electric field polarized in the x direction, while the dominant optical electric field in the TM modes is polarized in the z direction. $\underline{D}$ and $\underline{E}$ are parallel to each other for the dominant electric field in these two cases. Therefore, in the scalar approximation of the wave equation and for uniform F across the waveguide, the effective index of the TE modes can be calculated approximately by using n_o', while the effective index of TM modes can be calculated approximately by using n_e'.

A diffused waveguide on x-cut substrate is shown in Figure 8.1(b). For waveguides along the y direction on x-cut substrates, the TE modes have the dominant electric field polarized in the z direction, and the TM modes have the dominant electric field polarized in the x direction. $\underline{F}$ applied from the electrodes is predominantly in the z direction. In this case, the effective index of TE modes for uniform F can be calculated using n_e', and the effective index of TM modes can be calculated using n_o'.

In summary, in order to maximize the electro-optic effect, $\underline{F}$ is applied in the z direction to TE modes in x-cut $LiNbO_3$ and to TM modes in z-cut $LiNbO_3$. It is also clear that any change from these two cases, for example, an addition of F_x in addition to F_z may require us to find the x', y', and z' axes and then the new $\underline{D}_e$ and $\underline{D}_o$. The analysis of the effective index of the guided modes would then be much more complicated.

(b) The polymer waveguide

For polymer waveguides, the vertical direction in which the poling field is applied is usually defined as the z direction. The x and y axes are then in the plane parallel to the substrate. Material properties are symmetric in the x and y directions. The non-vanishing elements of the electro-optic tensor are $r_{13} = r_{23}$, $r_{42} = r_{51}$, and r_{33}[11]. The largest electro-optic coefficient is r_{33}. In order to maximize the electro-optic effect, $\underline{F}$ is usually applied in the z direction by the electrodes. The analysis of the electro-optic effect of TM modes is identical to that of the z-cut $LiNbO_3$ with $\underline{F} = F\underline{i_z}$ and a different r_{33} coefficient. On the other hand, the TE modes will not have any electro-optic effect. The value of the r_{33} coefficient will depend on the polymer material engineering. The reported r_{33} is much larger than that of $LiNbO_3$, making polymers very attractive for electro-optic applications. For example, $r_{33} = 130$ pm/V may be anticipated. In comparison, $r_{33} = 30.8$ pm/V in $LiNbO_3$. The challenge for polymer waveguide research is to obtain a material that has a high glass temperature, a low propagation loss, and a large electro-optic coefficient, simultaneously [12].

(c) The III–V compound semiconductor waveguide

GaAs or InP has $r_{41} = r_{52} = r_{63}$. All other r_{ij} are zero. In such a material with cubic crystalline symmetry, $n_x = n_y = n_z = n_o$. Therefore the equation of the index ellipsoid for all III–V compound semiconductor materials is:

$$\frac{X^2+Y^2+Z^2}{n_o^2}+2r_{41}(F_xYZ+F_yZX+F_zXY)=1 \tag{8.26}$$

In GaAs, $n_o = 3.6$ and $r_{41} = 1.1 \times 10^{-12}$ m/V at the 0.9 μm wavelength, and $n_o = 3.3$ and $r = 1.43 \times 10^{-12}$ m/V at 1.15 μm wavelength. Similar values of n_o and r_{41} have been reported in other III–V compound semiconductors. As an example, for a rf electric field, F, in the z direction, we obtain

$$\frac{X^2+Y^2+Z^2}{{n_o}^2}+2r_{41}F_zXY=1 \tag{8.27}$$

Let $z''=z$, $\sqrt{2}x''=x+y$ and $\sqrt{2}y''=-x+y$, then the index ellipsoid in x'', y'', and z'' is:

$$\left(\frac{1}{{n_o}^2}+r_{41}F_z\right)X''^2+\left(\frac{1}{{n_o}^2}-r_{41}F_z\right)Y''^2+\frac{1}{{n_o}^2}Z''^2=1 \tag{8.28}$$

For plane waves propagating along the y'' axis, the major and minor axes of the ellipse for $\underline{D}$ and $\underline{E}$ are the x'' and the z'' axes. Their n values are:

$$\text{For } \underline{D}//\underline{E}//z'' \text{ axis } \; n=n_o \tag{8.29}$$

$$\text{For } \underline{D}//\underline{E}//x'' \text{ axis } \; n=n=n_o-\frac{1}{2}{n_o}^3r_{41}F_z \tag{8.30}$$

For waveguides fabricated on z-cut substrates[1] and oriented in the y'' direction, as shown in Figure 8.2, the electric field is obtained by applying a electrical voltage across the i layer in a reverse biased p-i-n junction. Since the intrinsic layer is usually very thin, the electric field can be very high for a given voltage applied to the electrode. Let us assume again that the electric field is uniform in the intrinsic electro-optic layer. The effective indices of TE modes are found by perturbation analysis using $n=n_o-1/2\ {n_o}^3r_{41}F_z$ for the intrinsic layer. TM modes have no electro-optic effect. Since Eq. (8.26) is symmetric in x, y, and z, this result is applicable to x-cut or y-cut samples with electric field in the x or y directions.

No matter what material, $\underline{F}$, and waveguide configuration, are used, the electro-optic effect produces a $\Delta\chi'$ and Δn_{eff} of the guided-wave mode. After propagating a distance z, Δn_{eff} produces a phase shift $\Delta\varphi$ in the guided-wave mode where $\Delta\varphi=\Delta n_{eff}\,\omega\sqrt{\mu_o\varepsilon_o}z$. The relationship between $\Delta\chi'$ and F depends not only on the material properties, but also on waveguide and electrode configurations.

[1] Typically semiconductor waveguides are fabricated by epitaxial growth of the core and cladding layers that are parallel to the substrate surface. In order to apply the rf electric field most effectively, the core layer is usually an i layer sandwiched between n- and p-type semiconductor layers, and a reverse-biased voltage is applied to the p-i-n junction. Electrical voltage is applied across the ground and the signal electrodes. Thus $\underline{F}$ is usually in the direction of the cut of the sample. The channel ridge waveguide is often formed by etching.

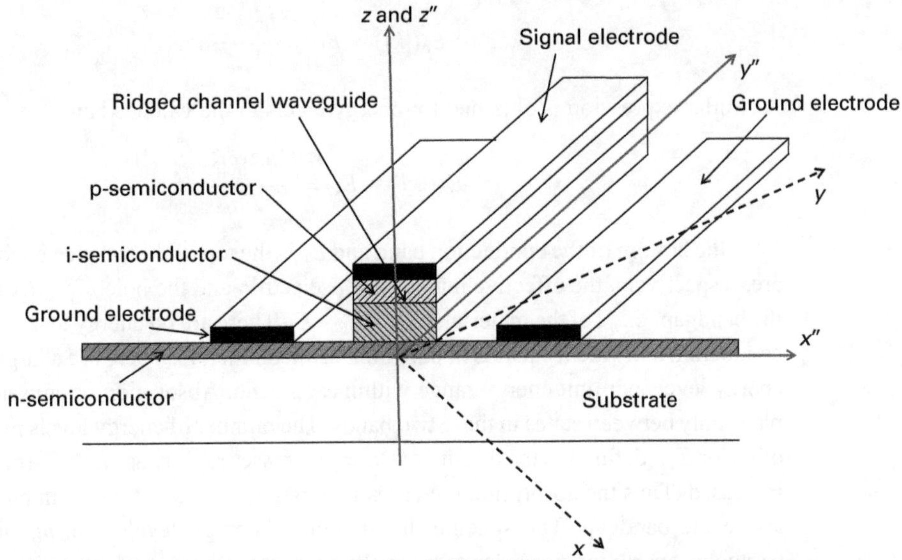

Figure 8.2 Examples of an electro-optic waveguides on III–V semiconductors. The ridged channel waveguide on the *z*-cut substrate is oriented along the *y*″ direction, which is 45° from both the *x* and the *y* axis. The high-index intrinsic core of the waveguide is sandwiched between a p-doped and an n-doped semiconductor. A reverse bias is applied from the electrodes to the i layer through the p-i-n junction.

8.2.2 $\Delta\chi''$ in semiconductors

In order to understand $\Delta\chi''$ in semiconductors, a brief discussion of the absorption properties of semiconductors is presented here.

(a) Stimulated absorption and the bandgap

Absorption of optical radiation in semiconductors can be understood most easily through the following abbreviated description of the process.

In semiconductors, electron and holes undertake stimulated emission and absorption. How such carriers are generated, transported, and recombined has been discussed extensively in the literature [13,14]. We note in particular that they occur in a periodic crystalline material. The energy levels of free electrons and holes are distributed in bulk crystalline semiconductors within conduction and valence bands. Within the conduction band and the valence band, each energy state has a wave function of the form

$$\Psi_C(\underline{r}) = u_{C\underline{k}}(\underline{r})e^{j\underline{k}\bullet\underline{r}}$$

where $u_{C\underline{k}}(\underline{r})$ has the periodicity of the crystalline lattice. The energy of electrons in the conduction band for a state with given $\underline{k}$ (in the parabolic approximation of the energy band structure) is

$$E_e(|\underline{k}|) - E_C = \frac{\hbar^2 |\underline{k}|^2}{2m_e} \tag{8.31}$$

A similar expression is obtained for energy levels in the valence band,

$$E_h(|\underline{k}|) - E_V = -\frac{\hbar^2 |\underline{k}|^2}{2m_h}. \tag{8.32}$$

E_C is the bottom of the conduction band and E_V is the top of the valence band. m_e and m_h are, respectively, the effective masses of the electron and the hole. $E_C - E_V$ is known as the bandgap, E_{gap}, of the material, $E_{gap} = E_C - E_V$. There are no energy levels between the conduction and valence bands in pure bulk semiconductors. There are a large number of energy levels per unit energy range within each band. Absorption or emission can take place only between states in these two bands. The number of energy levels per unit range of E_e (or E_h) (defined as the density of states) increases rapidly as "$E_e - E_c$ (or $E_v - E_h$)" is increased. Thus the absorption increases very rapidly for radiations with photon energy above the bandgap. The specific distribution of energy levels (i.e. m_e, m_h, and the parabolic approximation) depends on the material. Because of other issues, such as phonon interactions and electric-field-induced exciton effects, the variation of absorption near or below the band edge as a function of optical wavelength is rapid, but not abrupt.[2] In any case, whenever there is a change of bandgap (created by $\underline{F}$), there is a $\Delta\chi''$ for any radiation that has a photon energy just below the bandgap.

To maximize $\Delta\chi''$, the photon has an energy just below the bandgap. Since the absorption of a photon depends on the availability of energy states, the photon absorption depends on the changes in the bandgap and the energy states due to $\underline{F}$. This is known as the Frantz–Keldish effect in bulk semiconductors. Much research has been devoted to create material structures that will have energy states that provide a more rapid variation of the absorption of the phonon with the applied $\underline{F}$. The most effective method to do that is to obtain a material which has a quantum-confined Stark effect, QCSE.

(b) The quantum-confined Stark effect, QCSE

In order to understand QCSE, we must first understand quantum wells and their energy levels. We then need to understand the exciton absorption of these quantum wells, followed by how they can be utilized to obtain a rapid change in $\Delta\chi''$ by an applied electric field F.

Energy levels in quantum wells

A quantum well double heterostructure in semiconductors consists of a thin layer of material, called the well, that has a smaller bandgap, E_Γ, sandwiched between materials with a larger bandgap, *Eg*, called the barrier. These layers are typically III–V group semiconductors with different compositions that are grown epitaxially on a lattice matched to the GaAs or InP substrates. The thickness of the well L_W is typically 50 to 150 Å. The barrier is just thick enough (e.g. 50 to 100 Å) to isolate the wells.

[2] This is known in physics as the Erbach tail.

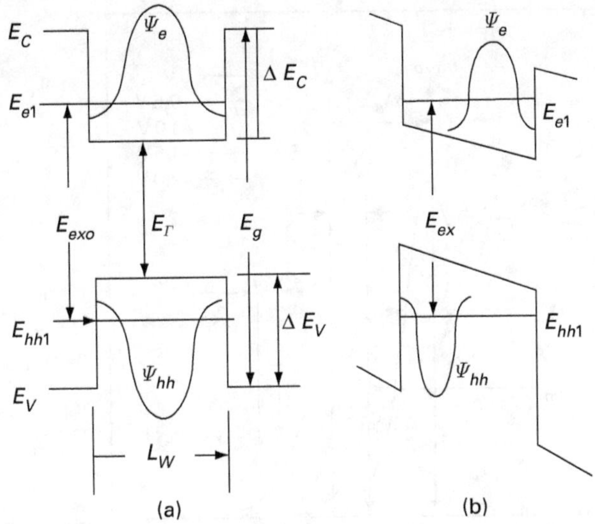

Figure 8.3 Potential energy diagrams, energy levels, and energy states in quantum wells. (a) At zero electric field. (b) At a bias electric field.

Figure 8.3(a) shows a typical one-dimensional potential diagram of the conduction and valence bands as a function of thickness position x at zero applied electric field. At the interface of the well and the barrier, there are discontinuities in the conduction band edge ΔE_C and the valence band edge ΔE_V, $\Delta E_C + \Delta E_V = E_g - E_\Gamma$. Quantum mechanical calculations of energy states in such potential wells yield discrete energy levels, E_e, for electrons in the conduction band and discrete energy levels, E_h, for holes in the valence band [15]. In the example illustrated in Figure 8.3(a), E_{e1} is the lowest-order energy level for electrons in the conduction band. The energy state for this energy level is illustrated as ψ_e. Some holes in the valence band have a heavier mass, called heavy holes, and some holes have a lighter mass, called light holes. The highest hole energy in the valence band is usually for heavy holes. Only E_{hh1} and its energy state ψ_{hh} are illustrated in Figure 8.3(a). Other higher electron levels and lower hole levels such as E_{e2} and E_{lh1} are not shown here. The energy states Ψ demonstrate that electrons and holes in a quantum well are confined in the thickness direction that we designated as the x direction. A multiple quantum well (MQW) structure consists simply of multiple of quantum wells separated by barriers.

Exciton transitions and absorption

E_e and E_h are the only energy levels of the electrons and holes in the thickness, x, direction. The total energy of electrons and holes is the sum of their energy in the x direction, i.e. $E_e + E_h$, and the energy of an electron–hole pair in the y–z pane, E_{yz}. In order to understand the energy of the electron–hole pair in the y–z plane, let us consider first the energy of an electron–hole pair in three dimensions in bulk semiconductors. When electron–hole pairs are created by absorption of a photon, they are initially close to each other. In bulk semiconductors, such electron and hole pairs will experience

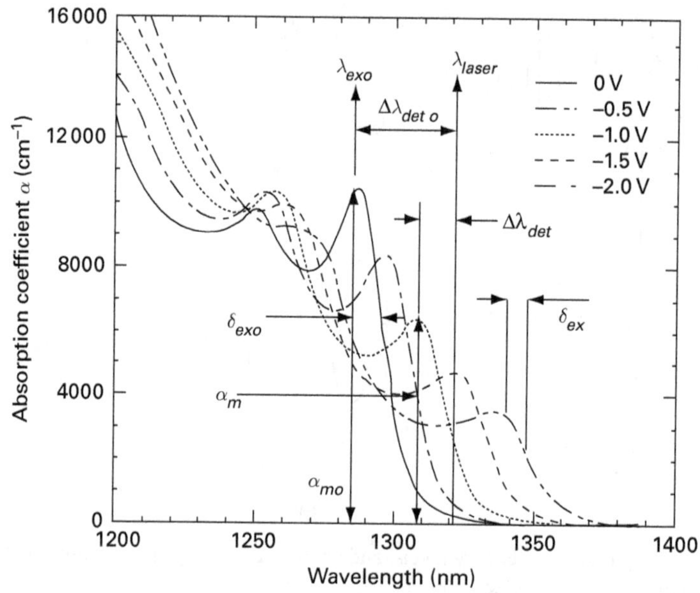

Figure 8.4 Absorption spectra of InAsP/GaInP multiple quantum well at different bias voltages.

mutual three-dimensional Coulomb forces similar to those present in a hydrogen atom. The energy of such an electron–hole pair is lower than the energy of free electrons and holes; this electron–hole coordination gives rise to a set of energy levels (called exciton levels) just below the bandgap. The exciton spectra in bulk materials have been directly observed only at very low temperatures. The situation is different in quantum wells. In the y and z directions of the quantum well, electrons and holes are also subject to periodic potentials and forces in a bulk crystal. The quantum confinement in the x direction increases the binding energy of the exciton. Therefore exciton transitions are directly observable at room temperatures.

Exciton absorption in quantum wells (i.e. $\Delta\chi''$) has been observed directly at room temperatures and under applied electric fields. The stimulated transition of heavy hole excitons takes place at photon energy just below $E_{e1} - E_{hh1}$. The solid curve in Figure 8.4 shows the TE polarized absorption spectrum of an $InAs_{0.4}P_{0.6}$(93 Å thick wells)/ $Ga_{0.13}In_{0.87}P$(135 Å thick barriers) multiple quantum well (MQW) at zero applied electric field [16]. The heavy hole exciton transition has a transition wavelength shown as λ_{exo} with a line width δ_{exo}. For this sample the half width at half maximum δ_{exo} is 6 meV. A second transition with a less distinct absorption peak due to a light hole can also be seen in this figure at $\lambda = 1.250$ μm.

Note that the absorption coefficient α will be dependent on the polarization of the electric field because the matrix element[3] for any induced transition between an electron and a hole is polarization dependent. For a TE guided wave in the y–z plane, its electric

[3] The transition probability of any two energy states is proportional to the matrix element of the applied radiation.

field is polarized in the y–z plane. Its absorption coefficient will be the same as α for a plane wave propagating in the x direction. For TM modes in a waveguide oriented in the y–z plane, its dominant electric field will be in the x direction, and it will have a different value of α.

The quantum-confined Stark effect (QCSE)

Under the application of an electric field in the x direction, the potential wells are tilted as shown in Figure 8.3(b). The quantum mechanical solution for the energy values in the quantum well indicates usually a reduction in $E_{e1} - E_{hh1}$. Therefore, the exciton absorption line at E_{exo} shifts normally toward longer wavelengths (i.e. the absorption peak is at smaller photon energy), known as a red shift. Occasionally the shift in a specific potential well configuration may be toward a shorter wavelength, known as a blue shift. This is the quantum-confined Stark effect (QCSE) [17, 18, 19]. Note also that as the potential wells are tilted, the wave functions of energy states for electrons and for holes are also shifted to the opposite side of the quantum well, as illustrated in Figure 8.3(b). Since the amplitude of the stimulated absorption between the two energy states depends on the matrix element of the electric dipole connecting ψ_e and ψ_{hh}, the shift of energy state function to the opposite side of the quantum well will produce a reduction of the exciton absorption as the electric field is increased.

The QCSE, the reduction in the absorption coefficient at the exciton peak α_m and the broadening of the exciton line width δ_{ex}, are clearly demonstrated experimentally for a specific sample in Figure 8.4 as the applied voltage is increased. In this case, the electric voltage shown in the figure is applied across a reverse-biased p-i-n junction that has an i-layer approximately 0.5 μm thick (containing 21 periods of quantum wells and barriers). Therefore the electric field F in units of V/cm applied to the quantum well is approximately 2×10^4 times the applied voltage divided by the thickness of the device. In this figure, the laser radiation at wavelength λ_{laser} is detuned from the exciton peak at $F = 0$ by $\Delta\lambda_{deto}$. As the QCSE increases, the absorption coefficient in the MQW for λ_{laser} shown in Figure 8.4 will first increase when $\Delta\lambda_{det} > 0$ and then decrease when $\Delta\lambda_{det} < 0$, as F is increased. When the electric voltage is changed from 0.5 V to 1.5 V, the change in absorption coefficient at λ_{laser} shown in the figure is $\Delta\alpha \cong 4000$ cm. Thus, $\Delta\alpha/\Delta F \cong 200 \times 10^{-3}$/V.

Figure 8.5 shows the measured QCSE and the calculated shift of $E_{e1} - E_{hh1}$ of the sample used in Figure 8.4. The discrepancy has been attributed to the variation of the exciton binding energy as the applied electric field is varied. Figure 8.6 illustrates $\Delta\alpha$ at different detuning energy and reverse biases that can be obtained in this sample. Note the importance of small δ_{exo} and appropriate choice of detuning energy and reverse bias in order to maximize $\Delta\alpha$ for a given ΔF.

QW structures became a reality because the epitaxy technology in material growth provided the means for control of the quantum well layer thickness and the smoothness of the interfaces up to atomic-level accuracy. Quantum wells and barriers are always parallel to the surface of the substrate. The direction of the applied electric field needs to be perpendicular to the substrate surface. The most effective way to apply such an electric field is by fabricating a p-i-n structure parallel to the substrate surface where the

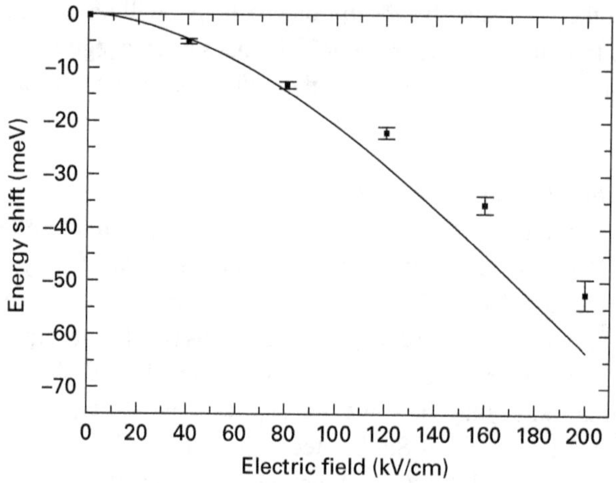

Figure 8.5 Quantum-confined Stark effect of the E_{e1}–E_{hh1} transition versus the electric field. The scattered signs are experimental data. The solid curve is calculated theoretically using the effective width model. Taken from ref. [15] with permission from X. Mei.

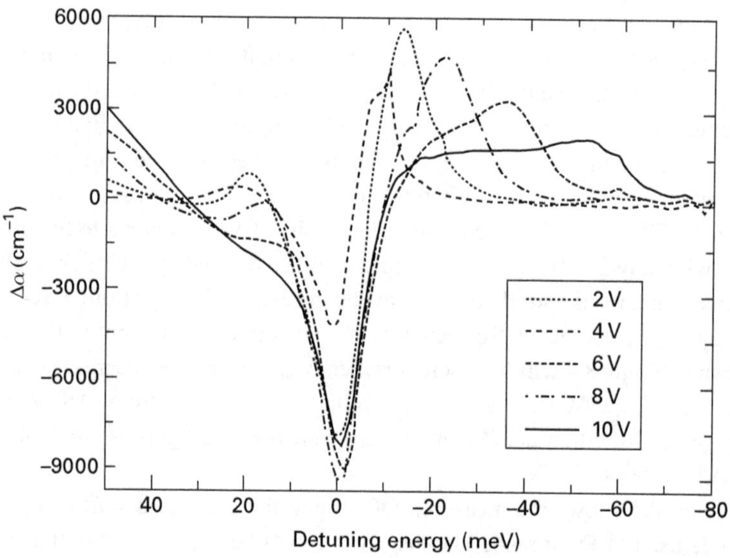

Figure 8.6 Change of absorption coefficient at different detuning energy and reverse biases. Taken from ref. [15] with permission from X. Mei.

MQW constitutes the i layer. In a reverse-biased p-i-n structure, the electrical field is predominantly perpendicular to and focused in the i layer. This is the way in which α is obtained in Figures 8.4, 8.5, and 8.6; it is α for the TE polarization. Note that the measured α is the averaged absorption coefficient of the entire MQW layer. For a given electric field, the actual absorption takes place only in the well, not in the barrier. Therefore $\Delta\alpha/\Delta F$ is increased by using a thinner barrier layer. The minimum barrier thickness will be governed by the decay of the energy state functions, ψ_e and ψ_h, in the barrier. The conventional guideline is that the barriers be thick enough so that the energy states of adjacent wells will not significantly interact with each other.

QCSE provided the largest $\Delta\chi''$ that can be obtained by a given $\underline{F}$. Figure 8.4 *demonstrated the change of absorption coefficient by QCSE in a specific sample for a specific photon energy. Quantum wells can also be designed to provide a more rapid change in the absorption coefficient for other photon energies. The goal of the quantum well material design is to maximize $\Delta\chi''/F$. The goal of the electrode design is to maximize F.*

In this section, several common physical mechanisms for obtaining $\Delta\chi$ have been presented. It is important to know how these mechanisms function so we can analyze the operation of active components. How to obtain maximum $\Delta\chi'$ or $\Delta\chi'$ for a given F and how to maximize F for a given rf signal power are clearly issues of individual design. Discussion of different designs is beyond the scope of this book. In the rest of this chapter, discussions of active components will be presented only in terms of a given $\Delta\chi'$ or $\Delta\chi''$.

8.3 Active opto-electronic devices

Commonly used electro-optical active components that will be presented here include the phase modulator, the directional coupler modulator/switch, the Mach–Zehnder modulator, and the electro-absorption modulator.

8.3.1 The phase modulator

The phase modulator is a very simple device. Phase modulation of the guided wave is obtained whenever an electric field is applied to a waveguide fabricated on electro-optic materials. $\Delta\chi'$ is created by the electric field.

Let there be a change of the material index,

$$\Delta n(x,y) = \Delta n_o g(x,y) \tag{8.33}$$

Δn creates Δn_{eff} for the guided wave; this guided wave propagating in a waveguide now has the phase variation,

$$A e^{-jn_{eff}\beta_o z} e^{-j\Delta n_{eff}\beta_o z} e^{j\omega t} \tag{8.34}$$

In Eq. (8.15), it has already been shown that

$$\Delta n_{eff,m} = \frac{n\Delta n_o}{n_{eff,m}}\Gamma_m, \quad \Gamma_m = \frac{\iint\limits_{\text{electro-optic region}} g(x,y)\underline{e_m}\bullet\underline{e_m}^* dxdy}{\int\limits_{-\infty}^{\infty}\int\limits_{-\infty}^{\infty} \underline{e_m}\bullet\underline{e_m}^* dxdy} \tag{8.35}$$

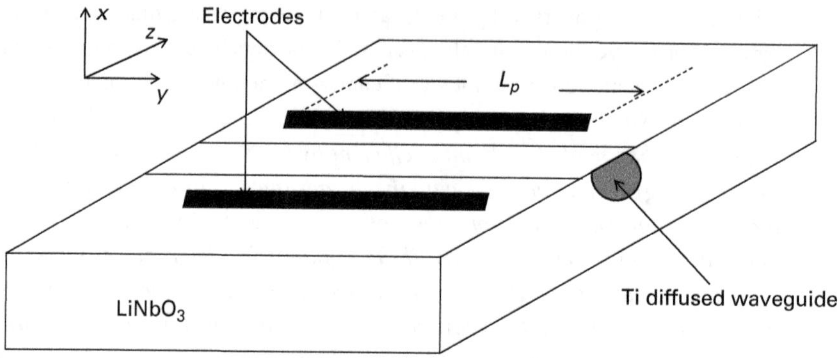

Figure 8.7 An x-cut $LiNbO_3$ phase modulator.

Therefore the electro-optic effect has created a phase shift, $\Delta n_{eff}\beta_o L$, where L is the length of the waveguide. Figure 8.7 illustrates a phase modulator on x-cut $LNbO_3$.

Note that the F that created the Δn is produced by the voltage, V, applied to the electrodes. A crude approximation to estimate F for a given V on the electrode is:

$$F_z = \frac{V}{d} \tag{8.36}$$

where d is the separation of the electrodes. In reality, calculation of F by V can be complex; nevertheless, F is proportional to V. Therefore, we can replace d by d_{eff}.

Note that the performance of a phase modulator will be measured by the voltage required to achieve a given $\Delta\varphi$. The larger Δn_o of the material and Γ, the larger is Δn_{eff}. The smaller d_e, the larger is F/V. The longer L, the lower the required F for a given Δn_{eff}.

The electrodes shown in Figure 8.7 are just a portion of the rf circuit, driving the modulator. In order to discuss the modulator performance as a function of the rf frequency, we also need to know how V is produced by the rf source through the electrical circuit. For our discussion here, let us consider the simplest case. The electrode is represented electrically by a capacitance, C. It is connected in parallel with a matching resistance, R, to match the rf signal source. The combination of R and C is driven by a current source that is represented by i_s in parallel with an internal resistance, R_s. Since the impedance of C is inversely proportion to frequency, V across the electrode drops to ½ from DC to ω_c when $\omega_c\left(RR_s/(R+R_s)\right)C = 1$. ω_c is known as the bandwidth of the modulator. In other words, the performance of the modulator is measured not only by the phase shift that can be achieved by a given rf input power, but also by its RC bandwidth.

8.3.2 The Mach–Zhender modulator

The Mach–Zhender (MZ) interferometer has already been discussed in Section 6.7.3. When there are shifts of phase, $\Delta\varphi_A$ of the guided wave in arm A and $\Delta\varphi_B$ of the guided-wave in arm B, there is a relative phase difference of the modes at the end of the arms,

$\Delta\varphi = \Delta\varphi_A - \Delta\varphi_B$. Note that as $\Delta\varphi$ is varied, the sum of the guided waves in the two arms at the input of the Y-branch coupler is

$$(\psi_a + e^{j\Delta\varphi}\psi_b)e^{-j\varphi_a} = [C(\psi_a + \psi_b) + D(\psi_a - \psi_b)]e^{-j\varphi_a}$$

The amplitude, C, of the symmetric mode $(\psi_a + \psi_b)$ is

$$C = \frac{1}{2}(1 + \cos\Delta\varphi - j\sin\Delta\varphi)$$

Only the symmetric mode is transmitted to the output of the Y-branch. The anti-symmetric mode is dissipated. Therefore, the amplitude of the guided wave in the output waveguide is C. As an example, when the relative phase shift is π, the amplitude of the output guided wave is 0; when the relative phase shift is 0, the output is 1. The output optical power is proportional to C^2.

Whenever an MZ interferometer is fabricated on an electro-optic material that yields Δn, each individual arm of the interferometer functions as a phase modulator, discussed in the previous section. The phase of the guided wave propagating for a distance L_p in the arm is $\Delta\varphi = \Delta n_{eff}\beta_o L_p$. Δn_{eff} is given in Eq. (8.35). It is proportional to the applied voltage, V. In single-arm modulators, an electrode such as that shown in Figure 8.7 is applied to just one arm. In push–pull modulators, electrodes are applied to both arms. In this case, the voltage applied to the two sets of electrodes is reversed, thereby doubling $\Delta\varphi$.

The performance of a Mach–Zehnder modulator is be measured by the voltage required to achieve a given depth of modulation.[4] The larger Δn_o of the material and Γ, the larger is Δn_{eff}. The smaller d_e, the larger is F/V. The longer L_p, the lower the required V to obtain a given $\Delta\varphi$. However, the capacitance, C, representing the electrical behavior of the electrode is proportional to L_p/d_e. It is larger for larger L_p and smaller d_e. The MZ modulator will have the same RC-limited bandwidth as the phase modulator.

8.3.3 The directional coupler modulator/switch

Directional coupler is discussed in Sections 6.3.4, 6.6, and 7.2.2, and illustrated in Figure 6.4. Let us consider a directional coupler that has two identical waveguides coupled together in the interaction region from $z = 0$ to $z = W$, as shown in Figure 6.4. In order to operate a directional as a modulator or switch, the waveguides are fabricated on electro-optical material. Electrodes such as those used in phase modulators are fabricated on waveguides in the interaction region to obtain Δn_{eff} in the waveguides. As shown in Eqs. (6.25) and (6.26), when the power is incident to one waveguide the output in the other waveguide P_{out} is proportional to

[4] Note that, although $\Delta\varphi$ is linearly proportional to V, the output power of an MZ modulator is not linearly proportional to $\Delta\varphi$. In digital applications, the desired modulation depth is determined by the required on/off ratio of optical power. In analog applications, the modulator may be biased at a specific V_{DC} in addition to the signal, V_{rf}. The desired modulation depth is determined by the rf-modulated optical output variation that satisfies the linearity requirement.

$$P_{out} \propto \sin^2\left(\sqrt{C_{BA}C_{AB} + \left(\frac{\Delta\beta}{2}\right)^2}\right) z, \qquad \Delta\beta = \beta_A - \beta_B = (n_{effA} - n_{effB})\beta_o \tag{8.37}$$

At different $\Delta\beta$ values, P_{out} varies from $P_{out}=0$, when $C_{BA}C_{AB} + \left((\Delta\beta/2)^2\right) = 0$ (or $n\pi$) to $P_{out}=1$ when $C_{BA}C_{AB} + \left((\Delta\beta/2)^2\right) = \pi/2$ (or $(2n+1)\pi/2$). Just like the phase modulators, n_{effA} and n_{effB} are proportional to the voltage applied to the electrodes on waveguide A and waveguide B. Reverse of the voltage will reverse Δn_{eff}.[5]

Such a directional coupler is a switch when the input is incident to one waveguide and the output is taken from the other waveguide. It is a modulator when the input power is incident to one waveguide and the output power is taken from the same input waveguide. Note that the electric field, F, and the voltage, V, on the electrode will vary as a function of the rf frequency in the same manner as the phase modulator discussed in Section 8.3.1. The smaller d_e, the larger is F/V, and the larger is its electrical capacitance, C. Similar to the phase modulator, there is an RC-limited bandwidth of the directional coupler or switch.

8.3.4 The electro-absorption modulator

Figure 8.8 illustrates an EA waveguide modulator. It shows a ridged waveguide on an InP substrate, where the waveguide core consists of a quaternary InGaAsP layer sandwiched

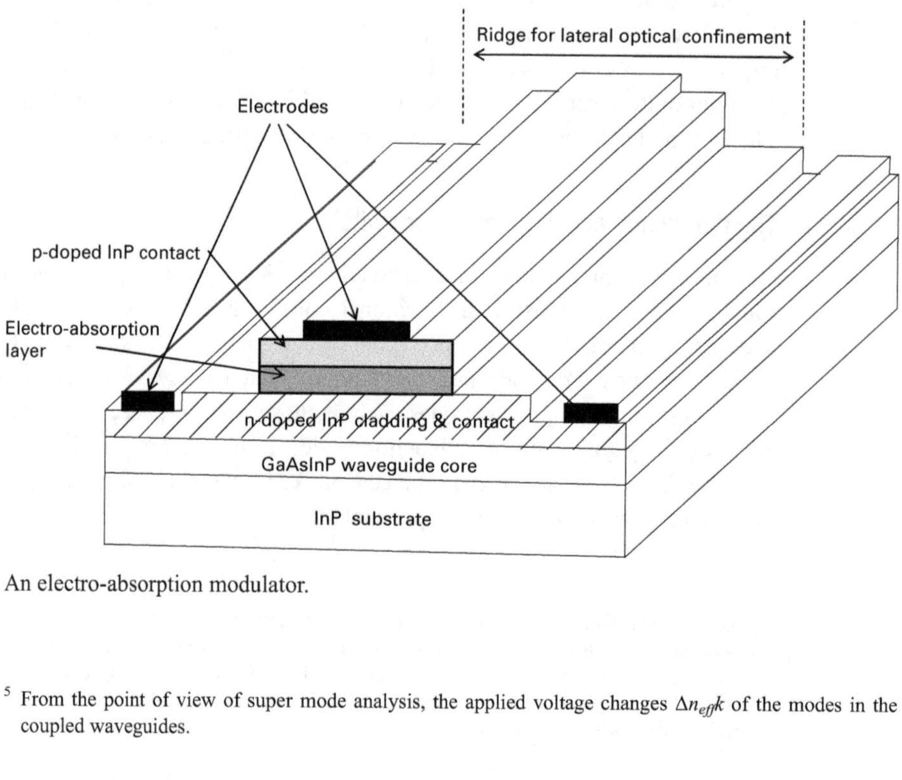

Figure 8.8 An electro-absorption modulator.

[5] From the point of view of super mode analysis, the applied voltage changes $\Delta n_{eff}k$ of the modes in the coupled waveguides.

among lower-index InP layers in the vertical direction. A ridge is etched on the top cladding layer to provide the mode confinement in the lateral direction. Within the cladding layers above the core, there is an EA layer, which is a QCSE layer, discussed in Section 8.2.2. In order to provide a large electric field from a given applied voltage, the EA layer is the intrinsic layer of a reverse-biased p-i-n diode. In addition to the electrode on top of the EA layer, the n layer under the QCSE layer serves as the bottom electrode. It is connected electrically to the electrodes on the sides via the n layers.

As we apply a voltage, V, to the electrodes, it creates $\Delta\chi''(x,y)$ via QCSE. As shown in Eqs. (8.18) and (8.19), the amplitude of the guided wave is reduced by this $\Delta\chi''$. V consists of a DC bias voltage V_b and an rf voltage V_{rf}. Thus a DC attenuation of the guided-wave α_b is created by V_b in addition to the attenuation $\Delta\alpha_n$ created by V_{rf}.

Let the amplitude be a_n at $z = 0$, then

$$a_n(z = L) = a_n e^{-\frac{\Delta\alpha L}{2}} = a_n e^{-\frac{\Delta\alpha_b L}{2}} e^{-\frac{\Delta\alpha_n L}{2}} \tag{8.38}$$

$\Delta\alpha_b$ is the attenuation due to V_b, and $\Delta\alpha_n$ is the attenuation due to V_{rf}

$$\Delta\alpha_n (or\ \Delta\alpha_b) = \Gamma_n \Delta\alpha_o = \frac{\beta_o}{n_{eff}} \Gamma_n \Delta\chi''_o \tag{8.39}$$

The power carried by the nth mode of guided wave is reduced by $e^{-(\Delta\alpha_b+\Delta\alpha_n)L}$ at the output end of the modulator. This is the essence of an electro-absorption, EA, modulator.

The performance of the EA modulator is measured by the smallest ΔV_{rf} required to achieve a given modulation of optical power. In digital applications, it is the total $e^{\Delta\alpha L}$ for ΔV_{rf} that produces the on/off signal. For analog applications, it is the linear portion of $e^{\Delta\alpha n L}/\Delta V_{rf}$[6] that is used for transmission of analog signals [10]. Note again that the electrical circuit representation of the electrodes across a p-i-n junction is a capacitance in parallel with a junction resistance. Therefore, similar to the phase modulator, the MZ modulator, and the DC modulator, the frequency response is RC limited.

8.4 The traveling wave modulator

The discussions in Section 8.3 assume that $\Delta\chi$ in the device is the same at any given instant of time. When the wavelength of the rf signal is comparable or shorter than the length of the electrode, the voltage- and current-induced modulation electric field F is no longer the same across the device at any instant of time.

Usually the rf voltage is applied to the electrodes at the start of the electrode at $z = 0$. The electrode is designed as an electrical transmission line. The rf voltage propagates as a traveling wave on the electrode with[7]

$$V_{rf}(z,t) = V_o \cos(\omega_m t - n_{eff,m} kz) \tag{8.40}$$

In a phase modulator, it produces an instantaneous Δn_{eff} that varies locally as

[6] V_b is adjusted to maximize the linear variation.
[7] This assumes that the microwave is a forward-propagating wave without reflection.

$$\Delta n_{eff,rf}(z,t) = \frac{1}{2}\Delta n_{eff}^{o}[e^{j(\omega_m t - n_{eff,m}kz)} + \text{complex conjugate}] \tag{8.41}$$

In the meantime, the optical wave propagates in the z direction with propagation wave number $n_m k$. $\Delta n_{eff}(z,t)$ seen by the photons entering the waveguide in the mth mode at $z = 0$ is

$$\Delta n_{eff} = \Delta n_{eff}^{o}\cos[\omega_m t - (n_m - n_{eff,m})kz] = \Delta n_{eff}^{o}\cos(\omega_m t - \delta nkz) \tag{8.42}$$

where $\delta n = n_m - n_{eff,m}$. The electric field of the mth mode of the guided wave is:

$$\underline{E(x,y,z)} = \underline{Ae_m(x,y)}e^{-jn_{eff,m}z}e^{-j\left(\Delta n_{eff,m}{}^{o}\int_0^z \cos(\omega_m t - \delta nkz)dz\right)} \tag{8.43}$$

Therefore, the total phase shift $\Delta\varphi$ of the guided wave produced by the CW traveling wave rf signal for a distance of L_p is

$$\Delta\phi = \Delta n_{eff,m}{}^{0}\int_0^{L_p}\cos(\omega_m t - \delta nk)dz = \Delta n_{eff,m}^{o} L_p \frac{\sin\left(\frac{\delta nkL_p}{2}\right)}{\left(\frac{\delta nkL_p}{2}\right)}\cos(\omega_m t - \delta nkL_p) \tag{8.44}$$

Note that when the microwave equivalent index, n_m, matches the optical effective index $n_{eff,m}$, then $\delta n = 0$ and the $\Delta\varphi$ modulation is a constant at all rf frequencies. In that case, large $\Delta\varphi$ can be obtained with large L_p. However, $\Delta\varphi$ is sensitive to δn at large kL_p. When $\delta nkL_p/2 = 1.4$, $\Delta\varphi$ is reduced to $1/\sqrt{2}$ of its maximum value at $\omega = 0$.[8] Therefore, the bandwidth of traveling wave modulation for a given δn and L_p is given in the literature to be

$$L_p\Delta\omega_m = \frac{2.8c}{\delta n} \tag{8.45}$$

The smaller δn is, the longer L_p that can be used. The longer L_p, the smaller $\Delta n_{eff,m}^{o}$ required to yield a given $\Delta\varphi$, and the smaller is the required rf modulation voltage. There is no RC limitation of electrical bandwidth. However, the microwave is also attenuated as it propagates. The attenuation increases as the rf frequency ω_m is increased.[9] Thus the microwave attenuation limits further both the effectiveness of using large L_p and the bandwidth of the modulator.

Traveling wave electro-absorption modulators can be analyzed in a similar manner [10].

[8] When $\delta nkL_p/2 = 1.9$, $\Delta\varphi$ is reduced to ½ of its maximum value at $\omega_m = 0$. If we use this criterion to define bandwidth, we will get a slightly different answer, The bandwidth $\Delta\omega$ will depend on how large the maximum variation of $\Delta\varphi$ is allowed within the band.

[9] Typically α_{rf} is proportional to $\sqrt{\omega}$ in the microwave transmission line.

Chapter summary

Besides lasers and detectors, there are switches and modulators. They operate by an electro-optical change of susceptibility. They can be classified into two categories: the devices that utilize $\Delta\chi'$ and the devices that utilize $\Delta\chi''$. In order to discuss them together systematically, the general effects of $\Delta\chi'$ and $\Delta\chi''$ on plane waves and waveguide modes are discussed first. The various mechanisms that produce $\Delta\chi'$ and $\Delta\chi''$ are presented next. The operation of individual active devices using different physical mechanisms is then analyzed. The electrical and optical performances of these devices are evaluated. At low frequencies, the bandwidths of these devices are limited by the RC time constants of the electrical circuit. At high frequencies, traveling wave modulators need to be used.

References

[1] A. E. Siegman, *Lasers*, University Science Books, 1986.

[2] S. Shimoda and H. Ishio, *Optical Amplifiers and Their Applications*, John Wiley & Sons, 1994.

[3] D. P. Schinke, R. G. Smith, and A. R. Hartman, Photodetectors, in *Semicondictor Devices for Optical Communication*, 2nd edition, ed. H. Kessel, Topics in Applied Physics, Vol. **39**, Springer-Verlag, 1982.

[4] P. K. L. YU and Ming C. Wu, Photodiodes for high performance analog links, Chapter 8, in *RF Photonic Technology in Optical Fiber Links*, ed. W. S. C. Chang, Cambridge University Press, 2002.

[5] L. A. Codrin and S. W. Corzine, *Diode Lasers and Photonic Integrated Circuits*, John Wiley & Sons, 1995.

[6] G. P. Argawal, *Semiconductor Lasers*, AIP Press, 1995.

[7] R. J. Keyes, *Optical and Infrared Detectors*, Springer-Verlag, 1980.

[8] M. Born and E. Wolf, *Principles of Optics*, Pergamon Press, 1959.

[9] A. Yariv, *Quantum Electronics*, John Wiley & Sons, 1989.

[10] William S. C. Chang, *Fundamentals of Guided-Wave Opto-Electronic Devices*, Cambridge University Press, 2010.

[11] Y.M. Cai and A. K-Y Jen, Thermally stable poled polyquinoline thin film with very large electro-optic response, *Applied Physics Letters*, **67**, 299, 1995.

[12] T. Van Eck, Polymer modulators for RF photonics, Chapter 7 in *RF Photonic Technology in Optical Fiber Links*, ed. W. S. C. Chang, Cambridge University Press, 2002.

[13] S. M. Sze, *Physics of Semiconductor Devices*, John Wiley & Sons, 1981.

[14] B. G. Streetman, *Solid State Electronic Devices*, Prentice Hall, 1995.

[15] X. Mei, *InAsP/GaInP Strain-compensated Multiple Quantum Wells and Their Optical Modulator Applications*, Ph.D. thesis, University of California San Diego, 1997.

[16] X. B. Mei, K. K. Loi, H. H. Wieder, W. S. C. Chang, and C. W. Tu, Strain compensated InAsP/GaInP multiple quantum wells for 1.3 μm waveguide modulators, *Applied Physics Letters*, **68**, 90, 1996.

[17] D. A. B. Miller, D. S. Chemla, T. C. Damen, *et al.*, Electric field dependence of optical absorption near the bandgap of quantum well structures, *Physical Review B*, **32**, 1043, 1985.

[18] D. A. B. Miller, J. S. Wiener, and D. S. Chemla, Electric field dependence of linear optical properties in quantum well structures: waveguide electroabsorption and sum rules, *IEEE Journal of Quantum Electronics*, **QE-22**, **816**, 1986.

[19] D. S. Chemla, D. A. B. Miller, P. W. Smith, A. C. Gossard, and W. Wiegmann, Room temperature excitonic non linear absorption and refraction in GaAs/GaAlAs multiple quantum well structures, *IEEE Journal of Quantum Electronics*, **QE-20**, **25**, 1984.

[20] D. S. Chemla, D. A. B. Miller, P. W. Smith, A. C. Gossard, and W. Wiegmann, Room temperature excitonic non linear absorption and refraction in GaAs/GaAlAs multiple quantum well structures, *IEEE Journal of Quantum Electronics*, **QE-20**, **25**, 1984

Appendix Solution of the scalar wave equation: Kirchoff's diffraction integral

The Helmholtz equation plus boundary conditions for TEM waves with time variation $e^{j\omega t}$, in Eq. (3.4) is:

$$\nabla^2 U + k^2 U = 0 \tag{A.1}$$

where, $k = \omega/c = 2\pi/\lambda$. It is typically solved mathematically using Green's function.

In the following, we will first present Eq. (A.2), which defines Green's function, G. Then we will show how a solution of G will let us find U at any given observer position (x_o, y_o, z_o) from U and ∇U at the boundary. Using Green's function, we obtain Kirchoff's integral. For known U on a planar boundary, Kirchoff's integral can be further simplified.

The equation for Green's function

Green's function, G, is the solution of the equation,

$$\begin{aligned}\nabla^2 G(x, y, z; x_o, y_o, z_o) + k^2 G &= -\delta(x - x_o, y - y_o, z - z_o)\\ &= -\delta(\underline{r} - \underline{r_o}).\end{aligned} \tag{A.2}$$

Eq. (A.2) is identical to Eq. (A.1) except for the δ function. The boundary conditions for G are the same as for U. δ is a unit impulse function that is 0 when $x \neq x_o$, $y \neq y_o$, and $z \neq z_o$. It tends to infinity when (x, y, z) approaches the discontinuity point (x_o, y_o, z_o). δ satisfies the normalization condition,

$$\begin{aligned}&\iiint_V \delta(x - x_o, y - y_o, z - z_o)\, \mathrm{d}x\mathrm{d}y\mathrm{d}z = 1\\ &= \iiint_V \delta(\underline{r} - \underline{r_o})\, \mathrm{d}v\end{aligned} \tag{A.3}$$

where $\underline{r} = x\underline{i_x} + y\underline{i_y} + z\underline{i_z}$, $\underline{r_o} = x_o\underline{i_x} + y_o\underline{i_y} + z_o\underline{i_z}$ and $\mathrm{d}v = \mathrm{d}x\mathrm{d}y\mathrm{d}z = r^2 \sin\theta \mathrm{d}r\mathrm{d}\theta\mathrm{d}\varphi$. V is any volume that includes the observation point (x_o, y_o, z_o).

Finding *U* from Green's function, *G*

From advanced calculus, we learned that,

$$\nabla \cdot (G\nabla U - U\nabla G) = G\nabla^2 U - U\nabla^2 G. \tag{A.4}$$

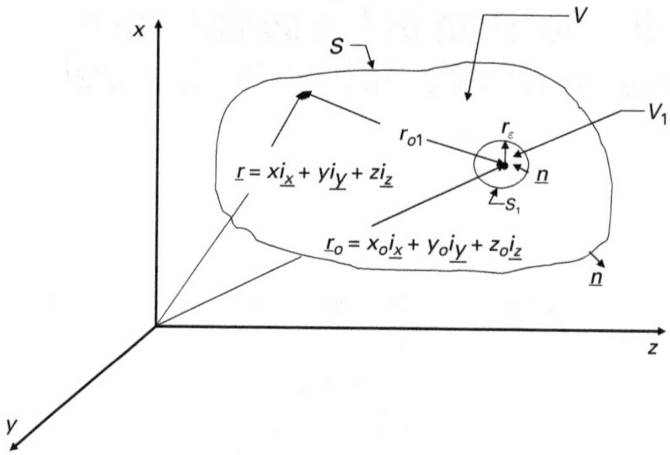

Figure A.1 Illustration of volumes and surfaces to which Green's theory applies. The volume to which Green's theory applies is V, which has a surface S. The outward unit vector of S is $\underline{n}$; $\underline{r}$ is any point in the x,y,z space. The observation point within V is r_o. For the volume V', V_1 around $\underline{r_o}$ is subtracted from V. V_1 has surface S_1, and the unit vector $\underline{n}$ is pointed outward from V'.

Applying a volume integral to both sides of the above equation and utilizing Eqs. (A.2) and (A.3), we obtain

$$\iiint_V \nabla\cdot(G\nabla U - U\nabla G)\mathrm{d}v = \iint_S (G\underline{n}\cdot\nabla U - U\underline{n}\cdot\nabla G)\,\mathrm{d}s$$
$$= \iiint_V [-k^2GU + k^2UG + U\delta(\underline{r} - \underline{r_o})]\,\mathrm{d}v = U(\underline{r_o}) \tag{A.5}$$

Here, V is any closed volume (within the boundary S) enclosing the observation point $\underline{r_o}$. $\underline{n}$ is the unit vector perpendicular to the boundary in the outward direction, as illustrated in Figure A.1.

Equation (A.5) is an important mathematical result. It shows that, when G is known, U at the position (x_o, y_o, z_o) can be expressed directly in terms of the values of U and ∇U on the boundary S, without solving explicitly the Helmholtz equation (A.1). Eq. (A.5) is known mathematically as Green's identity. The key is how to find G.

A general Green's function, *G*

A general Green's function, G[1] has been derived in many classical optics textbooks [2]. It is:

$$G = \frac{1}{4\pi}\frac{\mathrm{e}^{-jkr_{o1}}}{r_{o1}} \tag{A.6}$$

[1] There are different Green's functions, see [1].

where $r_{o1} = |\underline{r_o} - \underline{r}| = \sqrt{(x - x_o)^2 + (y - y_o)^2 + (z - z_o)^2}$. As shown in Figure A.1, r_{o1} is the distance between $\underline{r}_o$ and $\underline{r}$.

This G can be shown to satisfy Eq. (A.2) in two steps: (1) By direct differentiation, "$\nabla^2 G + k^2 G$" is clearly zero everywhere in any homogeneous medium except at $\underline{r} \cong \underline{r_o}$. Therefore Eq. (A.2) is satisfied within the volume V' which is V minus the volume V_1 (with boundary S_1) of a small sphere with radius r_ε enclosing $\underline{r_o}$in the limit as r_ε approaches 0. V_1 and S_1 are also illustrated in Figure A.1. (2) In order to find out the behavior of G near $\underline{r_o}$, we note that $|G| \to \infty$, as $r_{o1} \to 0$. If we perform the volume integration of the left-hand side of Eq. (A.2) over the volume V_1, we obtain:

$$\underset{r_\varepsilon \to 0}{Lim} \iiint_{V_1} [\nabla \cdot \nabla G + k^2 G] \mathrm{d}v = \iint_{S_1} \nabla G \cdot \underline{n} \mathrm{d}s$$

$$= \underset{r_\varepsilon \to 0}{Lim} \int_0^{2\pi} \int_{-\pi/2}^{\pi/2} \left[-\frac{\mathrm{e}^{-jkr_\varepsilon}}{4\pi r_\varepsilon^2} \right] r_\varepsilon^2 \sin\theta \, \mathrm{d}\theta \, \mathrm{d}\varphi = -1 \quad \text{(A.7)}$$

Thus, using this Green's function, the volume integration of the left-hand side of Eq. (A.2) yields the same result as the volume integration of the δ function. In short, G given in Eq. (A.6) satisfies Eq. (A.2) for any homogeneous medium.

From Eq. (A.5) and G, we obtain the well-known Kirchoff's diffraction formula,

$$U(\underline{r_o}) = \iint_S (G \nabla U - U \nabla G) \cdot \underline{n} \mathrm{d}s \quad \text{(A.8)}$$

Note that, in this format, we need to know both U and ∇U on the boundary in order to calculate its value at $\underline{r_o}$ inside the boundary.

For many practical applications, only U is known on a planar aperture, followed by a homogeneous medium with no additional radiation source. In this case, calculation of $U(\underline{r_o})$ *can be simplified.*

Green's function for known *U* in a planar aperture

Let there be an aperture on the planar surface $z = 0$. A known radiation U is incident on the aperture Ω from $z < 0$, and the observation point $\underline{z_o}$ is located at $z > 0$. As a mathematical approximation to this geometry, we define V to be the semi-infinite space at $z \geq 0$, bounded by the surface S. S consists of the plane $z = 0$ on the left and a large spherical surface with radius R on the right, as $R \to \infty$. Figure A.2 illustrates the hemisphere.

The boundary condition for a sourceless U at $z > 0$ is given by the radiation condition at very large R, as $R \to \infty$ [3],

$$\underset{R \to \infty}{\mathrm{Lim}} R \left(\frac{\partial U}{\partial n} + jkU \right) = 0 \quad \text{(A.9)}$$

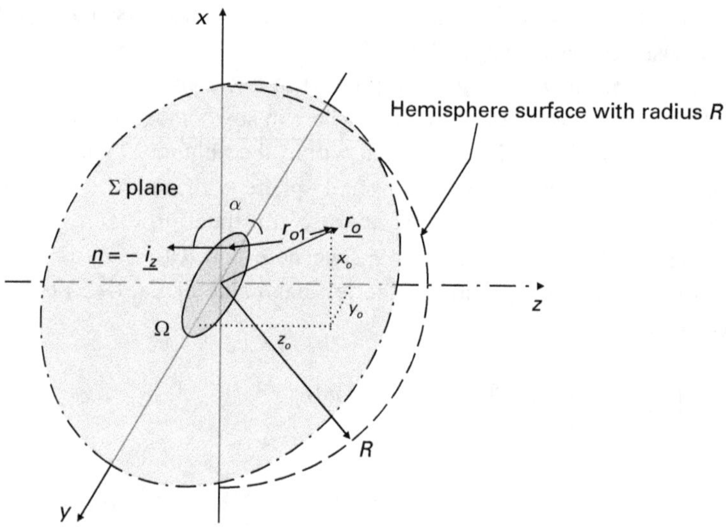

Figure A.2 Geometrical configuration of the hemispherical volume for Green's function, G. The surfaces to which Green's function applies consist of Σ, which is part of the xy plane, and a very large hemisphere which has a radius R, connected with Σ. The outward normal of the surface Σ and Ω is $-i_z$. The coordinates for the observation point $\underline{r_o}$ are x_o, y_o and z_o.

The radiation condition is essentially a mathematical statement that there is no incoming wave at very large R. Any U that represents an outgoing wave in the $z > 0$ space will satisfy Eq. (A.9).

If we do not know the ∇U term in Eq. (A.8), we need to find a Green's function that has G = 0 on the plane boundary (i.e. $z = 0$). However, we already know the incident radiation for $z < 0$. We need only apply Eq. (A.8) to the hemisphere S for $z > 0$. We note that any function, F, in the form of e^{-jkr}/r, will satisfy $[\nabla F + k^2 F = 0]$, as long as r is not allowed to approach 0. Thus we can add such a second term to G given in Eq. (A.6). This satisfies Eq. (A.2) for $z > 0$ as long as r never approaches 0.

To be more specific, let $\underline{r_i}$ be a mirror image of (x_o,y_o,z_o) across the $z = 0$ plane at $z < 0$. Let the second term be $e^{-jkr_{i1}}/r_{i1}$, where r_{i1} is the distance between (x,y,z) and $\underline{r_i}$. Since our Green's function will only be used for $z_o > 0$, r_{i1} for this second term will never approach zero for $z \geq 0$. Thus, as long as we seek a solution of U in the space $z > 0$, Eq. (A.2) is satisfied for $z > 0$. However, the difference is that the sum of the two terms is zero when (x,y,z) is on the $z = 0$ plane.

Let the Green's function for this configuration be

$$G_1 = \frac{1}{4\pi}\left[\frac{e^{-jkr_{01}}}{r_{01}} - \frac{e^{-jkr_{i1}}}{r_{i1}}\right] \tag{A.10}$$

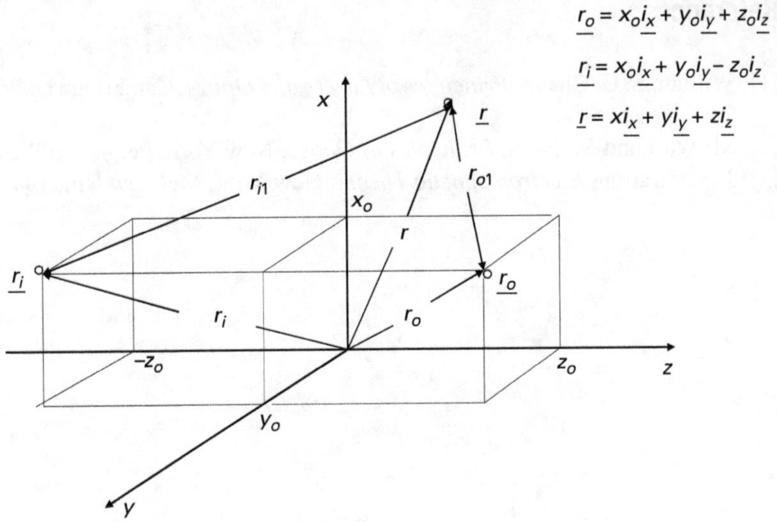

Figure A.3 Illustration of $\underline{r}$, the point of observation $\underline{r_o}$ and its image $\underline{ri}$, in the method of images. For G, the image plane is the x–y plane.

Here $\underline{r_i}$ is the image of $\underline{r}_o$ across the $z = 0$ plane. It is located at $z < 0$, as shown in Figure A.3. G_1 is zero on the x–y plane at $z = 0$. When this G_1 is applied to Green's identity, Eq. (A.8), we obtain

$$U(\underline{r_o}) = \iint_{\Sigma} U(x, y, z = 0)\, \frac{\partial G_1}{\partial z}\, \mathrm{d}x\mathrm{d}y \tag{A.11}$$

Here Σ refers to the x–y plane at $z = 0$. Because of the radiation condition expressed in Eq. (A.9), the value of the surface integration over the very large semi-sphere enclosing the $z > 0$ volume (with $R \to \infty$) is 0.

For most applications, $U \neq 0$ only in a small sub-area of Σ, e.g. the radiation U is incident on an opaque screen that has a limited open aperture Ω, or if the incident radiation has only a limited beam size Σ. In that case, $-\partial G_1/\partial z$ at $z_o >> \lambda$ can be simplified. We obtain

$$-\nabla G_1 \cdot \underline{i_z} = 2\cos\alpha\, \frac{\mathrm{e}^{-jkr_{o1}}}{4\pi r_{01}}(-jk) \tag{A.12}$$

α is illustrated in Figure A.2. Therefore, the simplified expression for U is:

$$U(\underline{r_o}) = \frac{j}{\lambda} \iint_{\Omega} U \frac{\mathrm{e}^{-jkr_{o1}}}{r_{o1}} \cos\alpha\, \mathrm{d}x\mathrm{d}y \tag{A.13}$$

This result is also known as Huygens' principle in classical optics.

References

[1] William S. C. Chang, *Principles of Lasers and Optics*, Cambridge University Press, 2005.
[2] M. Wolf and M. Born, *Principles of Optics*, New York, Pergamon Press, 1959.
[3] J. A. Stratton, *Electromagnetic Theory*, New York, McGraw Hill, 1941.

Index

www.ingramcontent.com/pod-product-compliance
Lightning Source LLC
LaVergne TN
LVHW062123200926
848484LV00007B/1747
* 9 7 8 1 1 0 7 0 7 4 9 0 3 *